Biomass Energy with Carbon Capture and Storage (BECCS)

Biomass Energy with Carbon Capture and Storage (BECCS)

Unlocking Negative Emissions

Edited by

Clair Gough
University of Manchester, UK

Patricia Thornley
University of Manchester, UK

Sarah Mander
University of Manchester, UK

Naomi Vaughan
University of East Anglia, Norwich, UK

Amanda Lea-Langton
University of Manchester, UK

Registered Office(s)
John Wiley & Sons, Inc., 111 River Street, Hoboken, NJ 07030, USA
John Wiley & Sons Ltd, The Atrium, Southern Gate, Chichester, West Sussex, PO19 8SQ, UK

Editorial Office
9600 Garsington Road, Oxford, OX4 2DQ, UK

For details of our global editorial offices, customer services, and more information about Wiley products visit us at www.wiley.com.

Wiley also publishes its books in a variety of electronic formats and by print-on-demand. Some content that appears in standard print versions of this book may not be available in other formats.

Library of Congress Cataloging-in-Publication Data

Names: Gough, Clair, editor. | Thornley, Patricia, editor. | Mander, Sarah,
   editor. | Vaughan, Naomi, editor. | Lea-Langton, Amanda, editor.
Title: Biomass energy with carbon capture and storage (BECCS) : unlocking
   negative emissions / edited by Clair Gough, Patricia Thornley, Sarah
   Mander, Naomi Vaughan, Amanda Lea-Langton.
Description: First edition. | Hoboken, NJ : John Wiley & Sons, 2018. |
   Includes bibliographical references and index. |
Identifiers: LCCN 2018001038 (print) | LCCN 2018007091 (ebook) | ISBN
   9781119237631 (pdf) | ISBN 9781119237686 (epub) | ISBN 9781119237723
   (cloth)
Subjects: LCSH: Carbon sequestration. | Biomass energy–Climatic factors. |
   Climate change mitigation.
Classification: LCC SD387.C37 (ebook) | LCC SD387.C37 B555 2018 (print) | DDC
   577/.144–dc23
LC record available at https://lccn.loc.gov/2018001038

Cover design by Wiley
Cover images © Pobytov/Getty Images

Set in 10/12pt WarnockPro by SPi Global, Chennai, India
Printed in Singapore by C.O.S. Printers Pte Ltd

10 9 8 7 6 5 4 3 2 1

Contents

List of Contributors

Gabrial Anandarajah
UCL Energy Institute
University College London
UK

Kevin Anderson
Tyndall Centre for Climate Change
Research, School of Mechanical
Aerospace and Civil Engineering
University of Manchester
UK

Gordon Andrews
School of Chemical and Process
Engineering
University of Leeds
UK

Amit Bhave
CMCL Innovations
Castle Park, Cambridge
UK

Bill Buschle
Institute for Energy Systems
School of Engineering
University of Edinburgh
UK

Hannah Chalmers
Institute for Energy Systems
School of Engineering
University of Edinburgh
UK

Olivier Dessens
UCL Energy Institute
University College London
UK

Temitope Falano
Tyndall Centre for Climate Change
Research, School of Mechanical
Aerospace and Civil Engineering
The University of Manchester
UK

Paul Fennell
Bone Building
Imperial College London
UK

Karen N. Finney
Energy 2050, Department of Mechanical
Engineering
University of Sheffield
UK

Dennis Gammer
Energy Technologies Institute
Holywell Building
Loughborough
UK

Clair Gough
Tyndall Centre for Climate Change
Research, School of Mechanical
Aerospace and Civil Engineering
The University of Manchester
UK

Geoffrey P. Hammond
Department of Mechanical Engineering
Institute for Sustainable Energy and the
Environment (ISEE)
University of Bath
UK

Alice Larkin
Tyndall Centre for Climate Change
Research, School of Mechanical
Aerospace and Civil Engineering
The University of Manchester
UK

Amanda Lea-Langton
Tyndall Centre for Climate Change
Research, School of Mechanical
Aerospace and Civil Engineering
The University of Manchester
UK

Mathieu Lucquiaud
Institute for Energy Systems, School
of Engineering
University of Edinburgh
UK

Leslie Mabon
Robert Gordon University
Aberdeen
UK

Niall Mac Dowell
Centre for Environmental Policy
Imperial College London
UK

Sarah Mander
Tyndall Centre for Climate Change
Research, School of Mechanical
Aerospace and Civil Engineering
The University of Manchester
UK

Will McDowall
UCL Institute for Sustainable Resources
University College London
UK

Alison Mohr
School of Sociology and Social Policy
University of Nottingham
UK

Geraldine Newton-Cross
Energy Technologies Institute
Loughborough
UK

Juan Riaza
Institute for Energy Systems,
School of Engineering
University of Edinburgh
UK

Nilay Shah
Bone Building
Imperial College London
UK

Raphael Slade
Centre for Environmental Policy
Imperial College
London
UK

János Szuhánszki
Energy 2050, Department of Mechanical
Engineering
University of Sheffield
UK

Richard H.S. Taylor
Bone Building
Imperial College London
UK

Patricia Thornley
Tyndall Centre for Climate Change
Research, School of Mechanical
Aerospace and Civil Engineering
The University of Manchester
UK

Andrew Welfle
Tyndall Centre for Climate Change
Research, School of Mechanical
Aerospace and Civil Engineering
The University of Manchester
UK

Naomi Vaughan
School of Environmental Sciences
University of East Anglia
Norwich
UK

Foreword

This book is essential reading for anybody interested in understanding the role that biomass energy with carbon capture and storage (BECCS) can play in the transition to a low-carbon economy and the challenge of achieving the targets embedded in the Paris Agreement. It is the first comprehensive assessment of the technical, scientific, social, economic, ethical and governance issues associated with BECCS.

There is no doubt that human activities are changing Earth's climate with adverse consequences for socio-economic sectors, ecological systems and human health. Recognising the danger associated with human-induced climate change, literally every government in the world signed the Paris Agreement (the United States has since withdrawn from the agreement), which called for the increase in global average temperature should be held to well below 2 °C above pre-industrial levels and to pursue efforts to limit the temperature increase to 1.5 °C above pre-industrial levels. To realise these targets requires a rapid transition to a low-carbon economy involving unprecedented global cooperation and dramatic energy-system transformations, and most models require global net negative emissions of greenhouse gases in the second half of this century through the widespread use of BECCS.

Key issues associated with BECCS such as the social and environmental sustainability of biomass production, sustainability of supply, state of pre- and post-combustion technologies, energy penalty due to carbon capture, investment and operating costs, scale required to deliver global net negative emissions, life-cycle analyses and accounting systems for greenhouse gas emissions and governance structures are all critically assessed in this book.

The analysis suggests that BECCS is a workable technology, but scaling up from the individual project level to the global scale requires a massive enhancement of the technical, biological, logistical, governance and social processes associated with BECCS. The analysis also suggests that the 1.5 °C target may be impossible to achieve without BECCS or other negative emissions processes and that without BECCS the carbon price in 2050 to achieve the 2 °C target could be doubled to about 300\$/tCO$_2$. Unfortunately, current uncertainties in technology development, enabling frameworks and policy development restrict quantification of the magnitude of global net negative emissions that can be delivered by BECCS, and hence its role in limiting human-induced climate change.

Given the potential role that BECCS can play in limiting the magnitude of human-induced climate change, it is imperative that governments and the private sector significantly enhance their research, development and demonstration agendas for this technology, and develop appropriate governance structures, transparent and reliable life-cycle accounting systems for greenhouse gas emissions and incentives, policies and mechanisms to stimulate and account for negative emissions on a global scale.

Director of Strategic Development at the Tyndall Centre
for Climate Change Research and
Chair of the Intergovernmental Science-Policy Platform
on Biodiversity and Ecosystem Services (IPBES)

Professor Sir Bob Watson
CMG FRS.

Preface

There is a growing and significant dependence on biomass energy with carbon capture and storage (BECCS) in future greenhouse gas emission scenarios in global integrated assessment models, and it has become central to the discourse around achieving a target of 2 °C global average temperature rise. This reliance on BECCS hinges on its potential to deliver so-called negative emissions in order to maintain a sustainable atmospheric concentration of CO_2 cost-effectively. As a young and untested group of technologies, there are many uncertainties associated with BECCS and there is strong imperative to understand better the conditions for and consequences of pursuing this group of technologies. BECCS technologies may offer a role in offsetting hard-to-abate sectors (e.g. agriculture and aviation) or may enable an 'overshoot' in reaching cumulative emissions budgets in the context of delayed action on mitigation. However, there is very little practical experience of implementing the technology in commercial applications and, indeed, relatively little research into its potential and the conditions for realising its deployment. This book aims to set out the technical and scientific parameters of delivering BECCS technologies within the wider social, economic and political context in which it sits, giving the reader a broad understanding of the key issues and potential consequences of pursuing BECCS.

To understand BECCS, what it can offer and how it might contribute to climate-change mitigation, it is essential to consider the variety of technical and non-technical constraints in a joined-up manner. Bringing together modern biomass energy systems with CCS not only presents technical and scientific challenges but also, to be delivered at scales large enough to contribute to substantive negative emissions, depends on other factors, such as geopolitics and supply chain integration and may have significant implications at a societal level. This book brings the issues together in a clear and accessible way that will support a more informed debate around the potential for this technology to unlock negative emissions.

List of Abbreviations/Acronyms

AFT	Adiabatic flame temperature.
ASU	Air separation unit.
BECCS	Biomass energy with carbon capture and storage.
BIGCC	Biomass integrated gasification combined cycle.
CCS	Carbon capture and storage.
CFBC	Circulating fluidized bed combustion.
CO_2	Carbon dioxide.
CV	Calorific value.
EOR	Enhanced oil recovery.
EJ	Exajoules $(1 \times 10^{18}$ J), unit of energy.
FGR	Flue gas recirculation/recycling.
Gha	Giga hectares $(1 \times 10^9$ ha).
HP	High pressure (steam turbine).
IAM	Integrated assessment model.
IGCC	Integrated gasification combined cycle.
IP	Intermediate pressure (steam turbine).
LCOE	Levelised cost of electricity.
LP	Low pressure (steam turbine).
LHV	Lower heating value.
MEA	Monoethanolamine.
NET	Negative emission technology.
NOAK	nth of a kind.
PJ	Petajoules $(1 \times 10^{15}$ J), unit of energy.
PM	Particulate matter.
PSA	Pressure swing adsorption.
RCP	Representative concentration pathway.
RFG	Reformulated gasoline.
SNG	Synthetic natural gas.
TRL	Technology readiness level.
UNFCCC	United Nations Framework Convention on Climate Change.

Part I

BECCS Technologies

1

Understanding Negative Emissions From BECCS

Clair Gough[1], Sarah Mander[1], Patricia Thornley[1],
Amanda Lea-Langton[1] and Naomi Vaughan[2]

[1]*Tyndall Centre for Climate Change Research, School of Mechanical Aerospace and Civil Engineering,*
University of Manchester, UK
[2]*School of Environmental Sciences, University of East Anglia, Norwich, UK*

1.1 Introduction

Changes in our climate are driven by human activity such as agriculture, deforestation and burning coal, oil and gas. The single most significant driver of climate change is the increase in the greenhouse gas carbon dioxide (CO_2) in our atmosphere from the combustion of fossil fuels. Efforts to limit the impacts of climate change focus on reducing the emissions of CO_2 and other greenhouse gases and adapting to live with the changing climate. In recent years, a third approach has gained significant attention: action to remove CO_2 from the atmosphere and store the CO_2 for long timescales (over hundreds of years). Recent negotiations under the UN Framework Convention on Climate Change (UNFCCC) delivered the 2015 Paris Agreement, which set a target of limiting global average temperature rise to 'well below 2 °C' (the 2 °C target having been agreed within the UNFCCC in 2010) while 'pursuing efforts to limit the temperature increase to 1.5°C' (UNFCCC, 2015). These are ambitious goals that will require immediate and radical emissions reductions if they are to be met. The idea of introducing 'negative emissions' is born out of the gap between the current trajectory in global emissions and the pathway necessary to avoid dangerous climate change. The most prominent proposal for achieving such negative emissions is to use biomass as a feedstock to generate electricity (or produce biofuels or hydrogen), capture the CO_2 during production and store it underground in geological reservoirs – biomass energy with carbon capture and storage, or BECCS for short. However, the negative emissions concept remains just that, a concept; in principle, technologies such as BECCS can deliver net CO_2 removal at a project scale, or potentially at a global scale sufficient to impact atmospheric concentrations of CO_2 and associated global average temperatures – but in practice, this potential has yet to be accessed at anything like a global scale. This book explores the challenges of unlocking negative emissions using BECCS.

Future climate change is most commonly explored using a suite of models that represent the Earth's climate system, the physical and socio-economic impacts of a changing climate and the greenhouse gases and other drivers generated by the global economy and energy systems. Integrated assessment models (IAMs) are used to create scenarios of future emissions that are used by climate and impact models. The growing and significant dependence on BECCS in future emissions scenarios in global IAMs has placed BECCS at the centre of the discourse around achieving targets of 2 °C global average temperature rise and, following the 2015 Paris Agreement, 1.5 °C. This reliance on BECCS hinges on its potential to remove CO_2 from the atmosphere in order to maintain a sustainable atmospheric concentration of CO_2 in a cost-effective manner.

There are many different technical options that could deliver negative emissions *via* BECCS and these vary in their technology readiness level (TRL). Some of the closer-to-market BECCS technologies are composed of component parts that have been proven and tested, but integration and deployment have not yet been demonstrated at commercial scale. Consequently, there remain significant uncertainties associated with BECCS performance and costs. Understanding the potential for, and implications of, pursuing BECCS requires an interdisciplinary approach. It has been suggested that BECCS could play a role in offsetting hard-to-abate sectors (e.g. agriculture and aviation) or enable delayed action on mitigation. While the atmospheric concentration of CO_2 continues to rise and policy objectives focus on limiting warming to 1.5 °C, it becomes increasingly likely that a means of delivering negative emissions will be required. Whether or not limiting warming to 1.5 °C is feasible without negative emissions remains unclear. In 2018, the IPCC will deliver a special report devoted to understanding the emissions pathways and impacts associated with 1.5 °C.

Despite its significance within the formal policy goals, there is very little practical experience of implementing the technology in commercial applications and limited research into the practicalities of implementation and conditions for accelerating deployment. Combining modern biomass energy systems with CCS not only presents technical and scientific challenges but, to be implemented at scales large enough to deliver global net negative emissions, also depends on other factors, such as geopolitics and supply-chain integration and may have significant societal implications. To understand BECCS, what it can offer and how it might contribute to climate-change mitigation, it is essential to consider the technical and non-technical constraints in a holistic manner.

This book aims to provide a comprehensive assessment of BECCS, describing the technology options available and the implications of its future deployment. While there is a rich literature relating to bioenergy and carbon capture and storage (CCS) separately, there is currently very little published research on the integration of these components. Our aim is to address this gap, bringing together technical, scientific, social, economic and governance issues relating to the potential deployment of BECCS as a key climate-change mitigation approach. The uniqueness of the book lies in bringing these subjects together and imposing order on the disparate sources of information. Doing this in a clear and accessible way will support a more informed debate around the potential for this technology to deliver deep cuts in emissions.

1.2 Climate-Change Mitigation

In its Fifth Assessment Report (AR5), the Intergovernmental Panel on Climate Change (IPCC, 2014) identified four so-called representative concentration pathways (RCPs), describing time-dependent ranges of atmospheric greenhouse gas concentration trajectories, emissions and land-use data between 2005 and 2100 (van Vuuren et al., 2011). Created by IAMs, each RCP is associated with emissions pathways that result in atmospheric concentrations correlated with different levels of radiative forcing; these are RCP2.6 (i.e. a radiative forcing of $2.6\,\mathrm{W\ m^{-2}}$), RCP4.5, RCP6, RCP8.5 (IPCC, 2014). Greenhouse gas concentrations within each RCP are associated with a probability of limiting temperature rise to below certain levels; only the lower concentrations within RCP2.6 are considered 'likely' (i.e. associated with a greater than 66% chance) to limit global atmospheric temperature rise to below 2 °C, or 'more unlikely than likely' (i.e. a less than 50% chance) for 1.5 °C (IPCC, 2014). The RCPs provide a consistent framework for analysis in different areas of climate-change research – for example, by climate modellers to analyse potential climate impacts associated with the pathways (including projected global average temperature rise) and in IAMs to explore alternative ways in which the emissions pathways for each RCP could be achieved (i.e. mitigation scenarios) under different economic, technological, demographic and policy conditions (IPCC, 2014; van Vuuren et al., 2011). The shared socio-economic pathways (SSPs) offer a further framework for IAMs to explore alternative emission pathways, by detailing different socioeconomic narratives that are consistent with the RCPs (O'Neill et al., 2017).

Climate-change mitigation policies are focused around limiting the increase in the global average temperature as described earlier. Achieving these targets is dependent on tight limits to cumulative emissions of CO_2 (and other greenhouse gases) in order to stabilise their atmospheric concentration. The cumulative emissions associated with a particular temperature goal is known as a carbon budget – the remaining budget for a 66% chance of keeping temperatures below a 2 °C increase is 800 Gt CO_2 (from 2017) (Le Quéré et al., 2016). With global emissions currently at about 36 Gt CO_2/year, this equates to about 20 years at current emissions rates before the budget is exceeded; until emissions are reduced to near zero, atmospheric CO_2 concentration will continue to rise (ibid).

In this context, by offering a route to delivering negative emissions, BECCS appears to be an attractive approach to potentially enabling mitigation costs to be reduced, more ambitious targets to become feasible than would otherwise be possible or allowing a delay to the year in which emissions will peak by enabling removal of CO_2 from the atmosphere in the future (Friedlingstein et al., 2011; van Vuuren et al., 2013). Typically, scenarios that are 'likely' to stay within 2 °C include such an overshoot in the concentration achieved through large-scale deployment of carbon dioxide removal (CDR) techniques (i.e. BECCS or afforestation) (IPCC, 2014).

A large majority of the pathways that deliver atmospheric CO_2 concentrations consistent with the 2 °C target (and indeed many of those associated with temperature increases up to 3 °C) require *global net* negative emissions by about 2070 (Fuss et al., 2014). The range of CO_2 removal through BECCS assumed in the IPCC scenarios is typically between 2 and 10 Gt CO_2/year by 2050 (Fuss et al., 2014; van Vuuren et al., 2013) with a median value of around 608 Gt CO_2 cumulatively removed by 2100 using BECCS (Wiltshire et al., 2015). Global net negative emissions are achieved when the amount of CO_2 removed from the atmosphere is greater than emissions from all other anthropogenic (i.e. resulting from or produced by human activities Allwood et al., 2014) sources (Fuss et al., 2014). When discussing the contribution of negative emissions from BECCS, it is useful to distinguish between three metrics that are typically used (Gough et al., 2018), as described in Figure 1.1:

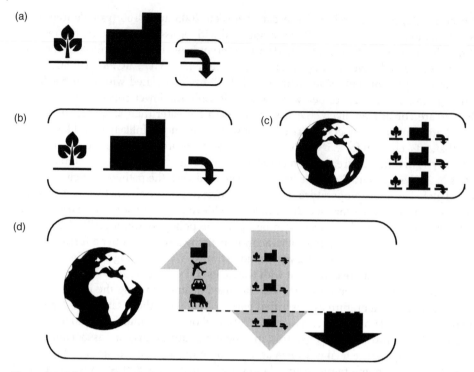

Figure 1.1 Schematic illustrating terminology for negative emissions from BECCS. (a) *CO_2 stored*, which can be calculated at a project, national or global level. (b) *Negative emissions*, which can be calculated at a project or national level, are achieved when the CO_2 removed from the atmosphere during biomass growth is greater than the CO_2 emissions from all the other processes in the supply chain. (c) *Global negative emissions*, the global sum of all negative emissions activity, this removes carbon dioxide from the atmosphere and can be used to 'offset' other anthropogenic CO_2 emissions. (d) *Global net negative emissions* occur when the global amount of negative emissions exceeds the CO_2 emissions from all other human sources, e.g. energy, transport and agriculture.

a) *CO_2 stored from BECCS systems*: this is the total CO_2 stored in a geological formation following capture in a CCS system and gives an indication of the geological storage capacity needed;
b) *Negative emissions from BECCS*: this is the net emissions from the BECCS supply chain, at a project scale, or cumulatively across all BECCS projects, accounting for system losses, emissions associated with land-use change and fossil fuel emissions;
c) *Global net negative emissions*: occur when the amount of global negative emissions (from all negative emissions approaches combined) exceeds the CO_2 emissions from all other human sources, e.g. energy, transport and agriculture.

Thus, global net negative emissions are used within the IAMs to offset outstanding anthropogenic emissions to deliver a 'net' emission trajectory in line with the RCP associated with a given temperature goal, such as 2 °C (Fuss et al., 2014). This is illustrated in Figure 1.2 to reveal how global net negative emissions are envisaged to contribute to keeping emissions within the carbon budget and the scale of the challenge (Anderson and Peters, 2016). Parties to the UNFCCC are required to submit Intended Nationally

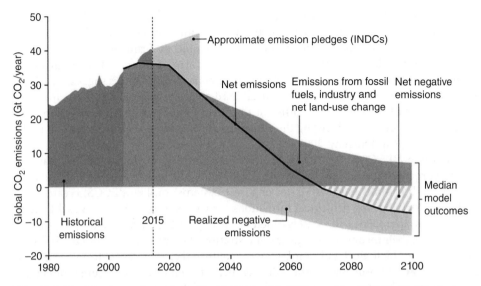

Figure 1.2 The role of negative emissions in relation to global CO_2 scenarios. Note that 'realised negative emissions' correspond to 'negative emissions described in Figure 1.1b and that 'net negative emissions in this figure correspond to 'global net negative emissions' described in Figure 1.1d. Source: Anderson and Peters (2016). (*See colour plate section for the colour representation of this figure.*)

Determined Contributions (INDCs); revised every 5 years, the INDCs are pledges for proposed emissions reductions at a national level. Figure 1.2 also shows that the current INDCs are not currently in line with the rapid decline in emissions necessary to stay within the budget – i.e. we are currently on track for an 'overshoot'. The further emissions rise above the 2 °C pathway, the greater the negative emissions required to stay within the total carbon budget. Results of the IAMs that describe alternative mitigation scenarios associated with achieving 2 °C include negative emissions from BECCS from around 2020 (Figure 1.1b,c), with global net negative emissions (Figure 1.1d) setting in around 2070 (Figure 1a in Fuss et al., 2014).

1.3 Negative Emissions Technologies

To date, climate-change mitigation has focused on ways to reduce the amount of greenhouse gases emitted to the atmosphere by reducing energy consumption or through the use of alternative technologies (such as renewable energy technologies) that emit less greenhouse gases. However, in the context of the rapidly diminishing carbon budgets described in Section 1.2, possibilities for geoengineering the climate have been raised; geoengineering is defined as 'deliberate large-scale manipulation of the planetary environment to counteract anthropogenic climate change' (Royal Society, 2009). Approaches for geoengineering include solar radiation management (reflecting the sun's energy back into space, for example, by cloud brightening or stratospheric aerosol injection) or removing CO_2 from the atmosphere (so-called CDR or negative emission technologies) (Royal Society, 2009; Vaughan and Lenton, 2011; McLaren, 2012). By removing CO_2

from the atmosphere and storing it underground, thus delivering 'negative emissions', BECCS can be considered part of this latter type of climate intervention. BECCS can be used to reduce emissions, i.e. mitigation, by, for example, co-firing biomass with coal in a power plant with CCS. Some consider it to be a form of climate-change mitigation rather than a geoengineering approach (Boucher et al., 2013). Others suggest BECCS is mitigation because it can be used to 'offset' hard-to-abate sectors. But there are those who consider BECCS at a scale sufficient to deliver global net negative emissions (see Figure 1.1d) to be geoengineering the climate, due to its large-scale intervention in the Earth system (Royal Society, 2009).

Later we set BECCS in the context of other CDR approaches. Solar radiation management approaches are not addressed in this book. CDR approaches seek to enhance existing carbon sinks (such as ocean or forest systems) or, like BECCS, create novel carbon sinks (i.e. using geological formations for CO_2 storage) (Vaughan and Lenton, 2011). CDR approaches can be classified according to the means of CO_2 capture or removal and their means of storage or sequestration. Tavoni and Socolow (2013) distinguish between biological and non-biological capture (i.e. 'drawdown' from the atmosphere) and storage processes. Others distinguish between whether capture is direct (i.e. chemical) or indirect (i.e. biological) and four types of storage (mineralised, pressurised, oceanic, biotic) (McLaren, 2012). Table 1.1 summarises the main families of negative emissions approaches currently considered in the wider literature. The purpose of this type of classification is to support understanding of the different types of approaches associated with delivering negative emissions. Although the approaches are highly heterogeneous, grouping them in this way may also help signpost where certain issues or challenges might be applicable to several quite different approaches, or identify potential conflicts between approaches (for example, BECCS and Direct Air Capture may both require large volumes of underground storage).

1.4 Why BECCS?

It is clear from Section 1.2 that the categorisation of negative emissions technologies covers a wide variety of extremely different approaches, the unifying feature being their potential to remove CO_2 from the atmosphere. This book focuses on the group of technologies that combines bioenergy feedstocks for energy conversion processes linked to industrial capture of CO_2 for subsequent geological storage – otherwise known as BECCS.

Assessments of key negative emissions approaches have considered technical status, potential capacity and limitations; these place BECCS at a TRL between 4 and 7, depending on the particular approach under consideration (Lomax et al., 2015; McLaren, 2012). TRLs, originally used by NASA but now widely applied, provide a means of comparing the maturity of different technologies, from the lowest level, TRL1, up to TRL9 when a system is mass deployed. Other negative emissions approaches may be considered to be at higher TRLs than BECCS but are more limited in their potential at sufficiently large scale (i.e. above 1 Gt CO_2/year); for example, those sequestering CO_2 in materials such as magnesium silicate cement (limited by demand for cement) or timber in construction (limited by demand for timber and availability of a sustainable supply of timber) and land-based approaches such as forest restoration and peatland

Table 1.1 Negative emissions approaches, classified according to the type of capture and storage processes involved.

Approach	Capture	Storage	Description
Biomass energy and carbon capture and storage (BECCS)	Biological Indirect	Geological Pressurised	Power generation using biomass feedstocks followed by CO_2 capture from flue gases for subsequent geological storage
Direct air capture (DAC)	Non-biological Direct	Geological Pressurised	Uses chemical sorbents to absorb and then release CO_2 for subsequent geological storage
Ocean fertilisation	Biological Indirect	Biological Oceanic	Addition of nutrients to enhance growth of photosynthesising organisms, which eventually fall to the ocean floor
Ocean liming	Non-biological Indirect	Non-biological Oceanic	Calcium or magnesium oxides (lime) dissolved in sea water to increase the ocean absorption of atmospheric CO_2
Enhanced weathering	Non-biological Indirect	Non-biological Mineralised	Finely ground silicate rocks are dissolved in oceans or soils, these absorb CO_2 and are eventually sequestered in shells of organisms
Biochar	Biological Indirect	Biological Biotic	Heat treatment of biomass, locking up the carbon and can be used as a soil improver
Afforestation	Biological Indirect	Biological Biotic	Establishing and maintaining forests that have not previously been forested for a given period (e.g. 50 years) (UNFCCC, 2015)

Source: McLaren (2012) and Tavoni and Socolow (2013). Reproduced with permission of Elsevier.

and wetland habitat restoration, which is limited by land availability and climate impacts (McLaren, 2012). Chapter 5 considers TRL for BECCS approaches in more detail.

BECCS has further advantages, compared to other CDR or negative emissions approaches, that electricity (or liquid biofuels or hydrogen) is generated during the process and it is broadly compatible with current energy and social infrastructures (McGlashan et al., 2012). Although it is often described as being the most mature or least costly of the alternative negative emissions approaches (Kriegler et al., 2013; McGlashan et al., 2012) and, given its prominence in the mitigation pathways described earlier, there is relatively little research into the challenges and viability of bringing BECCS into mainstream commercial deployment. There are around 15 pilot-scale BECCS plants across the world, and the first large-scale BECCS project is a corn-to-ethanol plant in Decatur, Illinois, storing 1 Mt CO_2/year in an onshore saline aquifer; an overview of the status of these projects may be found in GCCSI (2016). BECCS is therefore dominant among the options with the potential to deliver large-scale negative emissions over the next 10–20 years.

BECCS is featured as the dominant emissions technology in the IAM model runs, alongside, to a much lesser extent, afforestation (i.e. establishing and maintaining forests that have not previously been forested for a given period, e.g. 50 years) (UNFCCC, 2013). IAMs calculate cost-optimal pathways to satisfy carbon budget constraints; thus, a key feature of BECCS is its role in energy supply; allowing the production of a commodity (e.g. electricity) while delivering negative emissions makes it an attractive option in cost-optimised scenarios. However, although there is limited commercial experience of integrated BECCS systems, the component technologies and techniques are well developed. Although, at the time of writing, IAMs started to include other negative emissions technologies such as direct air capture (Detlef van Vuuren, personal communication) and enhanced weathering (Taylor et al., 2016), BECCS dominates the negative emissions scenarios; understanding both the technical and non-technical constraints on its potential is critical to the feasibility of these scenarios. This book aims to unpack the issues associated with developing integrated BECCS systems at scales sufficient to deliver global net negative emissions.

1.5 Structure of the Book

In the following chapters, we set out and consider the key issues associated with the challenge of using BECCS to deliver negative emissions. The book is divided into three sections, described later, covering: BECCS technologies (describing the various technologies involved in BECCS systems); BECCS system assessments (considering key system characteristics across the life cycle and the relative performance of technical systems) and BECCS in the energy system (including the role of BECCS as a mitigation approach, its role within integrated assessment modelling, governance and supply-chain accounting frameworks and its social and ethical implications). Thus, the book sections reflect increasing scales of analysis, moving from consideration of the component technologies (Part I) that may contribute to the overall operation of a BECCS system (Part II) to the role of BECCS within the wider energy, climate and societal systems (Part III).

As an edited book, the chapters reflect the specific viewpoints of the different authors; BECCS remains a controversial technology in relation to the current significance placed on it within global climate-change assessments and the implicit reliance on it within the Paris Agreement. Different perspectives anticipate different extents for its potential. Our aim here is to present some of the key arguments associated with the deployment of BECCS to deliver negative emissions from a variety of perspectives; in the final chapter, we aim to bring these arguments together to consider if and under what circumstances BECCS may provide a route to unlocking negative emissions on a global scale.

1.5.1 Part I: BECCS Technologies

This section describes the biomass energy and carbon capture components of BECCS technologies. The four chapters in this section present the issues related to the supply of biomass for use in energy systems (Chapter 2), the specific issues that the use of biomass feedstocks brings to CO_2 capture (Chapters 3 and 4) and the techno-economic performance of these systems (Chapter 5). In contrast, the processes that occur after CO_2 capture (i.e. those involved in compression, transport and geological storage) are the same as those used for fossil CCS and hence are beyond the remit of this book.

Chapter 2 provides an assessment of the overall extent of different forms of biomass at both the UK and international scales and the potential availability of these resources for the bioenergy sector, drawing upon current reports and published research. The use of these biomass resources should not impact the sustainability of other sectors, such as industries that directly compete for biomass resource, or the availability of land to meet demands for food. The interfaces between the supply of biomass for energy and the competing land and resource demands of industry are highlighted, providing an indication of the levels of biomass resource that may be used for energy end uses without causing adverse impacts to other sectors. Finally, the chapter draws on published research to provide an evaluation of the best uses of available biomass resource and to conclude whether there will be sufficient sustainable biomass for a future bio-CCS sector.

Chapter 3 first explores the different technologies and configurations used to attain CO_2 capture from biomass fuel sources, comparing post-combustion capture (based on air-firing) and oxy-fuel combustion capture techniques. For post-combustion capture, a range of separation technologies can be utilised to remove the CO_2 from the rest of the flue gas stream, including solvent-based (wet scrubbing) capture, which is assessed in detail, and membrane separation, which is overviewed briefly. The potential locations of steam extraction from the power plant to regenerate the capture solvents used for absorption are also considered. The relative merits of enriched-air firing and oxy-fuel options are evaluated, consisting of an assessment of flue gas recirculation configurations. As the nature of biomass is so dissimilar to fossil fuels in terms of composition and properties, the specific challenges associated with biomass utilisation under BECCS operating conditions are then outlined, focusing primarily on trace elements and impurities and their impacts on capture performance. The deployment potential of these various BECCS options is subsequently overviewed in light of these challenges, based on existing technical knowledge.

Chapter 4 gives an overview of the pre-combustion technologies used for the capture and storage of CO_2 for both power generation and chemical production processes. These techniques involve removal of CO_2 from a feedstock before combustion is completed, involving its conversion to a synthesis gas (syngas) containing predominantly hydrogen and CO_2; the CO_2 can then be separated from this gas mixture. The main technical features involved are detailed and the relevant chemistry is provided. The applications of this technology include uses for power *via* the integrated gasification combined cycle (IGCC) and uses in fuel refineries for syngas upgrading and liquid fuels. Applications based on conventional fuels are described and the opportunities and issues regarding their uptake into biomass systems are discussed, providing case studies and examples of ongoing research activity.

Chapter 5 describes a unique collaboration between industry, consultancies and academia in a study commissioned and funded by the Energy Technologies Institute (ETI) UK to explore the potential for BECCS technologies to be utilised to enable large-scale CO_2 removal from the air, at the same time producing electricity. The study compares an initial long list of 28 gasification or combustion technologies integrated with CCS (pre-, post- or oxy-combustion) and considers current progress towards market, likely future progress to market, cost, efficiency and feasibility, all with as many assumptions as possible harmonised across the investigated technologies. Detailed process modelling is conducted on a shortlist of eight of the technologies, which are considered to be most promising for wide-scale deployment by 2050. These technologies are co-fired power generation with amine scrubbing, oxy-fuel combustion, carbonate looping or utilising integrated gasification combined cycle (IGCC); and dedicated fully biomass fired power stations utilising amine scrubbing, oxy-fuel, chemical looping or IGCC. An important part of the results is an attempt to quantify the uncertainties associated with the parameters discussed, considering how far away from market they are.

1.5.2 Part II: BECCS System Assessments

This section explores the technical performance of various BECCS approaches and their component technologies at a project level. Quantifying the negative emissions from BECCS requires analysis across its entire supply chain from growing, harvesting, treating and transporting biomass energy to its combustion and through the CO_2 capture process and the subsequent compression, transport and storage of the CO_2. Currently, there is very limited experience of conducting such a life-cycle analysis for BECCS, and Chapter 6 sets out the requirements and challenges associated with conducting such an assessment. Looking in more detail at the performance of some of the component technologies, Chapter 7 focuses on the CCS elements and Chapter 8 explores the potential and performance of full-chain BECCS systems in the UK.

Chapter 6 examines the rationale for applying a supply-chain life-cycle assessment approach in order to assess potential emissions savings and critiques the importance of key decisions around choice of system boundary and assessment methods that directly affect results. There are many uncertainties with this emerging technology, including the possible greenhouse gas savings that have not been extensively verified. Key uncertainties are probed and recommendations made on methodological approaches that will deliver appropriately informed assessments of the actual emissions reduction potential associated with deployment of BECCS.

Chapter 7 presents parameters relating to the performance of power plants with carbon capture and storage (CCS), which can be characterised and evaluated according to a number of sustainability criteria; these can be based on well-established, often quantifiable, economic, energy-related and environmental (including climate change) criteria. However, CCS is at an early stage of research, development and demonstration (RD&D), and therefore many of the system performance characteristics need to be determined on a first of a kind (FOAK) basis. CO_2 capture facilities will hinder the performance of power plants and give rise to an energy penalty which, in turn, lowers the system (thermodynamic) efficiency. The levelised cost of electricity (LCOE) can then be used as an indicator of the impact of adding capture equipment on plant economics. Finally, the environmental performance of CCS developments can be assessed in terms of climate-change impacts (e.g. cost of carbon avoided or captured), as well as the effects on biodiversity, land use and water resources. Geological storage of CO_2 will, in many cases, have potential consequences for the marine environment. Such impacts vary as to whether they are on a global, regional or local scale and to which stage of the CCS life cycle they relate.

Chapter 8 presents the key findings from a recent assessment of the potential for BECCS to play a role in UK emissions reduction targets conducted by the Energy Technologies Institute UK (ETI). Bioenergy with CCS (BECCS) is a credible, scalable and efficient technology and is considered to be critical for the UK to meet its 2050 greenhouse gas emission reduction targets cost-effectively. Major advances in the fundamental science and technology development have been made by the ETI and others over the last 10 years, specifically in understanding: the costs, efficiencies and challenges of biomass-fed combustion systems with carbon capture; the opportunities for different bioenergy supply chains, based on particular feedstocks, to deliver negative emissions; the potential availability and sustainability of feedstocks relevant to the UK and the identification and assessment of suitable CO_2 storage sites around the UK and the infrastructure required to connect to them.

1.5.3 Part III: BECCS in the Energy System

In this section, chapters go beyond the project-scale analysis of BECCS to consider its role within the wider energy and climate mitigation system, and the issues associated with achieving global net negative emissions. We consider questions such as how the approach might fit within a cumulative emissions profile measured against a diminishing carbon budget in the context of 1.5 or 2 °C (Chapter 9); how it impacts mitigation scenarios developed within an IAM (Chapter 10); what policy challenges are associated with establishing, regulating and accounting for negative emissions at a global scale (Chapter 11) and what the social and ethical implications are of pursuing this path (Chapter 12).

Chapter 9 explores the role of BECCS technologies and the implications of their deployment for mitigating climate change. This chapter demonstrates how the need for negative emissions arises in order to reconcile the levels and rate of mitigation necessitated by the rapidly shrinking global carbon budget associated with the 2 °C ambitions of the Paris Agreement. BECCS is central to much of the integrated assessment modelling that meets the 2 °C target, and it is essential to understand the assumptions underpinning the modelling work informing climate policy; thus, we make explicit the assumptions within these models about bioenergy, carbon capture and storage and BECCS more specifically. The infrastructure challenges of developing the energy conversion technologies, CO_2 transport and CO_2 storage reservoirs are discussed, and given the current slow rates of deployment of CCS generally, we explore whether the required levels of negative emissions from BECCS can be achieved within the appropriate timescales.

Chapter 10 presents results analysed using the TIAM-UCL integrated assessment model to investigate the extent to which bioenergy with carbon capture and storage (BECCS) is critical for meeting global CO_2 reduction targets under different long-term scenarios out to 2100. The chapter also assesses the potential impacts of BECCS on mitigation costs under various scenarios at a global scale. Though previous work has suggested that BECCS can play a crucial role in meeting the global climate-change mitigation target, uncertainties remain in two main areas: the availability of biomass, which is affected by many factors including availability of land for biomass production; and the sustainability of bioenergy production, including consequences for greenhouse gas emissions. In order to assess the importance of these uncertainties, this chapter develops several scenarios by varying the availability of biomass (sustainability of the bioenergy production) and peaking year for greenhouse gas emissions under 2 and 1.5 °C climate-change mitigation targets at a global level.

Chapter 11 explores some of the themes around supply-chain rationales and scope of system, first established in Chapter 6, from a social policy and governance perspective. It dissects whether negative emissions are 'real' or simply a figment of the supply-chain boundaries and accounting procedures and considers applicable policy and incentive frameworks in an attempt to understand why there has been so little development to date. BECCS often emerges from IAMs as an essential component of the future energy system. This leads to a perceived need to implement BECCS systems at significant scale. However, making such a significant energy system transition is a huge undertaking that will require appropriate policy incentivisation. Existing global policy frameworks encourage low-carbon technologies, but often fail to take account of the human dimension and local impact trade-offs. BECCS involves mobilisation of local supply chains with impacts on relevant communities in different countries/regions. Combining these contributions to assess and verify net impact is challenging and not facilitated under current governance structures. This chapter explores the wider context for such incentives including regional and global implementation frameworks.

Chapter 12 explores some of the social and ethical issues relating to the use of BECCS to deliver negative emissions. It considers both the big questions relating to its potential role in a morally adequate response to climate change as well as the more specific social and ethical issues associated with deployment of the technology on the ground. The relationship between BECCS and the use of fossil fuels, how it sits relative to other mitigation options and in the context of other negative emissions approaches are also considered. The chapter identifies contexts in which BECCS might represent a sustainable as well as a just solution and how it might be received at a social and societal level. Reviewing current thinking on justice in the context of energy and climate change, paying particular attention to issues that are relevant to CCS, and specifically BECCS, we look in turn at distributional, procedural, financial and intergenerational aspects of justice. Results from an expert workshop convened to discuss issues of governance and ethics in CCS and BECCS are used to supplement the wider literature throughout the chapter.

1.5.4 Part IV: Summary and Conclusions

The final chapter (Chapter 13) of this book synthesises the key messages from across the chapters and identifies critical issues associated with moving from negative emissions at a project scale to global net negative emissions at a systems level. We identify four questions governing the potential for BECCS to provide a key to unlocking negative emissions: Do we need this technology? Can it work? Does the focus on BECCS distract from the imperative to radically reduce demand and transform the global energy system? How can BECCS unlock negative emissions?

References

Allwood, J.M., Bosetti, V., Dubash, N.K. et al. (2014). Glossary. In: *Climate Change 2014: Mitigation of Climate Change. Contribution of Working Group III to the Fifth Assessment Report of the Intergovernmental Panel on Climate Change* (ed. O. Edenhofer, R. Pichs-Madruga, Y. Sokona, E. Farahani, S. Kadner, K. Seyboth, A. Adler, I. Baum, S. Brunner, P. Eickemeier, B. Kriemann, J. Savolainen, S. Schlömer, C. von Stechow, T. Zwickel and J.C. Minx). Cambridge and New York: Cambridge University Press. Climate Change 2014 Synthesis Report Summary for Policymakers. Intergovernmental Panel on Climate Change.

Anderson, K. and Peters, G. (2016). The trouble with negative emissions. *Science* **354**: 182–183.

Boucher, O., Forster, P.M., Gruber, N. et al. (2013). Rethinking climate engineering categorization in the context of climate change mitigation and adaptation. *Wiley Interdisciplinary Reviews: Climate Change* **5**: 23–35.

Friedlingstein, P., Solomon, S., Plattner, G.K. et al. (2011). Long-term climate implications of twenty-first century options for carbon dioxide emission mitigation. *Nature Climate Change* **1**: 457–461.

Fuss, S., Canadell, J.G., Peters, G.P. et al. (2014). Betting on negative emissions. *Nature Climate Change* **4**: 850–853.

GCCSI (2016). *The Global Status of CCS*. Summary report. Canberra: The Global CCS Institute.

Gough, C., Garcia Freites, S., Jones, C. et al. (2018). Challenges to the use of BECCS as a keystone technology in pursuit of 1.5°C. *Global Sustainability*, In press.

IPCC (2014). *Climate Change 2014 Synthesis Report Summary for Policymakers*. Intergovernmental Panel on Climate Change.

Kriegler, E., Edenhofer, O., Reuster, L. et al. (2013). Is atmospheric carbon dioxide removal a game changer for climate change mitigation? *Climatic Change* **118**: 45–57.

Le Quéré, C., Andrew, R.M., Canadell, J.G. et al. (2016). Global carbon budget 2016. *Earth System Science Data* **8**: 605–649.

Lomax, G., Lenton, T.M., Adeosun, A., and Workman, M. (2015). Investing in negative emissions. *Nature Climate Change* **5**: 498–500.

McGlashan, N., Shah, N., Caldecott, B., and Workman, M. (2012). High-level techno-economic assessment of negative emissions technologies. *Process Safety and Environmental Protection* **90**: 501–510.

McLaren, D. (2012). A comparative global assessment of potential negative emissions technologies. *Process Safety and Environmental Protection* **90**: 489–500.

O'Neill, B.C., Kriegler, E., Ebi, K.L. et al. (2017). The roads ahead: narratives for shared socioeconomic pathways describing world futures in the 21st century. *Global Environmental Change* **42**: 169–180.

Royal Society (2009). *Geoengineering the Climate: Science, Governance and Uncertainty*. RS Policy Document 10/09, Spetmber 2009 RS1636.

Tavoni, M. and Socolow, R. (2013). Modeling meets science and technology: an introduction to a special issue on negative emissions. *Climatic Change* **118**: 1–14.

Taylor, L.L., Quirk, J., Thorley, R.M.S. et al. (2016). Enhanced weathering strategies for stabilizing climate and averting ocean acidification. *Nature Climate Change* **6**: 402–406.

UNFCCC (2013) *Afforestation and reforestation projects under the Clean Development Mechanism: A reference manual*, United Nations Framework Convention on Climate Change, http://unfccc.int/resource/docs/publications/cdm_afforestation_bro_web.pdf (accessed 13 December 2017).

UNFCCC (2015). *United Nations Framework Convention on Climate Change*. Adoption of the Paris Agreement FCCC/CP/2015/10/Add.1.

Vaughan, N. and Lenton, T. (2011). A review of climate geoengineering proposals. *Climatic Change* **109**: 745–790.

van Vuuren, D.P., Deetman, S., van Vliet, J. et al. (2013). The role of negative CO_2 emissions for reaching 2°C: insights from integrated assessment modelling. *Climatic Change* **118**: 15–27.

van Vuuren, D., Stehfest, E., den Elzen, M.J. et al. (2011). RCP2.6: exploring the possibility to keep global mean temperature increase below 2°C. *Climatic Change* **109**: 95–116.

Wiltshire, A., Davies-Bernard, T., Jones, C.D. (2015). *Planetary limits to Bio-Energy Carbon Capture and Storage (BECCS) negative emissions*. AVOID2 Report WPD2a. http://www.avoid.uk.net/2015/07/planetary-limits-to-beccs-negative-emissions-d2a/ (accessed 13 December 2017).

2

The Supply of Biomass for Bioenergy Systems

Andrew Welfle[1] and Raphael Slade[2]

[1]*Tyndall Centre for Climate Change Research, School of Mechanical Aerospace and Civil Engineering,*
The University of Manchester, UK
[2]*Centre for Environmental Policy, Imperial College, London, UK*

2.1 Introduction

The widespread deployment of BECCS technologies and their ultimate success in generating significant emissions reductions will be highly dependent on the growth of the global bioenergy sector. The energy strategies of many countries across the world are already targeted bioenergy as key low carbon renewable energy pathway for meeting their climate change targets. There are already concerns about whether there will be sufficient availability of sustainable biomass resource to balance global demands. The development and widespread deployment of BECCS technologies is only likely to intensify the global demand and competition for biomass resource. This chapter focuses on evaluating the current and growing demands for biomass resource and how biomass resource demand may be impacted by widespread deployment of BECCS technologies. A discussion is provided of how biomass resource potential forecasts are developed and the specific approaches used to evaluate the availability of key categories of biomass resource, including energy crops produced specifically for the bioenergy sector; wastes and residues from ongoing processes such as agriculture and industry; the biomass resource potential from forestry and waste resources from a wide range of sources. Many studies have been undertaken that forecast the potential availability of the different categories of biomass resource at different geographic scales and over different timelines. A summary of global biomass resource potential forecasts is presented, highlighting the range of resources potentially available for the bioenergy sector and explaining why there are such variations in the results of different biomass resource studies. The key global biomass resource supply and demand regions are highlighted and current expected future biomass trade routes are discussed. The chapter will go on to highlight the sustainability implications of large-scale biomass resource production/mobilisation and the potential risks and limitations that need to be considered.

Biomass Energy with Carbon Capture and Storage (BECCS): Unlocking Negative Emissions, First Edition.
Edited by Clair Gough, Patricia Thornley, Sarah Mander, Naomi Vaughan and Amanda Lea-Langton.
© 2018 John Wiley & Sons Ltd. Published 2018 by John Wiley & Sons Ltd.

2.2   Biomass Resource Demand

Energy from biomass currently contributes approximately 10% to overall global energy supply. Two-thirds of this bioenergy is generated within developing countries and the remainder within the industrialised world (IEA Bioenergy, 2013). Bioenergy is an attractive energy option for all stages of economic development due to its flexibility, potential for integration within broad-ranging development strategies and the general acceptance that greenhouse gas (GHG) emissions from biomass resources are lower than those from conventional energy pathways. The IEA/IRENA Global Renewable Energy Policies and Measures Database (IEA and IRENA, 2017) confirms that more than 60 countries currently have national targets or policies supporting renewable energy. In countries such as those in Europe, biomass is expected to play a major role in contributing over 50% towards their renewable energy targets (AEBIOM, 2013).

Despite widespread increasing global demand for biomass, some of the regions with the greatest demand have comparatively low potential of biomass resource availability (Berndes et al., 2003; Rokityanskiy et al., 2007; Smeets and Faaij, 2007; Chum et al., 2011). Trade therefore has an important role to play, biomass being described as 'the most important renewable energy carrier worldwide' (Thrän et al., 2010). Short-term trends show that regions with strong economies and development will increase their requirement for imported biomass resource, whilst developing countries will continue their development largely reliant on fossil fuels (Hillring, 2006).

As a result of recent energy policies, Europe has become the prime market for the trade of biomass for energy, with more than 30% of biomass resource consumed in the EU in 2012 imported, and demand is forecast to rise by almost 50% between 2010 and 2020 (European Academies Science Advisory Council, 2012). Global demand for biofuels is expected to rise sharply, driven in part by biofuel mandates of different countries. Demand for wood pellets is forecast to increase threefold by 2020, as governments similarly offer renewable energy subsidies with the aim of using bioenergy to meet their renewable energy and carbon-reduction targets (Bottcher et al., 2012).

In summary, many developed countries and regions are set on a trajectory of increased utilisation of bioenergy pathways to meet energy demands. As their bioenergy-reliant renewable energy strategies are implemented, ever greater quantities of biomass resource will have to be produced, mobilised and likely imported to generate this energy.

2.3   Resource Demand for BECCS Technologies

Large-scale bioenergy plants are targeted by many countries as the pathway of choice for achieving renewable energy generation and GHG emission reduction mandates. In addition, the co-firing of biomass with fossil fuels may now be regarded as a mature technology as it is increasingly being implemented at power plants and through transport fuel blending across the world. Large-scale bioenergy plants have correspondingly large fuel demands and emissions profiles and, therefore, represent key opportunities for large emission reductions if appropriate BECCS technologies can be proven. Therefore, the biomass resource demands of future scenarios with a major focus on BECCS will likely reflect those of scenarios focusing on large-scale bioenergy generation.

Table 2.1 Forecast global bioenergy demand by 2030.

Demand sector	2010 (EJ/yr)	2030 (EJ/yr)	Growth (% per yr)	Growth (EJ/yr)
Buildings (traditional)	27	12	−4.1	−0.8
Buildings (modern)	8	13	2.6%	0.3
Industry and manufacturing	8	21	4.9%	0.7
Transport fuels	5	31	9.7%	1.3
Power and district heat generation	5	31	10.0%	1.3
Totals	53	108	3.7%	2.8

Source: Adapted from IRENA (2014).

The Global Renewable Energy Roadmap (REmap) (IRENA, 2016) developed by the International Renewable Energy Agency (IRENA) provides a pathway for increasing renewable energy generation to 2030. The REmap 2030 analysis indicates that bioenergy will become the most important renewable energy technology by 2030 if the renewable energy strategies of key countries are implemented in full – bioenergy accounting for 60% of global renewable energy generation and 20% of global primary energy supply (IRENA, 2014). Table 2.1 presents the forecast of global bioenergy demand for key sectors by 2030 reflecting the full implementation of renewable energy strategies. These bioenergy demand forecasts have been largely developed without considering future significant deployment of BECCS technologies, so any future pathways with increased focus on BECCS will likely result in increases in bioenergy demand in addition to the already steeply rising forecasts for 2030. The amount of biomass resource required to be grown, harvested and mobilised to meet the bioenergy demands will be correspondingly large.

2.4   Forecasting the Availability of Biomass Resources

Forecasting the availability of resources has arguably been undertaken for as long as goods and resources have been in demand and traded. The control, extraction and use of resources have been at the heart of all human history. However, over recent decades, a consensus has grown that our demand for resources has increased to the extent that it is now considered to be a limiting factor and even a threat to the functionality of economies and society (Behrens et al., 2007). As a consequence, resource modelling has become a prominent tool used by governments, businesses and even individuals to inform decisions, management actions and strategies. Resource forecasts have the potential to influence decisions and policies that, in turn can, influence economic trends and activities that similarly can highly influence society and the environment.

However, the vast array of resource types and characteristics means that no default modelling methodology can be universally applied. Resource modelling forecasts vary in both accuracy and relevance depending on the resource, methodology applied and analysis scale (Lara and Doyen, 2010). Irrespective of the chosen modelling approach, the fundamental aim is typically to quantify resource reserves through the science of understanding supply versus demand – future resource supply and demand relying heavily on estimations of resource availability.

The key question for the developers of resource models is whether past trends and experiences can be used to provide a guide for the future (Dunham et al., 1978). Unforeseen technological uncertainties, political priorities and numerous other influences impacting resource supply can render future resource-modelling scenarios vastly different from historical experience; therefore, making simple predictions based on the extrapolation of past behaviours and trends may be highly suspect (Rowse, 1986). Nevertheless, many forms of resource modelling are widely undertaken, with current analysis indicating two distinct trends (Behrens et al., 2007):

- There is a global decoupling between economic growth and resource extraction and use – indicating that on a global scale economic performance is becoming less resource dependent.
- At the same time, resource extraction in absolute terms is increasing in all regions globally – reflecting a trend clearly in contradiction with the concepts of sustainable development.

2.4.1 Modelling Non-Renewable Resources

By their very definition, non-renewable resources are finite, and, therefore, the availability and abundance of these resources will drive supply and demand forecasts. The application of Hotelling's Rule has been successfully proven through time to predict pathways of non-renewable resource extraction. The rule states that 'extraction trends through time will follow the most socially and economically profitable pathways, determined by net revenues and rates of interest increase; thus ensuring the maximisation of resource stock value through time' (Hotelling, 1931). Essentially, the extent that a given resource is available is driven by the quantities of the resource that may be economically and socially acceptable to extract. When demand for a resource increases and therefore its value increases, it becomes economically and socially acceptable to extract more and more resources previously deemed uneconomical/antisocial to obtain.

This concept is demonstrated in Figure 2.1, where the ratio of global fossil fuel consumption compared to the forecast number of years of remaining fuel reserves are plotted through time (1980–2006). Fossil fuel reserve and supply/demand trends are typically not forecast based on scientific knowledge of global availability, but on economic parameters – consumption and price. Fuel reserve forecasts fluctuate in relation to global economic conditions, whilst 'proven reserve levels' fall when prices are too low for fossil fuels to be recovered economically and increase when the economics deem that fuels are economically recoverable.

Figure 2.1 shows that the ratio of oil and gas consumption to forecast reserves has been relatively constant for decades, suggesting that reserves have been increasing at the same rate as that of consumption. At the same time, the coal consumption to forecast reserve ratio has been intermittently falling for decades. In reality, the relatively constant oil and gas ratios reflect new resource reserves becoming economically viable through technological advancements reducing the cost of extraction; political/economic incentives and the disclosure of information through time relating to reserve

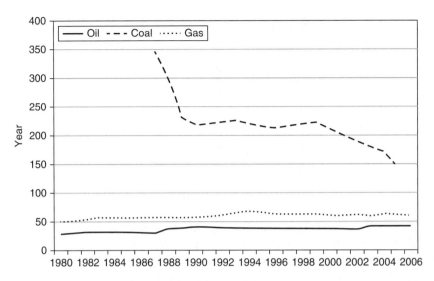

Figure 2.1 Ratio of global fossil fuel consumption to years of remaining reserve. *Source:* From Shafiee and Topal (2008). Reproduced with permission of Elsevier.

abundance and sizes in order to maintain maximised fuel prices (Shafiee and Topal, 2008). In contrast, the decrease in the ratio of coal consumption to reserves can be attributed to improvements in data rendering previous coal reserve projections inaccurate; changes in global coal demand, and thus the price of the fuel, influencing the levels of resource that may be economically extracted and the periods during which the ratio is shown to level off reflecting the introduction of technological advancements rendering more known reserves economically accessible.

2.4.2 Modelling Renewable Resources

The approaches and strategies for modelling renewable resources tend to differ from those for non-renewable resources, as a key additional dynamic exists – the resources are not finite. However, the focus of these models still revolves around projecting quantities of resource reserve and trends of supply and demand. Modelling biomass resources provides an excellent case study for the modelling of renewable resources.

2.4.2.1 Biomass Resource Modelling

> *The global biomass potential and its use to provide energy, cannot be measured; it can only be modelled.*
>
> Slade et al. (2011)

Biomass resource models, which typically aim to estimate the availability of biomass materials from a range of sources at a chosen geographic scale, vary greatly in complexity, but all aim to utilise data and information to estimate an aspect of the bioenergy sector's future development. Biomass availability is commonly categorised in terms of a hierarchy of 'potentials'; as Figure 2.2 illustrates, the size of the resource forecast

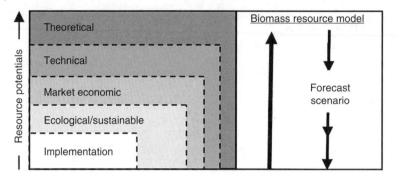

Figure 2.2 Biomass modelling resolution potentials. *Source:* Adapted from Welfle et al. (2014b).

progressively decreases as additional filters are applied. Depending on the type of analysis being undertaken and the choice of constraints applied, biomass resource forecasts may be extremely large or highly focused.

- *Theoretical modelling forecasts* reflect the quantity of biomass resource that can be grown/harvested/mobilised annually, limited only by fundamental physical and/or biological barriers. An example of this form of forecasting may be the maximum biological extent that different biomass crops may be grown in a given geography's climatic and land quality conditions. These forecasts are not typically useful for estimating realistic biomass production levels but can be used as comparative indicators of production potential.
- *Technical modelling forecasts* reflect the quantity of biomass resource that can be grown/harvested/collected based on technical constraints such as land area, land characteristic and agro-technological performance. An example of this form of forecasting may be the levels of straw resource that may be harvested from the land limited by the performance of the technology applied. These forecasts can be useful for evaluating how resource availability may change as a result of changing dynamics such as the introduction of technological advances.
- *Market economic modelling forecasts* reflect the quantity of biomass resource that can be grown/harvested/collected based on economic considerations, fundamentally driven by supply–demand curves. This is a highly variable modelling method as economic conditions and the markets relating to different biomass resource can fluctuate greatly over time. These forecasts are typically useful for evaluating how feasible bioenergy pathways may be in comparison to the cost of alternative energy pathways – largely driven by the cost of accessing the bioenergy feedstock.
- *Ecological or sustainable modelling forecasts* reflect the quantity of biomass resource that can be grown/harvested/collected based on ecological or sustainability constraints. These forecasts are typically useful to evaluate the levels of resource that may be sustainably mobilised without impacting the likelihood of sourcing more resource in the future. An example may be the levels of residue resource that may be sustainably harvested from a forest without causing impacts to the forest systems.

When comparing the outputs from different biomass modelling studies, it is important to consider the approach of the methodology and, specifically, the choice of constraints considered. Only the results derived from studies with comparable approaches can be realistically compared on a like-for-like basis.

A further important distinction that needs highlighting when comparing different biomass modelling studies are the concepts of 'biomass potential' and 'bioenergy potential'. 'Biomass potential' in many studies refers to a straightforward quantification of the amount (weight, volume, etc.) of a biomass resource – with further dynamics such as the moisture content (MC) of the biomass needing further consideration. Biomass potentials for different resources will typically be reported on a prescribed level of MC. At the same time, other studies may reference 'biomass potential' as the gross amount of energy contained within the biomass quantified, while the term 'bioenergy potential' typically reflects the extent that energy may be generated from a given quantity of biomass and the conversion efficiencies and losses of the chosen conversion pathway. These distinctions are not always clear, but should be investigated when comparing outputs from different biomass modelling studies.

2.4.3 Modelling Approaches – Bottom-Up versus Top-Down

As already discussed, there is no set design or approach for developing a biomass resource model; consequently, numerous methodological frameworks have emerged. The structure, analysis themes and overarching approach of the models play a crucial role in determining the types of output that the models generate and explain the (potentially high) variation in results from different models (Berndes et al., 2003). Biomass resource models can, however, be grouped into two distinct categories based on their analysis starting point. Models may be 'bottom-up' in that they are resource focused, aiming to analyse the availability of resources before determining the levels of bioenergy that these resource may be used to generate. Alternatively, models may be 'top-down' in that they have a target level of energy or bioenergy to be generated, before determining the levels of biomass resource that may be needed to generate this energy.

- *Bottom-up resource-focused models* typically aim to build an inventory of the different available biomass resources within a set geography, based on a function of supply. For example, this may reflect an evaluation of the extent that a certain area of land may be used for energy crops or forestry, or analysis of the state of industries and ongoing activities that may be providing wastes and/or residue resource. The results of resource-focused models are highly dependent on the boundary conditions, the methods of analysing biomass production systems and the number and types of resources modelled.
- *Top-down demand-driven models* focus on estimating the quantities of biomass required to meet various energy, biomass and renewable energy targets. Many demand-driven models are developed as part of wider energy-economic modelling studies as they can provide an indication of the competitiveness of bioenergy in comparison to conventional energy pathways. A key limitation of demand-driven models is that their assumptions are often aggregated and sometimes undefined – so although these studies can provide insights into the likelihood of increases in biomass utilisation, they often provide little insight into the size of the technical biomass potential.

2.5 Methods for Forecasting the Availability of Energy Crop Resources

Energy crops are those that are grown specifically as fuels for bioenergy and are typically produced within plantation systems that provide high resource outputs per hectare of land. Key categories of biomass grown specifically for bioenergy include varieties of short rotation coppice (SRC) such as willow; short rotation forestry (SRF) grown over short intervals such as eucalyptus; and high-yield energy grasses such as Miscanthus. The ultimate aim for the producers of energy crops is to maximise the growth output of energy crops from each harvest cycle, measured by either the crop yield per unit area of land or the quantities of bioenergy feedstock generated.

The extent that energy crops are produced is dependent on three key variables: the area of land available for energy crop production; the planting strategy determining how much of the land is utilised and what species are planted and finally the productivity yields achievable by the chosen species on the land. The extent that land is available for energy crops depends upon competing uses for the land, whilst the achievable crop productivity yields depend on many variables such as the fertility of the land and climatic conditions. The greatest competing use for land is for food production, whether it is arable land producing staple food crops, pasture land for livestock or land producing crops for animal feed. Therefore, changes to food systems can have large influences on the available land for alternative uses, such as for energy crop production.

Research has shown that technological improvements enhancing crop yields, population decreases and evolving diet preferences, such as reducing consumption of meat, has the potential for reducing the land required to balance food demands, therefore presenting opportunities for the bioenergy sector (Welfle et al., 2014b). There is also a series of interesting historical precedents; for example, during the 1990s around 67 Mha of arable land was removed from food production in the EU under the 'set-aside scheme' with the aim of improving the land (Boatman et al., 1999). This sudden abundance of available lands led to discussions in many European countries about the increased potential for energy crop production for bioenergy (Slade et al., 2011).

Much research has focused on the potential impacts and benefits of converting agricultural lands for energy crop production, in addition to the potential for using marginal and degraded lands, deforested/forest areas and extensive grasslands as used widely in the African savannah and Brazilian *Cerrado*. Questions remain related to the uncertainty in predicting future food needs and consequently what the priorities of different land uses should be – once land is converted into an energy crop plantation, the ease or even possibility of converting it back to its original uses may be uncertain.

Two modelling approaches are typically applied to evaluate the future availability of land. The 'availability factor' modelling approach focuses on categorising land in accordance with its current use and then developing an assumption of the fraction of each land category that may be suitable and/or available for energy crop production. These fractions may be determined by information such as agricultural surpluses, forestry and logging strategies or development plans. The availability factor approach can be simple and highly transparent but may not be able to capture all the dynamics of competing land demands and/or the spatial variations in yields. 'Land balance models' provide an alternative approach, where land areas are set aside depending on their suitability for different crops, the specific demands for food production and strategies and trends such as urbanisation and nature conservation – the remaining land area can then be identified as being potentially available for energy crop production. The land balance model approach can be highly sophisticated accounting for changing drivers such as food demand, climate change and demand for land. This approach has also been criticised for overestimating the availability of suitable land, and/or identifying lands that are already cultivated or have existing uses. A further land analysis approach increasingly used relies on mapping software such as geographic information systems (GIS) to generate maps identifying available land through overlaying of layers of excluded land categories. For example, Wicke (2011) combined data from the Global Land Cover Database with that of the World Database on Protected Areas in order to generate maps highlighting the areas and proportions of land that should be excluded from energy crop potential analyses.

Agriculture productivity yields are highly important assumptions when evaluating the potential for energy crops. Where yields can be increased agricultural lands may become increasingly available for energy crops; and similarly if energy crop species yields increase, then more resource may be produced by any given unit of land. Crop yields are essentially a function of the amount of sunlight received by the crop, the photosynthesis efficiency of the crop determining the amount of biomass that is produced and the proportion of the overall biomass that may be harvested (Monteith, 1977). The achievable yields for any given area of land will be determined by the complex interactions between agricultural management practices, local climatic conditions and plant physiology. A suitable balance of these variables is crucial for achieving high productivity yields; stresses to plants such as those caused by poor-quality soils or water scarcity can result in large reductions in the achievable yields. Productivity yields within energy crop potential analyses are often taken from databases that reflect typical crop yields achieved in different regions from different types of land. This yield data may be an extrapolation from measured case studies or sample plots, or may be model based, where complex empirical crop models have been developed to predict varying productivity yields.

2.6 Forecasting the Availability of Wastes and Residues From Ongoing Processes

Wastes and residues from ongoing processes such as agriculture or industry represent an abundant and robustly available resource for the bioenergy sector (Welfle et al., 2014a). These include resources such as straws from crop production; manure, slurry and litter from livestock and surplus organic material generated as part of industrial processes. These resources represent key opportunities for bioenergy as they sometimes have no alternative uses, for example, textile off-cuts from the clothing industry; or are produced in such volume that there is always surplus regardless of competing markets, as is the case in many circumstances where crop straw is generated.

Estimating the availability of these resources relies on developing an understanding of the resource's life cycle and how and why they are generated. The availability of residues for the bioenergy sector is typically a function of the amount of resource generated; the extent that the resource is harvestable/recoverable; the competition for the resource and the technical and economic feasibility of using the remaining available resource for bioenergy. Using crop straw as an example, it is possible to estimate the total amount of straw resource generated from an area of land through applying assumptions to calculate the quantity of straw produced in reflection of the quantity of food produced (for example, 1 t of usable wheat produced may generate a further 1 t of wheat straw). It is then possible to develop an assumption as to the extent that the total straw generated may be harvested from the land (it is not likely you will be able to collect every single stalk!). Of all the harvested straw, a proportion may be returned to the land/incorporated back into the soils, and a further proportion may have alternative uses such as for animal bedding. At the end of this calculation, if it is technically/economically feasible for the remaining straw to be used for bioenergy, this may be assumed to be the total available straw from a given area of land.

2.7 Forecasting the Availability of Forestry Resources

Biomass resource from forests for the bioenergy sector comes in many forms and from many sources. The majority of this resource is from virgin woods, classified as wood-based materials that have had no chemical treatments or the application of finishes. Their physical forms include barks, arboricultural residues, logs, sawdust, wood chips, pellets and briquettes. These resources typically have wide-ranging moisture contents (from oven dry to >60% in freshly harvested/green wood samples). Forestry resources may also include chemical contaminants taken up from the soil, water or air – although these will be generally quite low (Biomass Energy Centre, 2012). Resources harvested directly from forests have a wide range of physical shapes, sizes, characteristics and MC, so pretreatment is probably necessary to render the materials suitable for bioenergy systems.

Arboricultural residues (often called arisings) are a further key source of biomass potentially available for bioenergy. These residues include materials generated as a result of the management of municipal/private parks and gardens, tree surgery processes and the maintenance of infrastructure such as railways and road margins. Similar to forestry-sourced materials, virgin wood from arboricultural arisings will probably need pretreatment and processing, and due to the nature of the materials there may be a high proportion of brash material, bark and leaves. Chipping onsite is the typical processing method adopted.

Wood-based industries such as timber and paper mills and merchants provide another major potential source of forestry biomass for bioenergy. Again, these resources are typically highly variable including sawmill residues that are likely to include a high proportion of bark content; sawdust collected from different stages of processing with varying ranges of MC and all manner of other forms of wood residue generated at the different stages of industrial processes.

Harvesting biomass from forestry, whether it is through clear-felling an area, thinning sections of mature forests or collecting residues from the forest floor, can become a controversial topic. Tensions and differing opinions about using forestry resource for bioenergy stem from themes of biodiversity, conservation, carbon management and ecological economics. As a result, many biomass resource assessment studies exclude or severely limit the levels of natural virgin forests that may be used for biomass resource for energy.

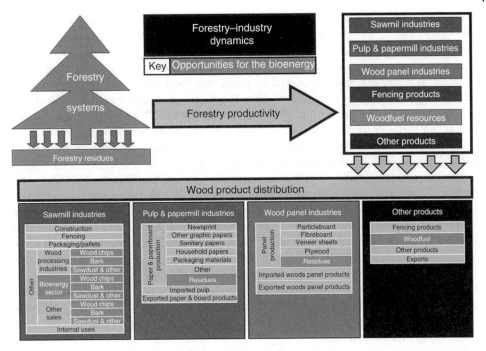

Figure 2.3 Example of the analysis boundaries and developed approach for evaluating forestry system and industry dynamics, as modelled for biomass resource assessment research (Welfle, 2014). (*See colour plate section for the colour representation of this figure.*)

Estimating the availability of biomass from actively managed forests within a given area depends on the complex dynamics between forestry systems and how they are managed in addition to the many industries that rely on forestry material. Forestry–industry dynamics and the flow of resource between them offer a wide range of opportunities for the bioenergy sector, both in terms of resources sourced directly from the forestry systems themselves and of residues resulting from industrial processes. Wood-based industry competes for resources, while generating residue resource opportunities for the bioenergy sector. These interactions may be modelled to evaluate the current state and future directions of forestry systems and their numerous links to industries. This process typically starts through analysing the current extent and productivity of forestry systems, followed by using sources such as the FAO's 'Global Wood Fibre Reports' (Williams and Duinker, 1997; FAO, 1998) to track how all resources produced by forests are used. National policies, strategies and targets can then provide an indication of how forestry systems and their management will evolve in the future. Figure 2.3 provides an example of the analysis boundaries and approach developed for modelling forestry system and industry dynamics within biomass resource assessment research (Welfle, 2014).

2.8 Forecasting the Availability of Waste Resources

Wastes represent a significant category of biomass resource that is increasingly being utilised as fuel for bioenergy. Key categories of waste for the bioenergy sector include municipal solid wastes (MSW) from our homes, places of work and businesses; food and other organic wastes and industrial wastes. Within any area, large levels of waste resource will be generated as part of everyday life. For example, within the food sector, wastes are generated at each step within the supply chain from growth/production through to disregarding of unwanted or unsuitable food at the end of the life cycle. Many categories of waste may be used directly within bioenergy systems such as anaerobic digesters or the generation of energy from waste (EfW) through incineration. Other categories of waste such as those from some industrial processes may need to be pretreated to become viable feedstocks or where there is high contamination and risk of hazard may be entirely unsuitable for bioenergy conversion.

The potential availability of different categories of waste resource for bioenergy can be forecast through modelling two key variables: waste generation trends and waste management strategies. Most countries and major municipal areas undertake studies and reporting procedures to develop an understanding of the types of waste generated in their area and will also have an idea of how these trends may evolve through time. Strategies are also likely to be developed that focus on the pathways for waste management, describing how the waste is managed, whether the focus is to recycle and reuse, landfill, export or on energy recovery. Through analysing any developed waste strategies and targets it is feasible to develop an understanding of the quantities of different types of waste that may be generated over time and the likely pathways for how different waste categories are managed.

Robust data on the generation and management of wastes are not always available. Assumptions can be developed, however, using top-down estimates of economic activity. The quantities of waste resource available for the bioenergy sector reflect: the level of economic activity; the assumed waste generation fraction for that activity; the waste recoverability fraction and the assumed level of waste that will be managed through an energy-recovered pathway.

2.9 Biomass Resource Availability

Many forecast energy scenarios indicate that bioenergy will have an increasingly prominent role in generating future heat, power and transport fuels for many countries at different stages of development. Continuing debate revolves around the question of whether there will be sufficient biomass resource to balance the demands of future bioenergy scenarios and significantly whether future biomass resource can be sustainably produced/mobilised. However, obtaining all-inclusive reliable estimates of total global biomass resource potential, let alone sustainable biomass resource potential, is currently a near impossible task. Unlike fossil fuel resources, no global companies currently undertake global assessments that gather up-to-date information on biomass production and consumption. Some general global estimates for biomass and bioenergy as developed by Boyle (2012) are shown in Table 2.2.

Estimates of biomass resource potential at different geographic scales are derived from biomass resource modelling research. Table 2.3 lists studies evaluating global biomass resource potential over short, medium and long forecast time frames. The modelling resolution and approach of these studies (as described in Section 2.4.2) is also listed, whilst their respective outputs are documented in Figures 2.4 and 2.5.

Table 2.2 Global biomass resource and bioenergy estimates.

Global biomass resource estimates	
Total mass of living matter	2000 billion tonnes (incl. moisture)
Total mass in land plants	1800 billion tonnes
Total mass in forests	1600 billion tonnes
Energy stored in terrestrial biomass	25 000 EJ
Net annual production of terrestrial biomass	400 000 Mt./year
Biomass energy consumption	56 EJ/year (1.6 TW)

Source: Data from Boyle (2012).

Table 2.3 Summary and characteristics of existing biomass resource modelling studies.

Time frame	Study references	Modelling resolution	Approach
2020	Bauen et al. (2004)	Theoretical + realistic	R-F
	FAO (2010)	Demand	D-D
2025	Sims et al. (2006)	Theoretical + technical	R-F
2030	IEA (2008)	Demand	D-D
	Moreira (2006)	Technical + economic	R-F
2050	Beringer et al. (2011)	Realistic	R-F
	Cannell (2003)	Theoretical + realistic	R-F
	de Vries et al. (2007)	Theoretical + technical + economic	R-F
	Erb et al. (2009)	Theoretical + technical + economic	R-F-L
	Field et al. (2008)	Theoretical	R-F-L
	Haberl et al. (2010), Berndes et al. (2003)	Geographic	R-F
	Hoogwijk (2004)	Geographic + technical + economic	R-F
	Dornburg et al. (2008)	Geographic	R-F
	Smeets et al. (2007), Thrän et al. (2010), WEA (2000), Wolf et al. (2003), Schubert et al. (2009), Hall et al. (1993), Fischer and Schrattenholzer (2001), Johansson et al. (1993)	Technical	R-F
2050– 2100	Hoogwijk et al. (2005)	Geographic + technical	D-D
	(Yamamoto et al., 2000)	Ultimate + realistic	R-F-L
2100	Yamamoto et al. (1999, 2001)	Ultimate + realistic	R-F-L
	Rokityanskiy et al. (2007)	Economic	R-F-L
Key	Approach	R-F	Resource-focused biomass models
		R-F-L	Resource-focused models driven by land-use assessments
		D-D	Demand-driven models

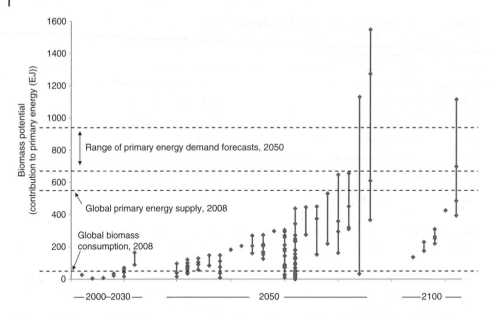

Figure 2.4 Range of global biomass resource estimates. *Source:* Adapted from Slade et al. (2011).

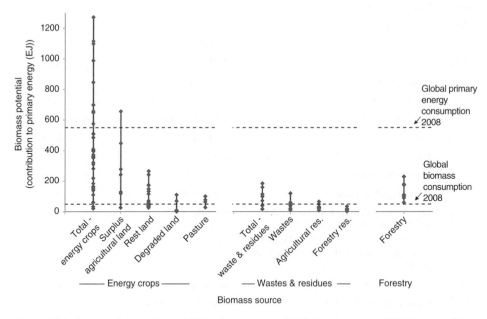

Figure 2.5 Indicative global ranges of different categories of biomass resource in 2050, from varying land uses. *Source:* Adapted from Slade et al. (2011).

Figure 2.4 presents the range of biomass resource potential forecasts for the studies listed in Table 2.3, each vertical line reflecting the range of resource potential estimates documented by a different study. The key message from Figure 2.4 is the large range of forecasts documented by different studies. For the 2050 time frame alone, a biomass resource potential forecast of zero is made by more than one study under their scenarios, whilst other studies have scenarios forecasting as much as >1.500 EJ/year by 2050 – reflecting approximately three times global primary energy supply in 2010 (British Petroleum, 2011). The wide range of biomass resource potential forecasts is a consequence of each study implementing different methodologies and assumptions and essentially each asking slightly different biomass resource questions – therefore, caution has to be taken when comparing outputs from different studies. A review of all the listed studies indicates that the majority of biomass resource potential forecasts fall between 50 and 300 EJ/year by 2050.

Figure 2.5 provides a further breakdown of the biomass resource potential forecasts into different categories of biomass resource. Again when interpreting and comparing the different studies' outputs, a degree of caution is required as key assumptions such as land-use categories are not consistently defined across the studies, and different constraints are applied when calculating the availability of different resource categories. However, Figure 2.5 highlights that energy crop resources present the greatest resource category globally by 2050; and potentially controversially, the majority of these will be produced on 'surplus agricultural lands', i.e. land that may otherwise be used for agriculture to produce food. Significant biomass resource potential compared to total global energy consumption can be provided by biomass wastes, agricultural residues and forestry.

2.10   Variability in Biomass Resource Forecasts

The variability in outputs from biomass resource assessments is a reflection of the specific assumptions and analysis methodologies adopted. There is value in developing and testing different analysis approaches when seeking to answer different bioenergy questions, while having broad-ranging output allows the identification of typical and outlier results. Biomass resource forecasts at the lower end of the spectrum may reflect scenarios where future biomass supply is comparable with (or less than) the current contributions from bioenergy; these may be restricted in the categories of resource analysed or the scale of geography assessed. At the upper end of the spectrum, forecasts may provide an indication of the upper technical limits of biomass resource potential to inform the limits of bioenergy potential. Mid-range results will likely reflect economic or sustainable levels of biomass resource that may be available to the future bioenergy sector without the need for extreme mobilisation/production strategies. Figure 2.6, developed by Slade et al. (2011), provides an example of analysis undertaken to evaluate the key variables that are indicative of the scale of biomass resource potentials developed by different studies.

All biomass resource models and assessments revolve around analysing the influence of different drivers. As such, the range of drivers listed within the literature that are identified as influencing biomass resource availability is extremely broad. Table 2.4 provides an overview of a number of these key drivers as identified and analysed in bioenergy research. Table 2.5 presents research undertaken by Welfle et al. (2014a)), where a series of supply chain drivers in the United Kingdom were analysed to evaluate their influence on determining the availability of different categories of biomass resource for the bioenergy sector.

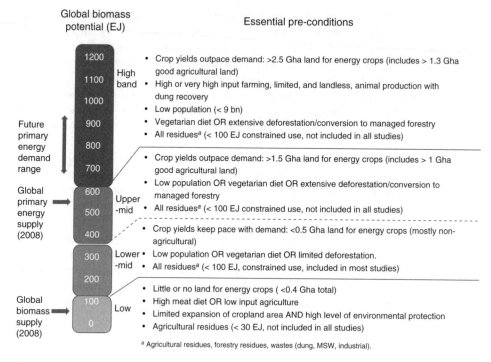

Figure 2.6 Typical assumptions driving variations in outputs from biomass resource assessments. *Source:* Adapted from Slade et al. (2011).

Table 2.4 Summary of key supply chain drivers influencing the availability of biomass resource for the bioenergy sector.

Categories	Supply chain drivers	
Economic and development drivers	● Population	● Industry productivity
	● Resource import/export	● Gross domestic product
	● Economic development	● Rural development
Infrastructure targets	● Energy system structure	● Supply chain development
	● Energy generation plant	
Physical and climate drivers	● Land-use change	● Flood protection
	● Climate change	● Soil degradation
	● Nature conservation	● Water availability
Food drivers	● Food demand and consumption	● Diet change
	● Calorie consumption	● Agriculture productivity yields
Resource mobilisation drivers	● Technological advances	● Residue generation
	● Forest system productivity	● Forestry residues collection
Resource demand drivers	● Resource use by industry	● Demand for wood fuel
	● Demand for round wood	● Competing demands
Policy drivers	● GHG emission targets	● Renewable and bioenergy targets
	● Energy efficiency targets	● Fuel security
	● Energy consumption targets	● Support policies and mechanisms

Source: Adapted from Welfle et al. (2014a).

Table 2.5 Evaluation of the influence of key drivers on determining the availability of biomass resource for the UK bioenergy section.

Ranking	Influencing drivers	
High ranking *Drivers and resources with the greatest influence*	• Focus of waste management strategies	• Extent of land dedicated to energy crop growth
Medium ranking *Drivers and resources with medium influence*	• Crop and agriculture productivity yields • Population change • Changes in built-up land area	• Waste generation trends • Utilisation extent of agricultural wastes and residues • Forestry expansion and productivity
Low ranking *Drivers and resources with the least influence*	• Food commodity imports • Food commodity exports • Wood-based industry productivity • Imports of forestry product	• Exports of forestry product • Utilisation extent of forestry residues • Utilisation of industrial residues • Utilisation of arboriculture arisings

Source: Adapted from Welfle et al. (2014a).

2.11 Biomass Supply and Demand Regions, and Key Trade Flows

There are multiple dynamics that collectively influence the biomass resource potential of a country or region, and further drivers that determine whether the country or region may be a net importer or exporter of resource. The key variables driving supply are typically the area and characteristics of the land and climate, types of industry present and political policies and mechanisms that promote the production/mobilisation of resource. Demand of countries and regions, on the other hand, is typically driven by the nature of policies and mandates that either target or avoid the use of bioenergy, in addition to the characteristics of competing industry with existing biomass demands.

The global trade of biomass resources for energy end uses is driven by the following key factors: the prices of fossil fuels (especially oil); the implementation of policy mandates aimed at increasing renewable energy generation and carbon reduction and financial mechanisms supporting bioenergy pathways (Junginger et al., 2011a). Figure 2.7 shows the major biomass trade flows around the world. Brazil is the major exporter of bioethanol, predominately to Europe, the United States and Japan. The United States, Argentina, Indonesia and Malaysia are the largest exporters of biodiesel, again mainly to Europe (Lamers et al., 2011). The major exporters of wood pellets are Canada, the United States, Russia and the Balkan States, with Europe being the largest importing region (Bradley et al., 2009; Spelter and Toth, 2009; Sikkema et al., 2011).

2.11.1 Trade Hub Europe

As a result of the renewable energy and carbon-reduction targets of its countries, Europe has emerged as the world's primary trading hub and demand region for each of the major categories of traded biomass resources. With all major biomass trade flows

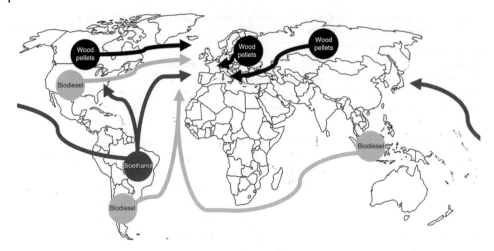

Figure 2.7 Biomass supply and demand regions and key biomass trade flows for energy end uses. *Source:* Welfle (2017). Reproduced with permission of Elsevier.

headed towards Europe, this may present both positive and negative issues for the bioenergy sectors of specific European countries. Being part of the central hub of biomass trade with increasingly established trade routes will present resource opportunities for the growth of the bioenergy sector. However, the competition for resource throughout Europe is only likely to intensify, with each European country having to increasingly compete with its neighbours who each have their own equally large and growing demands for resources (Welfle, 2017).

2.11.2   Bioethanol – Key Global Trade Flows

The United States and Brazil are the two leading producers of ethanol, accounting for over 85% of the global market. Brazil is the largest exporting country with the United States and the EU being the greatest importers, followed by Canada and Japan (Junginger et al., 2011a). Within Europe, the United Kingdom and Sweden are amongst the largest importers of ethanol, of which 32% in 2009 was estimated to be used to produce transport biofuels (Hewitt, 2011).

It has been estimated that between 40 and 51 PJ of ethanol fuels were traded globally in 2009 (Lamers et al., 2011). The uncertainty range of these data is due to the imprecise accounting methods for ethanol end uses. For example, traded ethanol may be accounted under multiple trade codes depending on whether it will be used within transport fuels, industrial processes or by beverage industries. The improper utilisation of trade codes and intra-trading and change of use once landed within a country also adds to the uncertainty (Junginger et al., 2011b).

2.11.3   Biodiesel – Key Global Trade Flows

The global production of biodiesel has been increasing exponentially, with 20 PJ annual global production in 2000, rising to around 565 PJ by 2009 (Lamers et al., 2011). Largely driven by renewable energy and carbon-reduction mandates, Europe now has the world's most developed biodiesel industry, accounting for two-thirds of global production in 2011 (Junginger et al., 2011a). Germany, France, Spain and Italy are the leading producers (EurObserv'ER, 2009), with rapeseed oil produced within Europe being the primary feedstock, providing two-thirds of total production. Imported feedstocks such as soybean oil, palm oil and, to a lesser extent, further rapeseed oil represent the final third (USDA, 2013). Other major biodiesel producers include the United States, Argentina and Brazil. More than 95% of global biodiesel exports are directed towards Europe (Carriquiry et al., 2010), almost half of which are consumed in Germany and France (Junginger et al., 2011b).

2.11.4 Wood Pellets – Key Global Trade Flows

The global wood pellet market has also been growing exponentially over recent years with levels currently comparable to those of both bioethanol and biodiesel in terms of traded volumes (Junginger et al., 2011a). Driven by mandates and incentives to increase renewable energy generation and reduce carbon emissions, Europe is the world's largest wood pellet market (USDA, 2013). Wood pellets are replacing, or being co-fired with, coal to either generate electricity (Bottcher et al., 2012) or to produce heating or cooling with reduced GHG emissions (USDA, 2013). To balance this increasing demand, Europe has become the largest importer of pellets with about 3.4 Mt imported in 2010, of which about half can be assumed to have been intra-traded inside the EU (Sikkema et al., 2009). The United Kingdom and specifically the United Kingdom's biomass–coal co-firing power stations are the destination for vast quantities of imported wood pellets (Woods et al., 2011). The European Biomass Association (AEBIOM) (AEBIOM, 2013) forecasts that Europe's consumption of wood pellets will continue its steep rise, estimating increases from 105 PJ in 2008 to as much as 837–1340 PJ in 2020. North American exports are targeted by Europe to provide the majority share of these near- and mid-term demands (Junginger et al., 2011a).

North America is the largest exporter of wood pellets, the majority of these going to Europe. Canada is the dominant exporter, also supplying pellets to the United States. In Scandinavian countries, despite there being significant domestic pellet production, increases in demand largely outpace production, leading to increasingly large volumes of pellets imported from the Baltic regions and Russia (USDA, 2013). Further minor biomass pellet trade flows to the EU are also expanding from Australia, Argentina and South Africa (AEBIOM, 2013).

2.11.5 Wood Chip – Key Global Trade Flows

In recent times the largest wood-chip-producing nations (based on 2013 data) have been Canada (25%), China (13%), South Africa (5%), Sweden (5%) and the United States (5%) (FAO, 2013). These countries, incidentally and relevantly, also have large paper and pulp production industries. This close relationship means that future wood-chip trade trends and opportunities will likely be influenced by the ongoing shift in the production of pulp and paper industries from the northern to the southern hemispheres (IEA Bioenergy, 2012).

For many years Japan has been the largest importer of wood chips, at times attracting over 50% of the global traded resource. Although the growing wood chip demands of industries in China is expected to see it take over as the predominant global importer within the next few years (Aguilar and Saunders, 2010). As a consequence of recent energy policies, price competitiveness and a strong forestry sector, Europe is also becoming an increasingly major market for the trade of wood-chip resource (Lamers et al., 2012). Europe produces large quantities of wood chips that are largely intra-traded within the continent. Sweden, Finland, Austria and Italy are current major importers of resource, whilst Germany, Latvia and Estonia are current major exporters. Future wood-chip-exporting markets are also expected to grow in Russia, Uruguay, Brazil and Canada (Alakangas et al., 2007).

2.12 Global Biomass Trade Limitations and Uncertainty

There are many limitations and uncertainties that need attention if we are to increase biomass production, processing and movement on a global scale with a view to developing global trade markets.

2.12.1 Technical Barriers

Technical barriers to the increased production and trade of biomass resource include specific performance characteristics of the traded biomass or processed fuels and the requirement that they meet potentially high specifications of the destination bioenergy system or end market. Technical performance levels essentially provide a benchmark for the suitability of available resources for trade for different markets. Technical standards are the key mechanisms applied to ensure that these minimum resource and fuel requirements are met. Adherence to these standards reduces the risks to potential importers, whilst also ensuring the safety of potential customers, as in the case of biofuels; for example, there are maximum proportions of either bioethanol or biodiesel that may be blended with fossil fuels to ensure that the imported biofuels comply with regulations and meet technical performance requirements (Junginger et al., 2011a).

Whilst these technical requirements may be an insurmountable barrier for developing suppliers needing to trade with new markets, they also provide an important constraint reducing the risk for importing markets. Technical barriers therefore reduce the overall extent that resources may be available for trade, but in the long term act as a driver for increasing the overall quality and performance of the resources traded globally.

2.12.2 Economic and Trade Barriers

The economic and financial mechanisms applicable to global trade markets are perhaps the most essential drivers of, and sometimes barriers to, the growth of biomass trade, both enabling and constraining resource imports, exports and utilisation (Junginger and Faaij, 2006).

Export subsidies or tariffs are a key example of a potential economic trade barrier that may impact the trade of biomass commodities, influencing the extent that countries develop export markets. The influence of export economic mechanisms on the global biomass markets can be seen in the case study of Argentina, where a differential in export tax was set with 20% tax for finished biodiesel products compared to 32% tax for the feedstocks of biodiesel production (CADER, 2009). This differential in tax incentivised the domestic production and export of finished biodiesel rather than the export of raw feedstocks for biodiesel generation elsewhere (CADER, 2009). This distortion, although generating growth in the Argentinean bioenergy sector, has created a disadvantage for other biodiesel-producing regions that may rely on the importation of raw feedstocks (EBB, 2009).

Import tariffs may either incentivise the importation of biomass resources or render it uneconomical. Biodiesel import tariffs are typically lower than for bioethanol (Steenblik and Simón, 2007), whilst countries typically have no tariffs for the importation of solid biomass such as pellets (Junginger et al., 2011a).

2.12.3 Logistical Barriers

A further potential major barrier to the growth of biomass trade markets are the logistics required to transport the resource around the world. If we are to develop mature global biomass trade markets and established supply chains, it is necessary to overcome these barriers. The majority of bioenergy systems require feedstocks with specific and uniform characteristics, such as moisture content, chemical composition or even physical size, whilst cost-efficient transportation and favourable energy characteristics are required to ensure the viability biomass trade (Junginger et al., 2011a). To achieve these fuel specification levels and to facilitate the actual transport of the resource, the development of infrastructure is required. A large proportion of resources come from key regions of the world where infrastructure lags far behind ambition (Junginger and Faaij, 2006).

2.12.4 Regulatory Barriers

Regulatory systems can be highly influential in determining both the types and the extent that different resources may or may not be imported from a given country's bioenergy sector. Regulatory systems can represent a major barrier or accelerator particularly for the development of the renewable energy sector (Bradley et al., 2010). Changing regulations that incentivise or prioritise different bioenergy pathways, and thus the resources required to fuel them, can strongly influence the resources traded on global markets. For example, changing the technology focus of feed-in-tariffs either incentivises the utilisation of particular resources in bioenergy pathways, or may otherwise render them uneconomical. Technical regulations requiring sanitary or phytosanitary measures, for example, can also place potential constraints on the trade and movement of resources across international borders. These are typically implemented to target the reduction risks associated with biological impacts, such as the spread of pests or pathogens around the world (Junginger et al., 2011a). Meeting these requirements is often straightforward through chain-of-custody tracking and assessments, although on occasion imported resources, especially from developing regions, are rejected by demand regions. For imports brought into markets such as the EU, thorough inspections are widespread, especially for under-bark untreated category imports (Steenblik and Simón, 2007).

Minimum sustainability criteria are also imposed on imported resources; compliance criteria include ecological, land-use, competing market and food system impacts, as well as embodied energy and GHG emission thresholds (Junginger and Faaij, 2006). The implications of sustainability barriers for the global trade of biomass are discussed further in the following sections.

2.12.5 Geopolitical Barriers

Geopolitical instabilities present risks and barriers to all international trade. The fluctuations in the price of fossil fuels following major global events or conflicts provide a perfect example of how the price of energy commodities can be highly influenced by security of supply. As many potential biomass-exporting regions of the world are at earlier stages of development than those where the resources are being traded to, there is always a risk of political instability, thus trading with these regions and having to invest in enabling infrastructure only adds further potential barriers to trade (Bradley et al., 2010). Global biomass markets are still developing, so the potential supply risks due to geopolitics events are deemed to be less extreme and more manageable than with fossil fuels, although they may represent major potential uncertainties and limitations as the markets mature (Carbon Trust, 2009).

2.13 Sustainability of Global Biomass Resource Production

Bioenergy differs from all other renewable and conventional energy pathways, as in many cases it is directly tied to the farms, forests and ecosystems from which biomass resources and feedstocks are produced and extracted. This close association within bioenergy systems and supply chains creates the potential for wide-ranging environmental and social impacts that can be both positive and negative. Many of these major potential impacts are discussed in the following sections.

2.13.1 Potential Land-Use Change Impacts

A close relationship exists between bioenergy and demand for finite land supplies. Therefore, the increased use of bioenergy and the requirement to produce more resources to balance demand represents a growing area of stress for land. Direct and indirect changes in land use and land management practices can potentially result in increased GHG emissions that could reverse any savings achieved through bioenergy utilisation in the first place (Institute for European Environmental Policy, 2011).

There is a growing consensus that in general terms the increased demand of resources for the bioenergy sector will likely accelerate intensification of agricultural productivity, leading to the widespread conversion of forestry and grassland systems to arable lands (Croezen et al., 2010). Land-use change emissions will be generated as forests, grasslands and other land categories are cleared in order to produce resources bound for bioenergy. Further emissions may be generated if agricultural lands are used to produce bioenergy feedstocks, and as a result food production is displaced to other areas where the land have to be first cleared – resulting in indirect land-use change (EESI, 2014). However, it is hard to identify where indirect land-use change is occurring, let alone its potential impacts. Impacts and GHG emissions arising from displaced lands may take place in well beyond the country or region where the direct land-use change initially took place, but needs to be accounted (Fritsche, 2012). Thus, impacts attributed to indirect land-use change can be difficult to quantify. Current methods for estimating indirect land-use change rely on complex analysis using macro-economic, econometric and biophysical models (IFEU, 2009).

2.13.2 The 'Land for Food versus Land for Energy' Question

Probably the greatest perceived issue for bioenergy is the conflict between using land for the production of food commodities and for the production of biomass resources for energy. By far the most heated debates around this issue have focused on the production of crops such as maize, wheat, sugarcane and soy to be used to produce biofuels to substitute for petrol and diesel. The complexity of monitoring this type of shift in production and use is increased by the fact that many traditional food commodities are also being widely used as bioenergy feedstocks (EESI, 2014).

In many countries, the production of biofuels is highly supported by financial mechanisms such as subsidy regimes; consequently, in the 'food versus fuel' debate, biofuels have been viewed increasingly negatively. Key arguments against biofuels are about increasing competition for land, changes in food prices and local food security.

Increasing the demand for agricultural land will also undoubtedly lead to increased pressure to convert pasture and forest lands to arable production. This land-use change may result in potentially significant GHG impacts, if, for example, newly exposed carbon-rich soils oxidise. These emissions have the potential to negate many of the potential environmental benefits provided as the rationale for supporting the development of biofuels in the first place. More extreme claims have been made that extensive production of feedstocks for biofuels has led to famine, destruction of ecosystems and soils and depletion of water resources and is responsible for local and regional spikes in food prices (Eide, 2008).

To place this issue within the context of overall global energy consumption, in 2011 biofuels provided ~3 EJ (<1%) of primary energy (IPCC, 2011), and the food versus fuel issue only represents one aspect of a much broader renewable energy and bioenergy debate. An alternative argument envisages that increases in agricultural productivity, boosted in part by research and development in biofuels, may increase the availability of land for energy crop production alongside for food commodities. Much research has also focused on increasing the productivity of energy crops on marginal and even fallow lands, reducing competition for land for food and limiting deforestation (Rathmann et al., 2010).

2.13.3 Potential Social Impacts

Potential societal impacts of bioenergy production systems are also frequently overlooked, with the most important issues related to food security. Land is a limited resource, and, therefore, the production of biomass itself may have impacts on a region's food productivity – this may present fundamental implications for local populations reliant on their region's land for food (Fritsche, 2012). Further localised impacts on food availability may have the effect of driving up the price of food commodities as demand for land increases. Negative food security impacts are especially prevalent for net food-importing countries, and specifically net food-purchasing households which do not produce food themselves are therefore highly vulnerable to any changes in the price of key staple food commodities (GBEP, 2011).

A key issue in many of the regions seeing an expansion in their biomass production as part of large-scale agro-industries is land ownership. This includes scenarios where land owners come under pressure to sell/lease their lands to large organisations wanting to produce biomass resource on a large scale or where individuals or communities are located adjacent to large land holdings producing biomass for the energy sector. The issue of land ownership and access becomes particularly acute in the context of marginal and degraded lands – these areas are often not privately owned, but are frequently extensively used by the rural poor (Schubert et al., 2009). Grazing lands are a perfect example – they may be productive during the wet season, but look barren, unmanaged and ownerless during the dry season, causing them to be classified as degraded (Ariza-Montobbio et al., 2010). From a central government perspective, working these lands and bringing them into production contributes to growing the economy; they may be therefore be allocated to/infringed upon by large agro-industry organisations, generating problems and hardship for the individuals previously reliant on the land.

2.13.4 Potential Ecosystem and Biodiversity Impacts

Increased use of land for any purpose has the potential to create stresses for biodiversity and ecosystems. Potential impacts from increased cultivation of land for biomass production can range from the complete transformation and potential destruction of ecosystems, to the extraction of selected resources leading to the gradual degradation in the health of the ecosystem (ESA, 2010).

In certain global regions where unsustainable practices occur, the loss of ecosystems continues to be the greatest impact of increased agriculture, unsustainable forestry management and increased biomass resource production (Fritsche, 2012). The protection of 'biodiversity zones' is now widely recognised internationally as being an insufficient mechanism for halting the destruction and degradation of ecosystems, with continuing exploitation at the fringes of these zones resulting in gradual impacts on the wider ecosystem. A further key potential impact of bad practices is on soil systems. Unsuitable and unmanaged cultivation practices potentially lead to erosion and loss of topsoil, in addition to the risk of soil compaction. Soil systems represent vast sinks of organic carbon, which, if released through degradation or erosion of the soil structures, will create large carbon emission impacts potentially far exceeding any benefits achieved through sustainable bioenergy generation (Fritsche, 2012).

It is therefore imperative that specific unsustainable activities related to biomass resource cultivation are curbed, especially the unregulated expansion of activities and use of unsustainable land management practices. In contrast, it is also acknowledged that biomass resource production can actually provide positive biodiversity benefits through appropriate planning and management of production systems (Institute for European Environmental Policy, 2011).

2.13.5 Potential Water Impacts

The potential impact of crop production on water systems is often overlooked in discussions. Globally, agriculture accounts for about 70% of all freshwater use (FAO, 2007);

therefore, in simplistic terms, agriculture around the world already places significant demands on water resources, and increasing the production of biomass risks inevitable further stresses. Given that some of the regions associated with greatest potential biomass production already have water availability concerns, anything that increases water stress is likely to exacerbate existing problems. It is essential to ensure that cultivation for biomass production remains in line with water availability and avoids the use of chemicals that can impact water quality if conflict with other demands for water are to be averted (Institute for European Environmental Policy, 2011). As with all forms of agriculture, there is a constant drive to enhance productivity yields. The close relationship between crop yields and transpiration means that upper levels of crop growth will only be achieved when there are no restrictions to water availability (Legg, 2005). When producing crops there is also an important distinction between surviving and thriving: if an annual crop survives a period of drought only to produce a negligible yield, then its survival may be of little benefit (Sinclair et al., 2004), although for perennial crops survival in times of drought may prove a considerable advantage in avoiding replanting over a longer time frame. Depending on the location and the character of the land and climate, it is important to identify carefully which types of crops to cultivate based on their suitability for the climate and whether they are likely to require additional irrigation beyond expected precipitation.

2.13.6   Potential Air-Quality Impacts

Potential impacts on air quality represent a further often overlooked sustainability issue for the bioenergy sector. There is potential for significant air pollution where biological materials are combusted without the use of appropriate treatment technologies to limit particulates (Institute for European Environmental Policy, 2011). Although particulate controls have vastly improved over recent years, in many places around the world small particulates remain a key concern for local air quality, especially where small-scale bioenergy systems or traditional combustion practices are utilised (DEFRA, 2007). Any air-quality issues resulting from increases in bioenergy generation through widespread BECCS deployment are still unknown so will require research and regulatory attention.

2.14   Conclusions – Biomass Resource Potential and BECCS

Future global demand for biomass resource for energy end uses is rapidly increasing and is expected to continue to grow as bioenergy forms a major constituent of many countries' energy strategies around the world. Much of the increase in resource demand is linked to large-scale bioenergy systems and, in particular, power plants constructed or converted for biomass co-firing with fossil fuels, which are likely to be the primary target for BECCS technologies. A large number of studies focusing on different geographic scales, time frames and categories have forecast the future availability of biomass resources. Variability in these forecasts is the result of different analysis methodologies and assumptions: each study essentially asks a slightly different 'biomass resource question'. Scenario outputs from some studies do highlight that there may be no or extremely low biomass resource potentially available for bioenergy by 2050, although the majority of scenario outputs from studies indicate significant resource potential. Energy crops, wastes, agricultural residues and forestry resources are highlighted as the key categories of biomass resource that will be available for the future bioenergy sector.

Many studies suggest that energy crops produced specifically for bioenergy from 'surplus agricultural lands' are the category with greatest global potential by 2050. This is a highly controversial scenario as the use of agricultural lands to produce anything other than food commodities will always come up against opposition. The widespread production/mobilisation of biomass resources will likely have many such sustainability limitations and implications, including risks associated with land-use change, water impacts, social impacts, conservation and biodiversity impacts. These sustainability impacts need to be analysed fully and steps taken to ensure that biomass resource produced/mobilised to meet growing future demands do not result in environmental, social and economic impacts in their source regions. Also that the targeted bioenergy pathways and use of BECCS technologies provide genuine emissions reductions when the whole life cycle of biomass resource and bioenergy conversion is taken into account.

References

AEBIOM (2013). *Forest Sustainability & Carbon Balance of EU Importation of North American Forest Biomass for Bioenergy Production*. Brussels: European Biomass Association.

Aguilar, F. and Saunders, A. (2010). Policy instruments promoting wood-to-energy uses in the continental United States. *Journal of Forestry* **108**: 132–140.

Alakangas, E., Heikkinen, A., Lensu, T., and Vesterinen, P. (2007). *Biomass Fuel Trade in Europe*. Jyväskylä, Finland: VTT Technical Research Centre of Finland.

Ariza-Montobbio, P., Lele, S., Kallis, G., and Martinez-Alier, J. (2010). The political ecology of *Jatropha* plantations for biodiesel in Tamil Nadu, India. *Journal of Peasant Studies* **37** (4): 875–897.

Bauen, A., Woods, J., and Hailes, R. (2004). *Bioelectricity Vision: Achieving 15% of Electricity from Biomass in OECD Countries by 2020*. London: Imperial College London.

Behrens, A., Giljum, S., Kovanda, J., and Niza, S. (2007). The material basis of the global economy worldwide patterns of natural resource extraction and their implications for sustainable resource use policies. *Ecological Economics* **64**: 444–453.

Beringer, T., Lucht, W., and Schaphoff, S. (2011). Bioenergy production potential of global biomass plantations under environmental & agricultural constraints. *GCB Bioenergy* **3** (4): 299–312.

Berndes, G., Hoogwijk, M., and Van Den Broek, R. (2003). The contribution of biomass in future global energy supply. *Biomass & Bioenergy* **25** (1): 1–28.

Biomass Energy Centre (2012). *Resource availability*. http://www.biomassenergycentre. uk/portal/page?_pageid=74,393180&_dad=portal&_schema=PORTAL (accessed 28 November 2017).

Boatman, N., Stoate, C., Gooch, R. et al. (1999). *The Environmental Impact of Arable Crop Production in the European Union: Practical Options for Improvement*. Loddington: Allerton Research and Educational Trust.

Bottcher, H., Frank, S., and Havlik, P. (2012). *Biomass Availability & Supply Analysis*. Laxenburg: International Institute for Applied Systems Analysis.

Boyle, G. (2012). *Renewable Energy*, 3rde. Oxford: Oxford University Press.

Bradley, D., Diesenreiter, F and Tromborg E (2009). *World Biofuel Maritime Shipping Study. IEA Bioenergy Task 40.*

Bradley, D., Hektor, B. and Schouwenberg, P. (2010). *World Bio-Trade Equity Fund Study. IEA Task 40 Bio-Trade.*

British Petroleum (2011). *British petroleum statistical review of world energy.* www. bp.com/sectionbodycopy.do?categoryId=7500&contentId=7068481 (accessed 28 November 2017).

CADER (2009). *International Production Rankings, Increasing Production Levels, & Ongoing International Commercial Disputes.* Buenos Aires: Cámara Argentina de Energías Renovables.

Cannell, M. (2003). Carbon sequestration & biomass energy offset: theoretical, potential & achievable capacities globally, in Europe & the UK. *Biomass & Bioenergy* **24** (2): 97–116.

Carbon Trust (2009). *Biomass heating - a practical guide for potential users.* www. carbontrust.co.uk/SiteCollectionDocuments/Various/Emerging technologies/Current Focus Areas/Biomass Heat/Biomass end user guide.pdf (accessed August 2011).

Carriquiry, M., Dong, F., Du, X. et al. (2010). *World Market Impacts of High Biofuel Use in the European Union.* Ames, IA: Iowa State University.

Chum, H., Faaij, A., Moreira, J. et al. (2011). *Bioenergy.* In: *IPCC Special Report on Renewable Energy Sources & Climate Change Mitigation*, 46–55. Geneva: Intergovernmental Panel on Climate Change.

Croezen, H., Bergsma, G.C., Otten, M.B.J., and van Valkengoed, M.P.J. (2010). *Biofuels: Indirect Land use Change & Climate Impact.* Delft, Netherlands: CE Delft.

DEFRA (2007). *UK biomass strategy.* www.decc.gov.uk/assets/decc/what we do/uk energy supply/energy mix/renewable energy/explained/bioenergy/policy_ strat/1_20091021164854_e_@@_ukbiomassstrategy.pdf (accessed 28 November 2017).

de Vries, B., van Vuuren, D., and Hoogwijk, M. (2007). Renewable energy sources: their global potential for the first-half of the 21st century at a global level: an integrated approach. *Energy Policy* **35** (4): 2590–2610.

Dornburg, V., Faaij, A., Verweij, P. et al. (2008). *Biomass Assessment: Assessment of Global Biomass Potentials & their Links to Food, Water, Biodiversity, Energy Demand & Economy: Main Report.* Amsterdam: Netherlands Research Programme on Climate Change.

Dunham, K., Manners, G., Govett, M.H., and Govett, G.J.S. (1978). Resources policy conference 78 discussion report: session 1 – future supply & demand for resources. *Resources Policy* **4** (3): 210–211.

EASAC (2012). *The Current Status of Biofuels in the European Union, their Environmental Impacts & Future Prospects.* Brussels: European Academies Science Advisory Council.

EBB (2009). *Restoring a Level-Playing Field with Argentine Biodiesel Producers.* Brussels: European Biodiesel Board.

EESI (2014). *Biomass & Land use.* Washington, DC: Environmental and Energy Study Institute.

Eide, A. (2008). *The Right to Food and the Impact of Liquid Biofuels (Agrofuels).* Rome: http://publish.uwo.ca/~dgrafton/righttofood.pdf (accessed 30 November 2017.

Erb, K. et al. (2009). *Eating the Planet: Feeding & Fuelling the World Sustainably, Fairly & Humanely - a Scoping Study.* Vienna: Institute of Social Ecology, Klagenfurt University.

ESA (2010). *Biofuels: Implications for Land use & Biodiversity.* Washington, DC: Ecological Society of America.

EurObserv'ER (2009). *Biofuels Barometer.*

FAO (1998). *Global Fibre Supply Model.* www.fao.org/docrep/006/X0105E/X0105E00. HTM (accessed 30 November 2017).

FAO (2007). *Coping with Water Scarcity: Challenge of the Twenty-First Century.* Rome: Food and Agriculture Organization.

FAO (2010). *OECD-FAO Agricultural Outlook 2010–2019.* Rome: Food and Agriculture Organization.

FAO (2013). *FAO Statistics Forestry Datasets.* Rome. http://faostat.fao.org/site/626/ DesktopDefault.aspx?PageID=626#ancor (accessed 30 November 2017).

Field, B., Campbell, J., and Lobell, D. (2008). Biomass energy: the scale of the potential resource. *Trends in Ecology & Evolution* **23** (2): 65–72.

Fischer, G. and Schrattenholzer, L. (2001). Global bioenergy potentials through 2050. *Biomass & Bioenergy* **20** (3): 151–159.

Fritsche, U. (2012). *Sustainable Bioenergy: Key Criteria & Indicators.* Darmstadt: Oeko-Institut.

GBEP (2011). *The GBEP Sustainability Indicators for Bioenergy.* Rome: Global Bioenergy Partnership.

Haberl, H., Beringer, T., Bhattacharya, S.C. et al. (2010). The global technical potential of bio-energy in 2050 considering sustainability constraints. *Current Opinion in Environmental Sustainability* **2** (5–6): 394–403.

Hall, D., Rosillo-Calle, F., Williams, R.H. et al. (1993). *Biomass for Energy: Supply Prospects,* Chapter 4. Washington, DC: Island Press.

Hewitt, J. (2011). *Flows of Biomass to & from the EU: An Analysis of Data & Trends.* Brussels: FERN.

Hillring, B. (2006). World trade in forest products & wood fuel. *Biomass & Bioenergy* **30**: 815–825.

Hoogwijk, M. (2004). *On the Global & Regional Potential of Renewable Energy Sources.* Utrecht: University of Utrecht.

Hoogwijk, M., Faaij, A., and Eickhout, B. (2005). Potentials of biomass energy out to 2100 for four IPCC SRES land-use scenarios. *Biomass & Bioenergy* **29**: 225–257.

Hotelling, H. (1931). The economics of exhaustible resources. *Journal of Political Economy* **39**: 137–175.

IEA (2008). *World Energy Outlook 2010.* Paris: International Energy Agency.

Bioenergy, I.E.A. (2012). *Global Wood Chip Trade for Energy.* Utrecht: University of Utrecht.

IEA Bioenergy (2013). *Large Industrial Users of Energy Biomass.* Lappeenranta: Lappeenranta University of Technology.

IEA & IRENA (2017). *Global renewable energy policies & measures database.* www.iea.org/ policiesandmeasures/renewableenergy/ (accessed 30 November 2017).

IEEP (2011). *Securing biomass for energy - developing an environmentally responsible industry for the UK now & into the future.* http://www.woodlandtrust.org.uk/en/ about-us/publications/Documents/ieep-uk-responsible-bioenergy-2011-08-16-final.pdf (accessed 30 November 2017).

IFEU (2009). *Synopsis Current Models & Methods Applicable to Indirect Land-Use Change.* Heidelberg: Institut für Energie und Umweltforschung.

IPCC (2011). *Bioenergy IN IPCC Special Report on Renewable Energy Sources and Climate Change Mitigation.* Geneva: Intergovernmental Panel on Climate Change.

IRENA (2014). *Global Bioenergy Supply and Demand Projections*. Paris: International Renewable Energy Agency.

IRENA (2016). *Roadmap for a Renewable Energy Future*. Paris: International Renewable Energy Agency.

Johansson, T.B., Kelly, A.K.N., Williams, R.H. et al. (1993). *A Renewables-Intensive Global Energy Scenario (RIDGES) Appendix T*. Washington, DC: Island Press.

Junginger, M. and Faaij, A. (2006). *Biofuel Trade Issues*. Utrecht: Copernicus Institute, Utrecht University.

Junginger, M., van Dam, J., Zarrilli, S. et al. (2011a). Opportunities & barriers for international bioenergy trade. *Energy Policy* **39**: 2028–2041.

Junginger, H.M., Jonker, J.G.G., Faaij, A. et al. (2011b). *Summary, Synthesis & Conclusions from IEA Bioenergy Task 40 Country Reports on International Bioenergy Trade*. Utrecht: Copernicus Institute, Utrecht University.

Lamers, P., Hamelinck, C., Junginger, M., and Faaij, A. (2011). International bioenergy trade – a review of past developments in the liquid biofuel market. *Renewable & Sustainable Energy Reviews* **15** (6): 2655–2676.

Lamers, P., Junginger, M., Hamelinck, C., and Faaij, A. (2012). Developments in international solid biofuel trade - an analysis of volumes, policies, & market factors. *Renewable & Sustainable Energy Reviews* **16** (5): 3176–3199.

Lara, M. and Doyen, L. (2010). *Sustainable Management of Natural Resources: Mathematical Models and Methods*. New York: Springer Publishing.

Legg, B. (2005). Crop improvement technologies for the 21st century. In: *Yields of Farmed Species: Constraints and Opportunities in the 21st Century*, 31–50. Nottingham: Nottingham University Press.

Monteith, J. (1977). Climate and the efficiency of crop production in Britain. *Philosophical Transactions of the Royal Society of London* **281**: 277–294.

Moreira, J. (2006). Global biomass energy potential. *Mitigation & Adaptation Strategies for Global Change* **11**: 313–342.

Rathmann, R., Szklo, A., and Schaeffer, R. (2010). Land use competition for production of food and liquid biofuels: an analysis of the arguments in the current debate. *Renewable Energy* **35**: 14–22.

Rokityanskiy, D., Benítez, P.C., Kraxner, F. et al. (2007). Geographically explicit global modeling of land-use change, carbon sequestration, & biomass supply. *Technological Forecasting and Social Change* **74** (7): 1057–1082.

Rowse, J. (1986). Constructing a supply function for a depletable resource. *Resources & Energy* **10**: 15–29.

Schubert, R., Schellnhuber, H.J., Buchmann, N. et al. (2009). *Future Bioenergy & Sustainable Land use*. Berlin: German Advisory Council on Global Change.

Shafiee, S. and Topal, E. (2008). When will fossil fuel reserves be diminished? *Energy Policy* **37** (1): 181–189.

Sikkema, R., Steiner, M., Junginger, M. et al. (2009). *Development & Promotion of a Transparent European Pellets Market - Final Report on Producers, Traders & Consumers of Wood Pellets*. Vienna: HFA Holzforschung Austria.

Sikkema, R., Steiner, M., Junginger, H.M. et al. (2011). The European wood pellet markets: current status & prospects for 2020. *Biofuels, Bioproducts & Biorefining* **5** (3): 250–278.

Sims, R., Hastings, A., and Schlamadinger, B. (2006). Energy crops: current status & future prospects. *Global Change Biology* **12** (11): 2054–2076.

Sinclair, T., Purcell, L., and Sneller, C. (2004). Crop transformation and the challenge to increase yield potential. *Trends in Plant Science* **9** (2): 70–75.

Slade, R., Saunders, R., Gross, R., and Bauen, A. (2011). *Energy from Biomass: The Size of the Global Resource.* London: UK Energy Research Centre.

Smeets, E. and Faaij, A. (2007). Bioenergy potentials from forestry in 2050: an assessment of the drivers that determine the potentials. *Climatic Change* **81**: 353–390.

Smeets, E., Faaij, A.P.C., Lewandowski, I.M., and Turkenburg, W.C. (2007). A bottom-up assessment & review of global bio-energy potentials to 2050. *Progress in Energy & Combustion Science* **33** (1): 56–106.

Spelter, H. and Toth, D. (2009). *North America's Wood Pellet Sector.* Madison, WI: U.S Department of Agriculture, Forest Service, Forest Products Laboratory.

Steenblik, R. and Simón, J. (2007). *Biofuels - at What Cost? Government Support for Ethanol & Biodiesel in Switzerland.* Geneva: International Institute for Sustainable Development.

Thrän, D., Seidenberger, T., Zeddies, J., and Offermann, R. (2010). Global biomass potentials — resources, drivers and scenario results. *Energy for Sustainable Development* **14** (3): 200–205.

USDA (2013). *Biofuels Annual Report: Brazil.* Washington, DC: U.S Department of Agriculture.

WEA (2000). *World Energy Assessment (WEA): Energy & the Challenge of Sustainability.* New York: United Nations Development Programme.

Welfle, A. (2014). *Biomass Resource Analyses & Future Bioenergy Scenarios.* Manchester: Tyndall Centre for Climate Change Research, University of Manchester.

Welfle, A. (2017). Balancing growing global bioenergy resource demands - Brazil's biomass potential and the availability of resource for trade. *Biomass & Bioenergy* **105**: 83–95.

Welfle, A., Gilbert, P., and Thornley, P. (2014a). Increasing biomass resource availability through supply chain analysis. *Biomass and Bioenergy.* **70**: 249–266.

Welfle, A., Gilbert, P., and Thornley, P. (2014b). Securing a bioenergy future without imports. *Energy Policy* **68**: 1–14.

Wicke, B. (2011). *Bioenergy Production on Degraded & Marginal Land: Assessing its Potentials, Economic Performance, & Environmental Impacts for Different Settings & Geographical Scales.* Utrecht: University of Utrecht.

Williams, J. and Duinker, P. (1997). *Implications of Sustainable Forest Management for Global Fibre Supply.* Rome: Food and Agriculture Organization.

Wolf, J., Bindraban, P.S., Luijten, J.C., and Vleeshouwers, L.M. (2003). Exploratory study on the land area required for global food supply & the potential global, production of bioenergy. *Agricultural Systems* **76** (3): 841–861.

Woods, J., Rosillo-Calle, F., Murphy, R.J. et al. (2011). *The Availability of Sustainable Biomass for Use in UK Power Generation.* London: LCAworks.

Yamamoto, H., Fujino, J., and Yamaji, K. (2001). Evaluation of bioenergy potential with a multi-regional global-land-use-and-energy model. *Biomass & Bioenergy* **21** (3): 185–203.

Yamamoto, H., Yamaji, K., and Fujino, J. (1999). Evaluation of bioenergy resources with a global land use & energy model formulated with SD technique. *Applied Energy* **63** (2): 101–113.

Yamamoto, H., Yamaji, K., and Fujino, J. (2000). Scenario analysis of bioenergy resources & CO_2 emissions with a global land model. *Applied Energy* **66** (4): 325–337.

3

Post-combustion and Oxy-combustion Technologies

Karen N. Finney[1], Hannah Chalmers[2], Mathieu Lucquiaud[2], Juan Riaza[2], János Szuhánszki[1] and Bill Buschle[2]

[1] *Energy 2050, Department of Mechanical Engineering, University of Sheffield, UK*
[2] *Institute for Energy Systems, School of Engineering, University of Edinburgh, UK*

3.1   Introduction

Achieving negative net carbon emissions from biomass power generation is possible using a variety of different technologies. In order for carbon dioxide emitted from electricity generation and other industrial processes to be stored, it needs to have high purity, as impurities in the outlet stream can lead to difficulties in compression, transportation and sequestration (or utilisation) phases. The most common method is post-combustion capture, which burns the biomass fuel in a more or less conventional manner and then uses a separation process to remove the biogenic CO_2 from the other components in the flue gases. Air firing of biomass coupled with post-combustion capture requires minimal modifications to the base power plant, with the capture plant retrofittable; thus, existing plants may already be capture ready. CO_2 separation from the flue gas can be achieved with various methods, as explored in Section 3.2. Oxy-fuel combustion, the focus of Section 3.3, can also be retrofitted to some existing boilers as well as be applied to newly built power stations units.

Due to the possibility of retrofits, both of these technologies have the potential to address future CO_2 emissions that are already locked into existing fossil fuel plants (Florin and Fennell, 2010). The primary focus of carbon capture research has largely been on decarbonising energy generation based on fossil fuels (mainly coal), and consequently, thus far, only limited work has been conducted on the assessment of coupling or integrating biomass firing with carbon capture and storage (CCS). Co-firing – simultaneously burning fossil fuels and biomass – combined with CCS has so far garnered significantly more interest compared to CCS from dedicated biomass power. This is most likely due to small incremental changes from the known (coal CCS) to the unknown biomass energy with carbon capture and storage (BECCS) being required to secure investment.

A key reason for the lack of BECCS deployment to date may be the number of specific technological challenges associated with the use of biomass, which often become more

Biomass Energy with Carbon Capture and Storage (BECCS): Unlocking Negative Emissions, First Edition.
Edited by Clair Gough, Patricia Thornley, Sarah Mander, Naomi Vaughan and Amanda Lea-Langton.
© 2018 John Wiley & Sons Ltd. Published 2018 by John Wiley & Sons Ltd.

apparent or intensified when combined with carbon capture, as assessed in Section 3.4. The fuel composition of most commonly used biomass fuels highlight the significant trace elements content, which can have negative impacts throughout the power and capture plants – from the furnace to spent solvent storage. These can take the form of slagging, fouling, deposition and corrosion in the boiler (caused by the presence of high levels of alkali metals and halogens, such as K, Na, Cl), as well as solvent degradation initiated and exacerbated by acidic elements (Cl, S, N) and transition metals (Fe, Cu) – all species carried over in the flue gases from the initial biomass materials. Due to the diversity of elements present in the flue gases from such combustion and their impacts on capture effectiveness, there may be a need for specific solvent management options that are specifically designed to operate under these regimes – i.e. biomass-specific solvent processes.

3.2 Air Firing with Post-combustion Capture

The combustion process used for post-combustion carbon capture is or can be the same as conventional combustion, where the biomass fuel is burned in air, with little or no modification required to the burner or furnace. The same configurations of the combustion chamber are used here, where the fuel and/or air can be staged, usually with excess oxidant provided, both to aid burnout. As with oxy-fuel combustion (Section 3.3), the biomass fuel would usually be pulverised when used on a large scale. Smaller plants operating on a distributed scale are more likely to utilise biomass via fluidised beds and moving or reciprocating grates (fixed bed); smaller-scale facilities are less likely to be used for initial CCS applications due to the volumes of CO_2 produced.

The post-combustion capture process takes place after combustion (see Figure 3.1), when the carbon dioxide is separated from the rest of the flue gas stream, before they are released into the atmosphere and the carbon dioxide can be compressed and sent to a suitable storage location. The most common method for CO_2 separation is through some form of wet gas scrubbing process with CO_2 absorbing into a chemical solvent. Despite the extensive knowledge and research into solvent-based capture, other techniques could potentially supersede the conventional post-combustion amine scrubbing method in the future. These include, but are not limited to, the different post-combustion capture systems explored in Sections 3.2.2 and 3.2.3, as well as oxy-combustion processes, also outlined subsequently Section 3.3. Although solvent-based capture may not necessarily appear to be the most cost-effective option compared to more novel

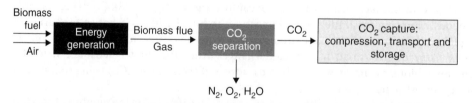

Figure 3.1 Overview of post-combustion carbon capture from biomass, showing the key stages and processes used to separate the CO_2 from the remaining flue gas components that go to the atmosphere.

second- and third-generation technologies where cost and technology-readiness uncertainties still remain large, cost reductions occurring through learning-by-doing are likely to be key considerations. Until significant cost reductions are observed for other capture systems, wet solvent scrubbing may continue to be the most favourable overall.

3.2.1 Wet Scrubbing Technologies: Solvent-Based Capture Using Chemical Absorption

The flue gas from the biomass combustion plant may undergo some preprocessing before entering the capture system; this can include particulate removal, as well as flue gas desulphurisation. The primary aim of these processes is to comply with environmental regulations, but they serve the additional purpose of removing specific species from the flue gases that could prove problematic for capture (see Section 3.4), as well as other CO_2 removal technologies. For gas scrubbing, the solvent solution first captures the carbon dioxide from the biomass flue gas through chemical absorption in an absorber or separation column and then releases it from the solvent via a thermal regeneration process in a stripper column (or desorber/recovery tower) – this usually involves heating the solvent with steam from an integrated reboiler to produce a high-purity stream of CO_2. The absorber and stripper columns, with their process flows of CO_2-rich and regenerated solvent, are shown in Figure 3.2. This regeneration allows the solvent to be reused (recycled), but also releases the CO_2 in a pure stream ready for sequestration, once water is condensed out. The regeneration of the solvent utilises thermal energy – typically by extracting steam from the power generation steam cycle

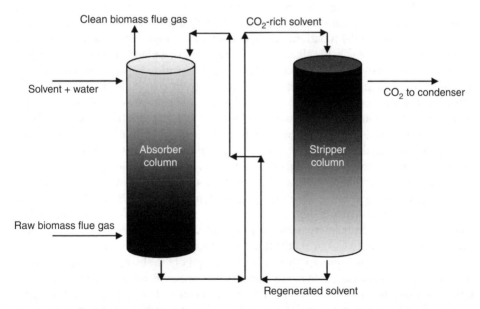

Figure 3.2 Simplified flow diagram of solvent scrubbing for post-combustion carbon capture from biomass. The solvent is added initially from a separate system and then recirculated upon regeneration in the stripper column.

that would otherwise be used for power generation – and as such imposes an electrical output penalty on the system.

Although this is now a commercially available and relatively mature technology, there have been no large-scale demonstration projects on biomass to date; however, fossil-fuel capture projects of various scales based on this technology are currently being deployed, such as the SaskPower Boundary Dam Carbon Capture Project in Saskatchewan, Canada – the world's first post-combustion CCS project on a coal-fired power station (SaskPower, 2016).

Post-combustion capture is an excellent CCS technology to retrofit for biomass plants, as little adaptation of the original combustor and power facility will be needed. Consequently, all new biomass-fired power stations can be expected to be made carbon capture ready, either via regulatory requirement, as it is the case in the United Kingdom for stations with an output greater than 300 MW_e, or via commercial drivers.

The range of solvents that can be used in the scrubbing process are divided into physical solvents (which predominantly use physical adsorption processes to separate CO_2 from the flue gas) typically used for high-pressure streams encountered in gasification systems, and chemical solvents (which largely use acid-base neutralisation chemical reactions to separate out CO_2), typically used for atmospheric capture from air-fired boilers – or intermediate solvents, which do some of each. Aqueous amine chemical solvents are often chosen for flue gas scrubbing applications because of their ability to absorb selectively CO_2 gas and their relatively low thermal energy requirement for regeneration. Biomass-specific capture solvents are considered in further detail in Section 3.4.1.

3.2.1.1 Amine-Based Capture

As noted earlier, at present, the most common solvents are amine based, composed primarily of water and alkanolamine (those containing both hydroxyl (–OH) and amino (–NH_2, –NHR and -NR_2) functional groups) chemical compounds. These solvents are designed on either one specific alkanolamine chemistry or a blend of chemistries to give a variety of benefits – the desired combination of absorption capacity, thermal regeneration energy requirement, operational life and solvent cost. Solvents utilising sterically hindered amines and advanced amine blends have been specifically designed for CCS purposes, although not necessarily for biogenic CO_2 capture (Blomen et al., 2009). In addition to amines, other solvents, such as amino-acid or ionic liquids, are being developed (Mumford et al., 2015). Monoethanolamine (MEA) is commonly used in baseline studies for comparative purposes, due to its low cost and long history of use in natural gas sweetening applications. The use of solvents has advantages and disadvantages compared to membrane-based CO_2 separation (see Section 3.2.2). Chemical solvents are less dependent on the partial pressure of CO_2, which is beneficial at low concentrations, and thus low partial pressures, of CO_2. Moreover, chemical solvents can thoroughly separate CO_2 from the flue gas stream; 95% CO_2 separation and capture can be economically possible. Further, since the design of the absorption column means the capture process can occur at nearly atmospheric pressure, no significant pressurisation of the flue gas is required for effective CO_2 separation. However, thermal energy is required for regeneration to break the bonds in the CO_2-rich solvent and reverse the chemical reaction between the solvent and the carbon dioxide; this energy could have been used for power regeneration, hence the energy penalty considered above.

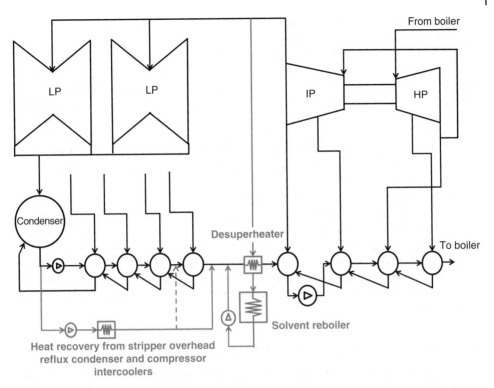

Figure 3.3 Steam turbine configuration of a thermal steam plant retrofitted with a post-combustion capture and compression process; the grey lines indicate the additions to the steam cycle when capture is implemented. HP, IP and LP are the high-, intermediate- and low-pressure steam turbines, respectively. Adapted from Lucquiaud and Gibbins (2011).

3.2.1.2 Steam Extraction for Solvent Regeneration

For biomass power stations fitted with post-combustion carbon capture, it can be reasonably expected that thermal energy for solvent regeneration may be supplied by extracting steam from the power cycle, as is routinely proposed for thermal power plants. Steam flow rates are typically of the order of 30% to 50% of the flow rate entering the low-pressure turbine of a steam cycle without capture. In practice, this implies that access is limited to crossover pipes between the turbines. Since amine solvents are typically being regenerated at temperatures around 100–150 °C, low-pressure steam around 3–5 bar is extracted prior to the inlet of the low-pressure turbine and then condenses by transferring heat to the solvent. The condensate is then pumped back to the steam cycle. This process is illustrated in Figure 3.3.

3.2.2 Membrane Separation

In addition to solvent scrubbing, membrane separation can be used to extract CO_2 from the biomass flue gas in a post-combustion set-up. For post-combustion capture applications, such devices allow only CO_2 to pass through; however, selectivity and efficiency of different materials can vary greatly. (In pre-combustion capture, for example, H_2-selective membranes, where H_2 passes through the membrane leaving behind a CO_2 stream to be transported to sequestration, are also under consideration.)

Significant progress has been made in developing membrane materials that are suitable for post-combustion CO_2 capture applications. For example, Merkel et al. (2010) report materials developed with support from the US Department of Energy that have potential to provide significant breakthroughs in reducing the energy penalty associated with post-combustion CO_2 capture. It is, however, challenging to identify materials that are able to achieve sufficient recovery and purity of CO_2 in one stage. A significant strand of current research is, therefore, focused on identifying suitable configurations of stages and modules to achieve simultaneously sufficient recovery and purity of CO_2 (e.g. Merkel et al., 2010; Zhao et al., 2010).

Since membranes for CO_2 capture applications are, in general, at an earlier stage of development than wet scrubbing technologies, further work is required to fully understand the specific issues that are likely to arise in BECCS application of membrane separation. One challenge is to ensure that materials are exposed to flue gases that contain a realistic mixture of components (which are typically difficult to obtain at lab scale) but without requiring significant volumes of material to be manufactured. One solution to this problem is to test materials on slipstreams of larger flows of the flue gas (e.g. Watson et al., 2013).

3.2.3 Brief Overview of Other Separation Methods

In addition to the aforementioned methods, a wide range of other post-combustion technologies are available for separating CO_2 from the other flue gas components, and they too can be applied to biomass combustion systems. These include using solid absorbents or adsorbents (such as in rotating wheel systems or static beds), gas-phase separation, cryogenic fractionation, low-temperature distillation and hybrid systems as an alternative means of flue gas decarbonisation. Regeneration of sorbents, as with the solvents above, can be via temperature or pressure swing.

There are also many other capture methodologies and techniques that are applicable to biomass-fuelled systems, including pre-combustion and chemical looping discussed in Chapter 4.

3.3 Oxy-Fuel Combustion

Oxy-fuel combustion, which has long been a common process in the glass, cement and steel industries, is also among the leading CCS technologies. It was first proposed for pulverised fuel power plants by Abraham et al. (1982), with the aim of producing CO_2 for enhanced oil recovery (EOR). In oxy-fuel combustion, the main difference from conventional air firing is that the fuel is burned in a mixture of O_2 and recycled flue gas. The O_2 is produced by an air separation unit (ASU), which removes the atmospheric N_2 from the oxidiser stream. By removing the N_2 upstream of the process, a flue gas rich in CO_2 and water vapour is produced, which eliminates the need for a post-combustion capture plant. The water vapour can be removed by condensation, leaving a product stream of relatively high-purity CO_2 which, after subsequent purification and dehydration, can be pumped to a geological storage site.

In between conventional air firing and full oxy-fuel mode is combustion in an enriched-air environment – using air plus additional O_2 as the oxidiser to the system; these both are considered in more detail in the subsequent sections.

3.3.1 Oxy-Combustion of Biomass Using Flue Gas Recirculation

Oxy-fuel operation in a CO_2–O_2 environment, as depicted in Figure 3.4, utilises flue gas recirculation (FGR) to achieve high concentrations of CO_2 in the exhaust stream. In addition to the condenser for water vapour removal, other treatment processes are required to remove trace contaminants (such as oxygen, nitrogen and those outlined in Section 3.4.1) from CO_2 prior to transportation. FGR is necessary here as burning fuels, such as biomass, in pure O_2 (without N_2 from air acting as a diluent) would produce furnace temperatures that could not be withstood by conventional furnace materials. Oxygen concentrations at the combustor inlet typically vary between 21% and 35% (Álvarez et al., 2014).

Circulating fluidised bed combustion (CFBC) is an alternative technology that could be used for oxy-fuel combustion. CFBC has some advantages over pulverised fuel boilers for oxy-fuel combustion that may make it a better choice for CO_2 capture in some cases. For example, better temperature control allows fluidisation with higher oxygen concentrations and so operation is achieved with lower flue-gas recycle ratios. In comparison, pulverised oxy-fuel combustors must operate at relatively high FGR rates. CFBC systems also have the benefit of being able to combust a wide variety of biomass fuels. However, costs of this type of boiler are usually higher.

The major cost associated with oxy-fuel combustion derives from the O_2 production in the ASU. At present, on a large scale, cryogenic air separation is the most viable option. Darde et al. (2009) found that the efficiency of cryogenic ASUs has more than doubled since 1971, and they project a continued increase due to technological advances. The operational cost of the ASU also depends on the purity of O_2 required. Shah (2006) found that 95–97.5% purity is optimal and also recognised the need for minimising air ingress to the system. Jordal et al. (2004) noted that this may be particularly challenging and ultimately a limiting factor for retrofit purposes of this technology for boilers where the level of air leakage into the boiler or downstream may already be significant.

Oxy-fuel technology can lead to increased efficiencies associated with the reduced heat losses from the flue gases, as the overall volume of the flue gas exhausted from the stack is lower, and the latent heat of water vapour is recovered during the condensation

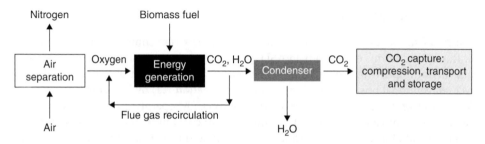

Figure 3.4 Overview of oxy-fuel combustion for carbon capture from biomass, showing the key processes and stages; some purification is also likely to be required at the dehydration stage.

step. Furthermore, due to significantly lower total NO_x (oxides of nitrogen – i.e. NO and NO_2) production, costly de-NO_x plants are often omitted from the cost estimates of oxy-fuel plants.

Several demonstration projects, together with a number of small- and medium-scale studies, have contributed significantly to our current understanding of the oxy-fuel process. Although the individual components of the process are all commercially available, they have never been integrated together on a large scale. There was, therefore, a requirement for full-scale plant experience to assess the plant's availability and load-following capability when using high-pressure steam cycles, viable with unit sizes of typically above 250 MW_e (Davidson and Santos, 2010). A number of demonstration projects has been successfully implemented in order to provide information for scaling up, which include Doosan Babcock's 40 MW OxyCoal™ system at Renfrew, UK; Vattenfall's 30 MW plant at Schwärze Pumpe, Germany; CS Energy's Callide 'A', a 30 MW retrofit plant in Queensland, Australia and CIUDEN's 20 MW plant (along with a 30 MW CFBC unit) in Spain (Davidson and Santos, 2010; MIT, 2015) – although as seen, these are largely coal-based projects.

In order to maximise the chance that the large-scale plants will succeed (and thus demonstrate the technological and economic feasibility of the process), the fundamentals of oxy-fuel combustion technology have to be understood fully. It is clear that replacing N_2 with recirculated CO_2 changes the fundamentals of the combustion process. This is because the thermophysical properties of CO_2 differ significantly from N_2. The differences include the following:

- CO_2, being a tri-atomic gas, actively participates in thermal radiation thus impacting the in-furnace heat-transfer profiles.
- CO_2 has a higher heat capacity than that of N_2, acting as an increased heat sink, which can lower the adiabatic flame temperature (AFT) and reduce flame speeds.
- The increased CO_2 concentration in oxy-fuel combustion results in a lower O_2 diffusion rate than in conventional cases, hindering O_2 diffusion to the char particle, and may also increase the importance of the Boudouard gasification reaction, influencing char burnout.

For recent reviews of the technological developments and overall status of the field, see Yin and Yan (2016), Chen et al. (2012), Scheffknecht et al. (2011), Davidson and Santos (2010), Toftegaard et al. (2010) and Wall et al. (2009). Specific challenges for progressing oxy-fuel combustion for biomass fuels are considered further in Section 3.4.

3.3.2 Enriched-Air Combustion

The use of enriched-air combustion – increasing the oxygen content of the combustion air through additions of O_2 from an ASU – has garnered a certain amount of interest as a means of improving the capture performance. This approach may lead to benefits over conventional post-combustion capture (see Section 3.2) by lowering the volumetric flow rate of flue gas, requiring post-combustion CO_2 separation through eliminating some N_2 and concentrating the CO_2. Compared to conventional oxy-fuel combustion, benefits would be expected to include reductions in the sizes of the ASU and capture systems of the integrated BECCS plant, which may lead to reduced capital costs of the plant. It is possible, however, that costs (capital and operating) and maintenance of two significant additional plant components could outweigh the overall benefits of this, compared to installing one plant with more conventional post-combustion or oxy-fuel capture only.

3.4 Challenges Associated with Biomass Utilisation Under BECCS Operating Conditions

There are many technical challenges associated with the use of biomass as a fuel source – both in general and in the context of BECCS. Whilst there is a number of power plant issues associated with combusting biomass feedstocks, adding a capture process, either post-combustion or oxy-fuel, means that many of the key technological challenges are also passed further downstream, impacting on the post-combustion capture solvent and/or altering the combustion process. As a result, biomass-specific capture methods are considered, such as solvents which are designed to be more able to deal with the variations in flue gas composition. The composition, particularly, the trace element content, of biomass is by nature unlike that of conventional fossil fuels and could negatively impact on either post-combustion or oxy-fuel capture, as assessed in the subsequent sections.

3.4.1 Impacts of Biomass Trace Elements on Post-combustion Capture Performance

The issue with biomass trace elements is that they are often distinctly different from those found in coal, and furthermore, the trace species and contaminants present in different biomass fuels also vary greatly. Biomass combustion therefore already differs from that of fossil fuels, and, consequently, capture methods originally designed for flue gas produced by combustion of coal (or natural gas) will therefore most likely need to be adapted to ensure suitability for efficient and effective capture from biomass. These will need to consider new control measures and technology options to compensate for these different trace elements. In this different environment, there are a range of species that can have detrimental impacts across both power and capture plants, including alkali and transition metals, acidic elements and particulate matter (PM); these are considered in turn below as the causes of a range of operational issues throughout the BECCS system, comprising slagging, fouling, deposition, corrosion and problems with environmental control systems, including various forms of solvent degradation in post-combustion capture plants. Control measures are outlined for each.

3.4.1.1 Alkali Metals

Biomass fuels often have elevated concentrations of alkali metals, namely potassium and sodium, which help to catalyse thermal conversion reactions in the combustor (Jones et al., 2007; Saddawi et al., 2012; Jiang et al., 2015), but also lead to the operational issues of deposition in the form of slagging and fouling in the boiler and heat-exchanger components (Nutalapati et al., 2007; Garcia-Maraver et al., 2016; Niu et al., 2016). These alkali-induced deposits build up over time, leading to the need for regular (and costly) removal and cleaning; otherwise they will lead to extensive damage of the system materials and often substantial reductions in heat flux. These problems can be exacerbated by corrosion under the deposits, caused by acidic elements, considered further below, that react with the metal surfaces. As noted, trace elements are found in significantly lower concentrations in fossil fuels, such as coal, compared to biomass. Due to these differences in the composition and properties of biomass fuels compared to conventionally used coal, adaptations may need to be made to both the combustion process and to the CCS technologies employed. Alternatively, removing some of the pollutants described here through fuel pretreatment methods could be an option to ensure that they do not affect the operation of combustion and capture plants. Washing often takes place simply with water and can remove alkali and other problematic ash-forming metals, as well as acidic elements, considered below, with relatively high efficiencies; alternative solvents, namely acidic or alkali solutions, can also be utilised to dissolve these and other unfavourable species in the biomass prior to firing. A considerable amount of research has investigated this form of biomass treatment, as it is a relatively cheap and efficient means of altering the fuel input composition (e.g. Deng et al., 2013; Gudka et al., 2015). High-temperature processes, for instance, torrefaction, are also common; this is where the biomass is heated in an inert/non-oxidising atmosphere to improve the energy density and other properties (Arias et al., 2008). Hydrothermal-based treatments, such as carbonisation, also involve a washing stage, followed by the use of wet conditions at high temperatures and pressures – essentially wet torrefaction. Metals (alkalis in particular), as well as a range of other inorganics, are washed out into the liquid to varying degrees (depending on the operating conditions), leaving a relatively clean bio-coal type fuel (Kambo and Dutta, 2014, 2015; Smith and Ross, 2016). Both washing and hydrothermal treatments require a drying phase, however, and can thus be energy intensive, although they may still remain a cheaper option than dealing with the issues in the plant if trace elements are not removed before combustion. Instead of fuel pretreatments, additional flue gas conditioning stages may need to be integrated into the capture plant prior to the CO_2 separation phase to eliminate or at least minimise problematic species.

3.4.1.2 Transition Metals

In addition to the alkali metals outlined, transition metals, especially Fe and Cu, also have implications for post-combustion capture. These too are commonly found in a range of biomass fuels used for energy generation and BECCS. These primarily impact the capture media, by dissolving into aqueous solvents, which act as an oxygen carrier and contribute to oxidative solvent degradation, especially where temperatures are elevated, such as in the stripper column used for regeneration and the lean-rich cross heat exchanger (Bedell, 2009; Ciftja et al., 2012). Such elements can react directly with amines, but act to catalyse primarily the oxidation reaction between the solvent molecules and excess oxygen in the flue gases, forming stable complexes that remain in the solvent (Bedell, 2009). If degradation products are allowed to build up in the solvent, it affects the capacity of the solvent to remove CO_2 and to transport it to the stripper. When capture efficiency drops or the energy use increases, the solvent needs to be treated or replaced, which can be expensive and also leads to a waste stream that needs to be disposed of safely. Furthermore, degradation products can also cause additional operational issues by changing the physical properties of the solvent, including its viscosity and surface tension. Further, the increased ion concentration in the solvent can increase its corrosivity, limiting the life of wetted plant components, such as pump heads, heat exchangers and pipework. Corrosion inhibitors can be added to solvents to reduce reaction rates with oxygen by increasing their activation energy and minimising degradation product formation (Voice and Rochelle, 2011).

3.4.1.3 Acidic Elements

Corrosion initiation is a particular issue when there are notable concentrations of acidic elements (Cl, S, N) found within the initial biomass feedstock that form inorganic anion compounds, such as NO_x and SO_x (oxides of nitrogen and sulphur), which remain in the flue gas. Acidic elements can therefore be problematical throughout the BECCS system as well. Moreover, these species may also cause solvent poisoning through the formation of heat-stable salts, although there was, at the time of writing, very little experience with carbon capture solvents exposed to biomass flue gas. Solvent poisoning is reversible through a process known as solvent reclaiming, although this is relatively expensive and energy intensive. Instead, removal of these compounds, either fully or partially, specifically halogens and sulphur, can also be achieved using washing and hydrothermal carbonisation fuel pretreatment processes described above for alkali metal elimination. Additionally, flue gas treatments prior to the post-combustion capture process can be employed to remove such species after combustion but before capture. For example, techniques such as flue gas desulphurisation are already common in power stations and are generally expected to be employed in the BECCS environmental control system for the flue gas prior to the capture plant, although some systems can have these integrated into their design so separate treatments are not required.

3.4.1.4 Particulate Matter

Particle carryover also has implications for capture solvents – again possibly causing solvent degradation. Fine fly ash particles, often sub-micron in size, can frequently penetrate particulate removal devices, and considering the high levels of metal contaminants here, their impact on capture performance can be significant. Transition metals, as delineated above, can be particularly problematic, especially as these tend to be concentrated in the smaller fly ash particles as opposed to the bottom ash PM (Chandan et al., 2014). This can be most challenging when using capture on solid fuels, especially pulverised fuel combustion, as most large-scale biomass applications would be. Highly efficient and extensive particle removal systems for small and potentially sub-micron PM, for instance, fabric filters and electrostatic precipitators, as well as pre-separator devices such as cyclones or wet scrubbers, would be required on BECCS plants to minimise such issues, as well as to meet with emissions control limits.

3.4.1.5 Biomass-Specific Solvents for Post-combustion BECCS

This section aims to address several key questions concerning capture solvents in this context. In particular, do conventional post-combustion capture solvents aimed traditionally at fossil fuels work effectively when applied to biogenic CO_2 in biomass flue gases? Also, do biomass-specific solvents for carbon capture exist and is there a need for them? Whilst there is very little experimental information available pertaining to post-combustion capture from biomass co-firing with coal – academic or industrial – there is even less on dedicated biomass plant operation with integrated solvent-based post-combustion capture, which is clearly yet to be explored empirically in a comprehensive manner. Studies in this area tend to be wholly theoretical and are typically either techno-economic studies (e.g. Domenichini et al., 2011; Koornneef et al., 2012; Al-Qayim et al., 2015) or modelling and process simulations (e.g. Berstad et al., 2011) – and sometimes both of these (e.g. Spath and Mann, 2004; NETL, 2012). Even in this literature, research tends to largely focus on co-firing rather than dedicated systems (such as Benetto et al., 2004; Khorshidi et al., 2013, 2014; Schakel et al., 2014; Fogarasi and Cormos, 2015). Furthermore, there is limited technical output concerning life cycle or environmental impact assessments for full BECCS systems based on post-combustion technologies due to the lack of information elsewhere from which to source input data. Since there is lack of understanding of the impacts that biomass-generated flue gases may have on solvent-based capture, it cannot really be accurately determined whether conventional capture solvents usually used for CCS from fossil fuels are able to capture the biogenic CO_2 cost-effectively from biomass flue gases without the additional challenges brought about by the differing fuel composition. It is worth noting though that the energetics of the process can easily be extrapolated from previous work on coal flue gas. Extensive laboratory and pilot–/ large-scale biomass combustion trials are needed, integrated with solvent-based post-combustion capture, to assess their suitability when exposed to potentially significant levels of alkali and transition metals and biomass-derived PM. These could then provide insights into managing solvent inventory, building on existing experience derived from coal and developing practical approaches in terms of degradation and other solvent issues. ZEP (2012) – the EU Zero Emissions Platform – highlights a number of key challenges specific to BECCS deployment in their publication on a way forward for European bio-CCS. Although they outline issues such as cost assessments and life cycle analyses, which appear to be being addressed (see above), other areas such as determining the 'effect of the composition of biogenic CO_2 on the CCS value chain in power plants (corrosion, effect on amine/ammonia solvents etc.)' and the 'specific storage properties for biogenic CO_2, i.e. biogenic impurities in the CO_2 stream' are dealt with less well. It would seem that there were, at the time of writing, no examples of biomass-specific solvents for such applications or any evidence of their development. So, why not? Based on the impurities within the fuel and the potential degradative issues caused by these in the capture plant – including corrosion and oxidative solvent degradation – as well as the strategy outlined by ZEP (2012) for accelerating BECCS deployment, this is something that should be, but is currently not being, researched and investigated further. This is most likely to entail adapting existing solvents and their operating conditions to make them more applicable and acceptable in this environment. Although there does not seem to be any evidence that conventional solvents are not suitable, there is limited confidence that they are – and based on the discussion at the start of Section 3.4.1, it would appear likely that such impurities would negatively impact on capture efficiencies through solvent degradation via a range of pathways.

3.4.2 Biomass Combustion Challenges for Oxy-Fuel Capture

Many of the key challenges of BECCS implementation using oxy-fuel processes are associated with combustion process. The presence of the flue gas recycle and the additional process plants (O_2 production and CO_2 processing units) further complicate the process and, as a result, oxy-fuel combustion differs from conventional air-fired combustion in a number of ways, including flame characteristics, feedstock reactivity, heat transfer, emissions characteristics, corrosion and process control.

3.4.2.1 Fuel Milling

Particle size distribution required for an efficient burnout will depend on the reactivity of the biomass, type of reactor and burner design. Fuel pelletisation processes are a very relevant stage, as most crushing in mills gives similar size distributions to the original biomass by breaking the pellets back into their source material. Higher-volatile-content biomasses require mill temperatures to be kept sufficiently low to control the risks of fires and potential explosions. An important consideration in assessing combustion safety at power plants burning pulverised biomass under oxy-fuel conditions is, therefore, the potential for fuel ignition from a fire in the mill and further explosion development during the milling stage. The primary recycle is introduced in the mills for drying and transporting the fuel into the boiler. Trabadela et al. (2014) identified the primary recycle O_2 content as one of the main variables for the risk of ignition, finding lower tendencies for ignition in oxy-fuel with the same O_2 levels than under similar conditions in air.

3.4.2.2 Flame Temperature

In pulverised fuel flames, flame temperature is considered to be the most important factor influencing particle heating, ignition and burnout (Dhungel, 2010). At the same O_2 concentrations, the adiabatic flame temperature (AFT) of an oxy-fuel flame is lower than that of an air-fired one. This is due to differences in the thermal properties of the mixture, which is characterised by product density and specific heat and is referred to as heat sink (Shaddix and Molina, 2009). However, by increasing the O_2 concentration, the AFT of the air-firing application can be matched. This is typically achieved at 27–30% O_2 concentration, which is controlled by varying the amount of the recycled flue gas, requiring approximately 60% recycle (Wall, 2007).

3.4.2.3 Heat Transfer

In pulverised fuel furnaces, radiation is the principal mode of heat transfer, which is dependent on the flame temperature and the radiative properties of the gas mixture and the particles. As emissivity of CO_2 and H_2O is higher than that of N_2, oxy-fuel combustion results in higher radiative heat fluxes, and therefore for retrofit applications lower AFTs are required to match the heat transfer to the radiative section. The increased heat transfer in the radiative section will, though, result in lower gas temperatures in the convective section. This heat transfer between the two sections will have to be optimised for retrofit applications. Smart et al. (2010, 2010) and Smart and Riley (2011) investigated the effect of the recycle ratio on the furnace radiation profiles and found that overall the radiative heat transfer matched air-firing cases at 72–74% recycle ratios.

3.4.2.4 Particle Heating, Ignition and Flame Propagation

Increasing the O_2 concentration also enhances reaction and heat release rates and consequently increases the rate of devolatilisation, lowers the time required for ignition and increases the flame propagation speed, all of which are reduced by substituting N_2 with CO_2; thus by O_2/CO_2 ratio selection, it is possible to match ignition times and volatile flames of oxy-fuel furnaces to conventional flames (Shaddix and Molina, 2009).

However, the additional degree of flexibility that comes from controlling the O_2/CO_2 ratio allows for a wider range of fuels to be used.

3.4.2.5 Burnout

Toftegaard et al. (2010) noted that char combustion is typically controlled by both reaction kinetics and diffusion, and as diffusivity of O_2 in CO_2 is 0.8 times its diffusivity in N_2 (at 1127 °C), at the same O_2 partial pressures, the lower diffusivity is expected to lower the rate of char combustion. Similarly, as diffusivity of light hydrocarbons is also lower in CO_2 atmospheres, the rate of volatile combustion is also negatively affected by the increased presence of CO_2. However, when the AFT is matched, the increase in O_2 concentration results in an overall enhanced burnout of char and volatiles. The enhanced burnout may also be explained by the increased importance of the Boudouard reaction at a high CO_2 partial pressure: $C + CO_2 \rightarrow 2CO$. An additional factor, which is expected to improve burnout for oxy-fuel combustion, is char gasification, which becomes more important in the later stages of combustion when the O_2 concentrations are low and the gas temperatures are still high. This is because these reactions are being slower than the char–O_2 reaction, and thus kinetically controlled (Chen et al., 2012).

3.4.2.6 Emissions

Numerous studies have been conducted on the impact of oxy-firing on pollutant formation (see for example Shaddix and Molina, 2009; Smart et al., 2010). The consensus view is that oxy-fuel combustion results in significantly lower NO_x and SO_x (oxides of nitrogen and sulphur) emissions per unit of energy produced than conventional air firing. The concentration of pollutant species in the flue is, though, significantly increased due to the cumulative effect of the recycle, which leads to a gradual build-up of the pollutants. The reduction of NO_x is attributed to a number of factors, including the lower thermal NO_x formation, due to the low nitrogen concentration (under oxy-firing conditions N_2 in the oxidiser is replaced by CO_2) and lower AFT, and especially the reburn mechanism where the recycled NO is mostly destroyed by hydrocarbon radicals at the root of the flame (Davidson and Santos, 2010). Furthermore, Toftegaard et al. (2010) concluded that the reduction of the partial pressure of N_2 to very low levels causes oversaturation of the furnace atmosphere with NO, which reverses the Zeldovich mechanisms reducing the NO already formed to N and N_2. Higher concentration of water vapour in the case of wet recycling of the flue gas could also interact with the nitrogen compounds, reducing the NO in the flue gas (Álvarez et al., 2011). The reburn mechanism is important in the fuel-rich regions of the flame, where the recycled NO is attacked by the hydrocarbon radicals and converted back to intermediate nitrogenous species. Okazaki and Ando (1997) concluded that the NO reburn by CH_x radicals contributes to 50–80% of the recycled NO reduction when reformulated gasoline (RFG) is used. The decrease in SO_x emissions is attributed to higher retention rates of S in the fly ash particles due to higher in-boiler SO_x concentrations. Studies by Dhungel (2010) and Szuhánszki (2014) indicated that there are no significant issues in the control of CO emissions when moving to oxy-fuel combustion.

3.4.2.7 Corrosion

Jordal et al. (2004) noted that high concentration of acidic species, such as CO_2, sulphur and chlorine, which also impact air-combustion with post-combustion capture as discussed above, increases the corrosive nature of the flue gases. Fleig et al. (2009) found that as the partial pressure of SO_2 increases, the percentage of SO_3 conversion also increases compared to conventional air-blown combustion. This poses additional challenges for controlling corrosion (Wall et al., 2011), especially in the regions of the furnace, e.g. the primary (fuel carrier) stream, where temperatures below that of the acid dew point are present. To mitigate this, only dry recycle is used to blow the pulverised fuel in the furnaces, and the rest of the oxidiser, which may be wet, is typically kept above 250 °C (well above the acid dew point of the flue gas). In addition, the use of more advanced alloys might be required for heat exchanger surfaces in order to minimise corrosion.

3.5   Summary and Conclusions: Synopsis of Technical Knowledge and Assessment of Deployment Potential

The two CO_2 capture technologies introduced and outlined in this chapter for power plants and other industrial sites producing CO_2 – air firing with (primarily solvent-based) post-combustion capture and oxy-fuel combustion capture – are well understood, with extensive research and development leading to their current and imminent deployment on a global scale – predominantly, however, for coal and other fossil fuels. Nevertheless, our knowledge of these can be borrowed and adapted to suit biomass fuels and thus BECCS-specific applications. Both technology groups are suitable for large-scale deployment and are relevant for BECCS implementation, although there is scope for further work to address technical challenges, which would be expected to improve technical performance and most likely also reduce costs.

Whilst there are still a number of remaining challenges regarding BECCS operating conditions, including various forms of solvent degradation, slagging, fouling, deposition and corrosion, as explored here, which often apply to both air-firing and oxy-fuel regimes, options are available for minimising and mitigating these. Pretreatments of fuels, through washing, hydrothermal carbonisation and torrefaction can remove problematic species (acidic elements/inorganic anions, such as N, S and Cl and their compounds, and alkali and transition metals, such as K, Na, Fe and Cu) before combustion and additional stages in the emission control systems are able to eliminate or at least minimise these before the flue gases enter the carbon capture plant (for example, particulate matter).

State-of-the-art post-combustion capture processes using chemical-based wet scrubbing technology can be implemented with air-fired biomass combustion gases to generate negative net carbon products, e.g. electricity or heat. The process itself and its expected energetic performance can be extrapolated from previous work on coal flue gas. Yet there is a lack of experience and understanding of the impacts that biomass-generated flue gas impurities, such as acidic species, alkali/transition metals, inorganic anions and particulate matter, may have on long-term viability of capture solvents. It is not presently possible to determine accurately whether conventional solvent processes and chemistries, for instance, using MEA, would face additional challenges brought about by the differing fuel composition, in terms of reversing the formation of heat-stable salts and dealing with yet-unknown specific corrosion and degradation mechanisms.

Other options for deploying post-combustion capture are typically further away from commercialisation than wet scrubbing technologies. For example, membrane separation using CO_2-selective membranes is under development for post-combustion capture applications. As these technologies progress towards commercialisation, it will be important to identify and resolve any challenges related to applying these options in BECCS projects. For example, use of slipstream testing (e.g. Watson et al., 2013) as part of upscaling these options could provide a cost-effective approach to identifying and resolving any important issues that arise as these options develop.

Oxy-fuel combustion, either via pulverised fuel or CFBC, operates with flue gas recycling to enhance the flue gas CO_2 concentration and limit high temperatures as an alternative option for BECCS. Whereas post-combustion capture has the energy penalty downstream of the power plant (for CO_2 separation from the rest of the flue gas components in the capture unit), oxy-fuel has an upstream energy penalty for O_2 production in the ASU prior to combustion. Whilst both technologies are retrofittable, air leakage into the boiler in a retrofitted oxy-combustion plant can be considerable and prove to be a limiting factor here. Somewhere between air firing and oxy-combustion is enriched-air combustion, where the O_2 content of the combustion air is augmented by additions from an ASU. Although this has led to some limited interest in improving the capture performance through reductions in the total flue gas volumetric flow rate, the capital and operational costs of running both an upstream air separation system and a downstream CO_2 capture facility may outweigh the overall benefits of either of these capture systems separately.

Ways to improve these systems are often generic to capture, such as better thermal integration of power and capture plants, especially for solvent-based systems to effectively use minimal amounts of extracted steam to reduce the energy penalty. Others such as the development of solvents designed for biomass flue gases are clearly BECCS specific. Both, however, will be required to ensure faster deployment and exploitation of CCS in general and BECCS in particular. In addition to extensive laboratory and pilot–/large-scale biomass combustion trials, both air and oxy firing, demonstration projects that incorporate power generation and capture from biomass are therefore needed to support the further development of solvent-based post-combustion capture for BECCS applications and the downstream dehydration and other treatment processes for oxy-fuel capture. This will in turn lead to demonstration projects that are more effective, and thus, the roll-out of the technologies on a larger scale can be timelier.

In the light of the issues outlined in this chapter and throughout the preceding chapters of the book, the deployment potential of the capture options above needs to be considered in terms of the applications of these technologies for biomass fuels. Overall, although much research has been conducted in the broad field of BECCS, only a limited amount has focused either on post-combustion or oxy-combustion capture from biomass fuels – the primary concerns here. Consequently, there are clear openings for much additional work in this area to be undertaken to make the potential of net negative emissions from BECCS a global reality. Based on comments from the likes of the Intergovernmental Panel on Climate Change, the UK Committee on Climate Change, the EU Energy Roadmap 2050, the European Biofuels Technology Platform and the Zero Emission Platform, carbon-negative solutions such as BECCS need to be deployed widely to meet carbon budgets and reduce climate change in a cost-effective manner.

As post- and oxy-combustion capture technologies are the most advanced, from fossil and biomass fuels, it seems most likely and most practical to exploit these technologies at present to meet stringent emission limits and mitigate climate change as best we can, although there are a significant number of other avenues and research opportunities for future development and deployment. All these net negative emissions options will need to be utilised to achieve maximum potential carbon savings.

References

Abraham, B.M., Asbury, J.G., Lynch, E.P., and Teotia, A.P.S. (1982). Coal-oxygen process provides CO_2 for enhanced oil recovery. *Oil and Gas Journal* **80**: 68–75.

Al-Qayim, K., Nimmo, W., and Pourkashanian, M. (2015). Comparative techno-economic assessment of biomass and coal with CCS technologies in a pulverized combustion power plant in the United Kingdom. *International Journal of Greenhouse Gas Control* **43**: 82–92.

Álvarez, L., Riaza, J., Gil, M.V. et al. (2011). NO emissions in oxy-coal combustion with the addition of steam in an entrained flow reactor. *Greenhouse Gases: Science and Technology* **12**: 180–190.

Álvarez, L., Yin, C., Riaza, J. et al. (2014). Biomass co-firing under oxy-fuel conditions: a computational fluid dynamics modelling study and experimental validation. *Fuel Processing Technology* **120**: 22–33.

Arias, B., Pevida, C., Fermoso, J. et al. (2008). Influence of torrefaction on the grind ability and reactivity of woody biomass. *Fuel Processing Technology* **89** (2): 169–175.

Bedell, S.A. (2009). Oxidative degradation mechanisms for amines in flue gas capture. *Energy Procedia* **1**: 771–778.

Benetto, E., Popovici, E.-C., Rousseaux, P., and Blondin, J. (2004). Life cycle assessment of fossil CO_2 emissions reduction scenarios in coal-biomass based electricity production. *Energy Conversion and Management* **45**: 3053–3074.

Berstad, D., Arasto, A., Jordal, K., and Haugen, G. (2011). Parametric study and benchmarking of NGCC, coal and biomass power cycles integrated with MEA-based post-combustion CO_2 capture. *Energy Procedia* **4**: 1737–1744.

Blomen, E., Hendriks, C., and Neele, F. (2009). Capture technologies: improvements and promising developments. *Energy Procedia* **1**: 1505–1512.

Chandan, P., Richburg, L., Bhatnagar, S. et al. (2014). Impact of fly ash on monoethanolamine degradation during CO_2 capture. *International Journal of Greenhouse Gas Control* **25**: 102–108.

Chen, L., Yong, S.Z., and Ghoniem, A.F. (2012). Oxy-fuel combustion of pulverized coal: characterization, fundamentals, stabilization and CFD modelling. *Progress in Energy and Combustion Science* **38**: 156–214.

Ciftja, A.F., Hartono, A., Grimstvedt, A., and Svensdsen, H.F. (2012). NMR study on the oxidative degradation of MEA in presence of Fe^{2+}. *Energy Procedia* **23**: 111–118.

Darde, A., Prabhakar, R., Tranier, J.P., and Perrin, N. (2009). Air separation and flue gas compression and purification units for oxy-coal combustion systems. *Energy Procedia* **1**: 527–534.

Davidson, R.M. and Santos, S.O. (2010). *Oxyfuel Combustion of Pulverised Coal*. IEA Clean Coal Centre.

Deng, L., Zhang, T., and Che, D. (2013). Effect of water washing on fuel properties, pyrolysis and combustion characteristics, and ash fusibility of biomass. *Fuel Processing Technology* **106**: 712–720.

Dhungel, B. (2010). *Experimental investigations on combustion and emission behaviour during oxy-coal combustion*. PhD thesis. University of Stuttgart.

Domenichini, R., Gasparini, F., Cotone, P., and Santos, S. (2011). Techno-economic evaluation of biomass fired or co-fired power plants with post combustion CO_2 capture. *Energy Procedia* **4**: 1851–1860.

Fleig, D., Normann, F., Andersson, K. et al. (2009). The fate of sulphur during oxy-fuel combustion of lignite. *Energy Procedia* **1**: 383–390.

Florin, N. and Fennell, P.S. (2010). Assessment of the validity of "Approximate minimum land footprint for some types of CO_2 capture plant" provided as a guide to the Environment Agency assessment of Carbon Capture Readiness. In: *DECC's CCR Guide for Applications under Section 36 of the Energy Act 1998*.

Fogarasi, S. and Cormos, C.-C. (2015). Techno-economic assessment of coal and sawdust co-firing power generation with CO_2 capture. *Journal of Cleaner Production* **103**: 140–148.

Garcia-Maraver, A., Mata-Sanchez, J., Carpio, M., and Perez-Jimenez, J.A. (2016). Critical review of predictive coefficients for biomass ash deposition tendency. *Journal of the Energy Institute* doi: 10.1016/j.joei.2016.02.002.

Gudka, B., Jones, J.M., Lea-Langton, A.R. et al. (2015). A review of the mitigation of deposition and emission problems during biomass through washing pre-treatment. *Journal of the Energy Institute* doi: 10.1016/j.joei.2015.02.007.

Jiang, L., Hu, S., Wang, Y. et al. (2015). Catalytic effects of inherent alkali and alkaline earth metallic species of steam gasification of biomass. *International Journal of Hydrogen Energy* **40**: 15460–15469.

Jones, J.M., Darvell, L.I., Bridgeman, T.G. et al. (2007). An investigation of the thermal and catalytic behaviour of potassium in biomass combustion. *Proceedings of the Combustion Institute* **31**: 1955–1963.

Jordal, K., Anheden, M., Yan, J., and Strömberg, L. (2004). Oxyfuel combustion for coal-fired power generation with CO_2 capture – opportunities and challenges. *GHGT-7 Conference: Vancouver, Canada*.

Kambo, H.S. and Dutta, A. (2014). Strength, storage, and combustion characteristics of densified lignocellulosic biomass produced via torrefaction and hydrothermal carbonization. *Applied Energy* **135**: 182–191.

Kambo, H.S. and Dutta, A. (2015). Comparative evaluation of torrefaction and hydrothermal carbonization of lignocellulosic biomass for the production of solid biofuel. *Energy Conversion and Management* **105**: 746–755.

Khorshidi, Z., Ho, M.T., and Wiley, D.E. (2013). Techno-economic study of biomass co-firing with and without CO_2 capture in an Australian black coal-fired power plant. *International Journal of Greenhouse Gas Control* **21**: 191–202.

Khorshidi, Z., Ho, M.T., and Wiley, D.E. (2014). The impact of biomass quality and quantity on the performance and economics of co-firing plants with and without CO_2 capture. *Energy Procedia* **37**: 6035–6042.

Koornneef, J., van Breevoort, P., Hamelinck, C. et al. (2012). Global potential for biomass and carbon dioxide capture, transport and storage up to 2050. *International Journal of Greenhouse Gas Control* **11**: 117–132.

Lucquiaud, M. and Gibbins, J. (2011). On the integration of CO2 capture with coal-fired power plants: A methodology to assess and optimise solvent-based post-combustion capture systems. *Chemical Engineering Research and Design* **89**: 1553–1571.

Merkel, T.C., Lin, H., Wei, X., and Baker, R. (2010). Power plant post-combustion carbon dioxide capture: an opportunity for membranes. *Journal of Membrane Science* **359**: 126–139.

MIT (2015). *Power plant carbon dioxide capture and storage projects.* http://sequestration.mit.edu/tools/projects/index.html(accessed 15 December 2017).

Mumford, K.A., Wu, Y., Smith, K.H., and Stevens, G.W. (2015). Review of solvent based carbon-dioxide capture technologies. *Frontiers of Chemical Science and Engineering* **9**: 125–141.

National Energy Technology Laboratory (2012). *Greenhouse Gas Reductions in the Power Industry Using Domestic Coal and Biomass – Volume 2: Pulverized Coal Plants, Final Report. DOE/NETL-2012/1547.*

Niu, Y., Tan, H., and Hui, S. (2016). Ash-related issues during biomass combustion: alkali-induced slagging, silicate melt-induced slagging (ash fusion), agglomeration, corrosion, ash utilization, and related countermeasures. *Progress in Energy and Combustion Science* **52**: 1–61.

Nutalapati, D., Gupta, R., Moghtaderi, B., and Wall, T.F. (2007). Assessing slagging and fouling during biomass combustion: a thermodynamic approach allowing for alkali/ash reactions. *Fuel Processing Technology* **88**: 1044–1052.

Okazaki, K. and Ando, T. (1997). NO_x reduction mechanism in coal combustion with recycled CO_2. *Energy* **22**: 207–215.

Saddawi, A., Jones, J.M., and Williams, A. (2012). Influence of alkali metals on the kinetics of the thermal decomposition of biomass. *Fuel Processing Technology* **104**: 189–198.

SaskPower (2016). *SaskPower CCS: boundary dam carbon capture project.* http://www.saskpower.com/our-power-future/carbon-capture-and-storage/boundary-dam-carbon-capture-project/ (accessed 15 December 2017).

Schakel, W., Meerman, H., Talaei, A. et al. (2014). Comparative life cycle assessment of biomass co-firing plants with carbon capture and storage. *Applied Energy* **131**: 441–467.

Scheffknecht, G., Al-Makhadmeh, L., Schnell, U., and Maier, J. (2011). Oxy-fuel coal combustion – a review of the current state-of-the-art. *International Journal of Greenhouse Gas Control* **5**: S16–S35.

Shaddix, C.R. and Molina, A. (2009). Particle imaging of ignition and devolatilization of pulverized coal during oxy-fuel combustion. *Proceedings of the Combustion Institute* **32**: 2091–2098.

Shah, M.M. (2006). Oxy-fuel combustion for CO_2 capture from PC boilers. *31st International Conference on Coal Utilisation and Fuel Systems. Clearwater, FL.*

Smart, J., Lu, G., Yan, Y., and Riley, G. (2010). Characterisation of an oxy-coal flame through digital imaging. *Combustion and Flame* **157**: 1132–1139.

Smart, J.P., O'Nions, P., and Riley, G.S. (2010). Radiation and convective heat transfer, and burnout in oxy-coal combustion. *Fuel* **89**: 2468–2476.

Smart, J.P. and Riley, G.S. (2011). On the effects of firing semi-anthracite and bituminous coals under oxy-fuel firing conditions. *Fuel* **90**: 2812–2816.

Smith, A.M. and Ross, A.B. (2016). Production of bio-coal, bio-methane and fertilizer from seaweed via hydrothermal carbonisation. *Algal Research* **16**: 1–11.

Spath, P.L. and Mann, M.K. (2004). *Biomass Power and Conventional Fossil Systems with and without CO2 Sequestration – Comparing the Energy Balance, Greenhouse Gas Emissions and Economics*. NREL technical report. *NREL/TP-510-32575*.

Szuhánszki, J. (2014). *Advanced oxy-fuel combustion for carbon capture and sequestration*. PhD thesis. The University of Leeds.

Toftegaard, M.B., Brix, J., Jensen, P.A. et al. (2010). Oxy-fuel combustion of solid fuels. *Progress in Energy and Combustion Science* **36**: 581–625.

Trabadela, I., Chalmers, H., and Gibbins, J. (2014). Oxy-biomass ignition in air and relevant oxy-combustion atmospheres for safe primary recycle and oxy-burner development. *Energy Procedia* **63**: 403–414.

Voice, A.K. and Rochelle, G.T. (2011). Oxidation of amines at absorber conditions for CO_2 capture from flue gas. *Energy Procedia* **4**: 171–178.

Wall, T.F. (2007). Combustion processes for carbon capture. *Proceedings of the Combustion Institute* **31**: 31–47.

Wall, T., Liu, Y., Spero, C. et al. (2009). An overview on oxyfuel coal combustion – state of the art research and technology development. *Chemical Engineering Research and Design* **87**: 1003–1016.

Wall, T., Stanger, R., and Santos, S. (2011). Demonstrations of coal-fired oxy-fuel technology for carbon capture and storage and issues with commercial deployment. *International Journal of Greenhouse Gas Control* **5**: S5–S15.

Watson, R., Lucquiaud, M., and Gibbins, J. (2013). Development of a portable facility for in-situ testing of carbon capture technologies. *2nd Post Combustion Capture Conference, Bergen, Norway (September 2013)*.

Yin, C. and Yan, J. (2016). Oxy-fuel combustion of pulverized fuels: combustion fundamentals and modeling. *Applied Energy* **162**: 742–762.

ZEP (2012). *Biomass with CO_2 Capture and Storage (Bio-CCS) – The Way Forward for Europe, European Technology Platform for Zero Emission Fossil Fuel Power Plants. ZEP/EBTP Bio-CCS JTF Report*.

Zhao, L., Riensche, E., Blum, L., and Stolten, D. (2010). Multi-stage gas separation membrane processes used in post-combustion capture: energetic and economic analyses. *Journal of Membrane Science* **359**: 160–172.

4

Pre-combustion Technologies

Amanda Lea-Langton[1] and Gordon Andrews[2]

[1] *Tyndall Centre for Climate Change Research, School of Mechanical Aerospace and Civil Engineering, University of Manchester, UK*
[2] *School of Chemical and Process Engineering, University of Leeds, UK*

4.1 Introduction

Post-combustion carbon capture and storage (CCS) is a mature technology which is in use in full-scale plants, whereas the use of pre-combustion technologies for CCS in power production is less established. However, many of the processes involved in pre-combustion CCS are well established in other industries such as refineries and ammonia production. Biomass energy and carbon capture and storage (BECCS) has the potential to reduce overall atmospheric carbon, but conventional CCS has been considered a transitional technology required until renewable sources can fully supply energy demand (Trapp et al., 2015).

There are currently at least 33 integrated gasification combined cycle (IGCC) plants operating with pre-combustion carbon capture, and in 2015 there were at least 36 more under construction (Schultz, 2016). Most of these plants are refinery-based and not grid-connected electricity plants; petroleum coke is the most common gasification fuel. In comparison, post-combustion CO_2 capture plants are in their infancy, with only a few plants operating around the world and none in the United Kingdom. CCS schemes for electric power generation have had a low uptake, inhibited by policies of governments around the world that have failed to categorise them as 'green' sources of electricity, and thus available for 'green electricity' subsidies. It is not primarily technical issues that have resulted in few combustion CCS plants being built, but an economic issue. Because the higher cost of electricity with pre- or post-combustion carbon capture cannot be recovered without a subsidy, the plants are not built. These extra electricity costs are reviewed in this chapter. In principle, if a renewable biomass is used as the feedstock for IGCC plants with carbon capture, then negative CO_2 emissions will occur as the biomass is 'green' and the CO_2 recovered gives an overall net reduction in CO_2 emissions.

Pre-combustion carbon capture encompasses gasification processes that convert carbon-based fuels into a hydrogen-rich synthetic fuel gas, which is known as syngas. Further reaction of the gaseous components with water via the water-gas shift (WGS)

Biomass Energy with Carbon Capture and Storage (BECCS): Unlocking Negative Emissions, First Edition.
Edited by Clair Gough, Patricia Thornley, Sarah Mander, Naomi Vaughan and Amanda Lea-Langton.
© 2018 John Wiley & Sons Ltd. Published 2018 by John Wiley & Sons Ltd.

reaction converts the carbon monoxide into carbon dioxide (CO_2), which can be separated from the fuel gas and sequestered. As the CO_2 concentrations in the syngas after WGS are higher than in the post-combustion flue gases, the CO_2 solvent capture is more efficient and uses less power. In the case of fossil fuels, undesirable species such as sulphur-containing species can also be removed during the process. Biomass is generally a sulphur-free feedstock but can contain other trace elements that would affect the process, such as alkali metals and chlorine.

4.2   The Integrated Gasification Combined Cycle (IGCC)

Solid fuels such as coal and biomass cannot burn directly in a gas turbine; they must first be gasified to syngas and the gas burnt in a combined-cycle gas turbine (CCGT). The combined gasification and CCGT is referred to as an IGCC. A CCGT has two electric power generators, one driven directly by the gas turbine power turbine and one driven by a steam turbine using the waste heat from the gas turbine exhaust. In a typical large combined-cycle coal plant, the gas turbine would generate 300 MW and the steam turbine 150 MW. Most current CCGT plants operate with natural gas (NG) as the fuel and for the latest technology the thermal efficiency is 62%. IGCC plants provide fuel flexibility and comparatively low emissions compared with pulverised coal plants, which makes them promising options for biomass combustion (Trapp et al., 2015).

Gasification of solid fuels involves fuel-rich combustion, where CO, hydrogen and CO_2 are the normal products of the reaction with some hydrocarbons present due to inefficient rich combustion. With air as the oxidant, the syngas has a large proportion of nitrogen from the air, as well as CO and H_2 and hence has a low calorific value. With pure oxygen gasification, there is no nitrogen in the syngas and so this has a higher calorific value. However, the separation of oxygen from nitrogen in the air in an upstream air separation unit (ASU) takes energy and this reduces the overall thermal efficiency. It takes 0.81 MJ to produce 1 kg of oxygen from air (Iki et al., 2009) and typically for a 262 MW net electrical power output the ASU power is 22 MW or 8.4% of the power output, which is an 8.4% loss in the plant thermal efficiency (Li and Wang, 2012). An air gasifier does not have this loss, but does have a higher mass flow due to the N_2 present, which increases the size and pressure loss of the gasifier.

The gas turbine combustor and fuel flow system for air separation should have larger fuel flow pipes than oxygen-based gasification, as the volumetric flow rate of the oxygen gasification process is much lower. This also requires different combustor designs, as the stoichiometric A/F is different for the two syngases, with the air-gasified syngas having a much lower stoichiometric A/F than for an oxygen-based syngas. Thus, an IGCC gas turbine plant has to have a combustor designed for syngas and this will have to have a different fuel supply system and different air flow splits in the combustor liner compared with that for natural gas.

IGCC power generation is ideal for solid or liquid biomass as biomass has a much higher volatile content than coal and normally lower ash content. The high volatile content makes biomass easier to gasify. A consideration for using biomass is nature and concentration of trace elements within the fuels, some of which are semi-volatile and can carry through to the gas phase. The trace inorganic elements can be measured as ash. If these are not cleaned from the syngas, they could potentially foul and erode the turbine blades.

Although syngas could be used in a conventional steam boiler for power generation, the higher thermal efficiency of CCGT, at 62%, makes this the preferred route. The latest coal-fired supercritical steam generation plants for power generation have a 50% thermal efficiency, which is much lower than for CCGT. This, combined with the lower power loss from the CO_2 extraction after the WGS plant, due to the higher CO_2 concentration, means that pre-combustion CO_2 capture should be the preferred route for CCS. Unfortunately, pre-combustion CCS with the IGCC plant is not a retrofit option, and new plants have to be built to exploit the latest CCGT and IGCC technologies. In contrast, post-combustion CCS can be retrofitted to existing plants, even if it does not make economic sense to do so, as most existing coal-fired plants are near the end of their design life.

4.3   Gasification of Solid Fuels

The generic process for oxygen-blown gasification and carbon capture of solid fuels at atmospheric pressure uses the four stages shown in Figure 4.1.

Stage 1. Oxygen is separated from nitrogen in an ASU or oxygen plant to produce a high-purity oxygen stream.

Stage 2. The oxygen and slurry of milled solid fuels mixed with water are simultaneously fed into a gasifier, where the fuel will partially oxidise to produce syngas as per Reaction (4.1).

$$4CH + 2H_2O + O_2 \rightarrow 4H_2 + 4CO \tag{4.1}$$

Fuel + water + oxygen → hydrogen + carbon monoxide.

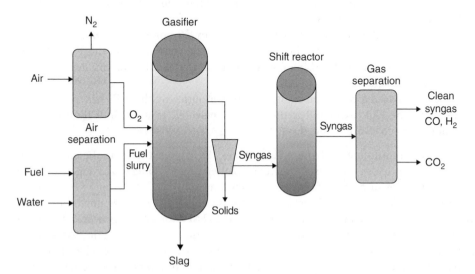

Figure 4.1 Generic oxygen-based gasification processes for solid fuels mixed with water as slurry feed with CO_2 separation.

The syngas is a mixture consisting predominantly of hydrogen, carbon monoxide, carbon dioxide and water. The exit from the gasifier usually includes a cyclone to remove any particulate matter from the syngas stream.

Stage 3. The syngas passes to a shift reactor where steam is added. Carbon monoxide and water react according to the WGS reaction to produce hydrogen and carbon dioxide as shown in Reaction (4.2).

WGS reaction

$$CO + H_2O \leftrightarrow CO_2 + H_2 \tag{4.2}$$

Carbon monoxide + water $\leftrightarrow$ carbon dioxide + hydrogen.

In the case of coal gasification, there is another process at this stage to remove sulphur from the gas stream (desulphurisation). This would not usually be needed for biomass, which has low fuel sulphur content; however, other contaminants would potentially be present.

Stage 4. CO_2 is separated from the gas stream, compressed and dehydrated prior to storage to leave a higher-purity hydrogen-rich fuel gas. The fuel gas can be combusted directly in a gas turbine as in the case of an IGCC plant, or can be further refined via the Fischer-Tropsch process to produce a liquid transportation fuel, or refined to higher-purity hydrogen for use in fuel cells.

Gasification with a gas turbine for power generation gives four options for the gasification process.

1) The gasifier can operate at atmospheric pressure on air or.
   (with an ASU) on oxygen. This will require a compressor to overcome the pressure loss of the ASU and a further compressor to raise the pressure of the syngas to above that of the GT compressor delivery pressure.
   The high mass flow of the gasified fuel for air gasification can lead to problems in a gas turbine as the turbine mass flow is greater than the compressor air mass flow. The alternatives are as follows:
2) Operate the gasifier at pressure with the air from the compressor outlet passing to the gasifier and the gases then being fed back to the combustor. In this case, the gas turbine compressor and turbine matching are not affected. Most models of IGCC use this approach.
3) For oxygen-based gasification using compressor outlet air, the ASU has to operate at pressure. The CO_2 extracted is then also at elevated pressure and the final CO_2 compression costs are reduced.

The energy requirement of the ASU reduces the overall thermal efficiency of the system, which is a loss not present in an air gasifier. Li and Wang (2012) calculated the electric power consumption of the ASU for oxygen supply as 37 MW for a net electrical power of 268 MW or 13.8% of the power. Majoumerd et al. (2015) estimated the ASU power as 50 MW from a net electrical power of 402 MW or 12.4%. However, the pressure loss of the ASU implies that the air has to be further compressed prior to the ASU. Giuffrida (2014) assessed the ASU power demand for a 500-MW net electric power CCGT generator plant as 17 MW for the additional air compressor and 12 MW for the ASU, a combined energy cost of 29 MW or 5.8%, which is much lower than that of other process models.

The syngas from an air gasifier is of a lower mass calorific value than for an oxygen-blown gasifier. The air-blown gasifier produces a gas of high volume flow rate due to the presence of nitrogen, and this needs a different combustion system and fuel supply system. Thus, the gasifier and gas turbine have to be designed together, taking into account whether oxygen or air gasification is being used.

The choice between the four gasification options is dependent on the differences in capital costs and the operational thermal efficiency of the overall system. The thermal efficiency of air- and oxygen-blown gasifiers will be compared in this chapter. Higher-pressure gasification is the most efficient, but requires a careful integration of various processes with the gas turbine combustor. Compressed-air-based systems have proved to exhibit operational difficulties, particularly, start-up difficulties, and many IGCCs do not use the more efficient high-pressure route (Parulekar, 2011). It is possible to have a mixed system with air and oxygen mixed together for the gasifier. This can reduce the cost of the ASU and improve the gasification reaction compared with air-only gasification.

Some typical compositions of syngas for oxygen- and air-blown gasifiers are shown in Table 4.1. Gasification reactor designs include fixed beds, fluidised beds and entrained flow designs. The most common gasification system in commercial use is entrained flow gasifier, as the coal ash is removed as slag from the gasifier and it has good fuel and oxidant mixing to give more uniform particle gasification. The variability in syngas composition for real and modelled gasifiers in Table 4.1 is due to differences in the gasifier design, operating conditions and coal composition. Bonzani (2006) showed that the syngas provided for the first commercial IGCC power plant in Italy is of variable composition. The feedstock for the gasifier was a refinery tar waste, which was converted into a hydrogen-rich gas using the oxygen-blown IGCC. Table 4.1 shows the variability in composition for a real gasification plant. The range in CO was 45–33%, which is similar to that for coal, and the range in H_2 was 45–33%, which is higher than that for coal. This range in compositions resulted in a lower heating value (LHV) range from 13 to 15 MJ/kg and a gas generator thermal efficiency of 67–78%, which is very similar to that for coal gasification. For comparison, the gasification of NG with a steam-methane reformer to produce hydrogen was reviewed by Corradetti and Desideri (2006) and a range of installations had gasification efficiencies from 71.6% to 78.9%. They also gave details of two NG gasification plants with CCS using pre-combustion CO_2 separation, and these had gasification efficiencies of 76–83%, which are a little higher than the values for coal IGCC with pre-combustion CCS, as shown below.

Table 4.1 shows that there is no difference in the predicted syngas composition for gasification at 1 bar and 24 bars. Most of the studies of air- and oxygen-based gasification for coal were carried out with pressurised gasification using the gas turbine compressor air, boosted to a higher pressure to overcome the gasifier pressure loss.

For air gasification of coal, the range of syngas compositions in Table 4.1 for CO is 20.6–30.5% for six different gasifiers and for hydrogen 10.1–12.9%. The range of LHV for air gasification in Table 4.1 is 3.47–4.64 MJ/kg$_{syngas}$ (four gasifiers). For oxygen gasification, the range of CO is 38.4–58.8% and for hydrogen 23.3–38.5% for seven gasifiers. For the oxygen-blown gasification gas calorific value (CV) the range in Table 4.1 is 9.49–12.9 MJ/kg$_{syngas}$ (five gasifiers). The efficiency of gasification in Table 4.1 (with the assumptions discussed above) ranges from 69.5% to 79.6% for air gasification of coal and from 51.6% to 82.7% for oxygen gasification of coal. Thus, it

Table 4.1 Composition percentage of gasified coal including water-gas shift for CO_2 removal (CCS in this table).

CV feed (MJ/kg)	Tar ~40	Coal dry 28.5	Coal dry 29	Coal slurry 29	Coal slurry 29	Coal dry 29	Coal dry 29	Coal slurry 29	Coal dry	Coal slurry		Bitumen coal 24.6	PRB coal biomass 18.1	Brown coal 10% H_2O 7.1	Coal dry 25.2	Coal dry 25.2	Coal dry 28.2	75% Coal 25% biomass 24.9
Gasifier	O_2	Air + 20% H_2O 1 bar MHI	Air 24 bar	O_2 1 bar	O_2 24 bar	O_2 24 bar	Air CCS 24 bar	O_2 CCS 24 bar	Air 20 bar	O_2 20 bar	Air 29.8 bar	Air CCS 32.5 bar	Air 25% O_2 25.3 bar	Air 25% O_2 25.3 bar	O_2 43 bar	O_2 43 bar CCS	O_2 Fluid bed 25 bar	O_2 Pressure ASU CCS
CO (%)	49–61	26.3	28.2	51.4	51.4	58.8	0.6	0.7	30.5	38.4	27.6	0.7	20.6	28.2	50.0	1.2	39.8	
H_2 (%)	45–33	10.1	12.9	33.3	33.3	32.2	40.5	58.2 (90 dry)	10.5	27.5	10.7	37.3	11.5	12.0	23.3	85.7	38.5	
CO_2 (%)	2.9–3.4	3.1	3.2	12.7	12.7	2.0	3.1	4.4	2.8	12.0	3.2	1.7	8.3	4.95	2.5	3.6	7.4	
CH_4 (%)	0.6–0.3	—	3.7	0.0	0.0	1.0	3.8	0.0	0.5	0.1	0.6	0.6	0.7	0.6	0.0	0.0	0.03	
H_2O (%)	0.2–0	2.8	0.3	0.3	0.3	0.3	0.3	35.1	5.0	20.0	0.3	0.3	0.3	0.3	16.4	0.0	11.3	
N_2 (%)	2.2–1.7	56.1	51.2	2.3	2.3	5.7	51.2	1.6	50.9	1.5	57.1	58.7	58.1	53.3	6.6	8.7	0.02	
Ar (%)	—	—	0.6	0	0	0	0.6	0	0.5	0.1	0.6	0.7	0.7	0.6	0.7	0.8	—	
MW	—	—	24.8	21.3	21.3	19.8	17.6	10.1										
LHV (MJ/m^3)													4.07	5.08				
LHV (MJ/kg$_{gas}$)	15.1–12.9		4.65	10.8	10.8	12.9	7.58	15.7			4.21	5.2	3.47	4.44	9.64	39.2	9.49	7.3
LHV (MJ/kg$_{fuel}$)	31.1–26.6		21.9	20.1	20.1	24.0	35.2	29.2			19.6	24.2	12.6	5.4	17.9	72.9	17.6	12.9
Efficiency (%)	78–67		75.4	69.3	69.3	82.7	122	100.7			79.6 (77)	98.3	69.5	75.9	71.2	289	62.6	51.6
Reference	Bonzani (2006)	Wang and Weng (2014)	Li et al., 2012	Li and Wang (2012)	Li and Wang (2012)	Li and Wang (2012)	Li and Wang (2012)	Li and Wang (2012)	Carbo et al. (2007)	Carbo et al. (2007)	Giuffrida et al. (2012)	Giuffrida et al. (2012)	Giuffrida (2014)	Giuffrida (2014)	Majoumerd et al. (2015)	Majoumerd et al. (2015)	Kawabata et al. (2012)	Galanti et al. (2009)

may be concluded that there is little advantage to oxygen gasification over air. It is clear from the above wide range in energy conservation from the fuel to the gas that there is a big difference between a well-designed system and some of those in the literature, particularly for oxygen-blown gasification.

Table 4.1 shows the low LHV on a MJ/kg$_{syngas}$ basis that results for air gasification compared with oxygen gasification. The LHV comparison in terms of MJ/kg$_{syngas}$ is not the best comparison, as the mass of syngas is significantly higher for air-blown gasifiers compared with oxygen-blown ones. It would be much more sensible if the LHV was expressed as MJ/kg$_{fuel}$ so that the proportion of the energy in the original fuel that was delivered in the syngas could be determined, which is gasification efficiency. Unfortunately, this is rarely done in the literature, but Giuffrida (2014a) called this 'cold gas efficiency', which is the thermal efficiency of the process with the input fuel and output gas at the same ambient temperature. The values of Giuffrid for different coals are shown in Table 4.1, and they agree with the calculations made in this chapter for other gasifiers. To convert the syngas LCV in the units of MJ/kg$_{syngas}$ into MJ/kg$_{fuel}$ in Table 4.1, the gasifier equivalence ratio has been assumed to be 3 (Wang and Weng, 2014). For coal, the stoichiometric A/F has been taken as 11 and 6 for biomass, unless the reference gave the actual value. For conversion MJ/kg$_{syngas}$ is multiplied by the ratio of the mass of syngas to the mass of the fuel, which is (1 + A/F). For oxygen-based gasification, the stoichiometric O$_2$/fuel ratio is the stoichiometric A/F for air times 0.234, the proportion of oxygen in air by mass. Table 4.1 shows the resultant calculations of the LHV in MJ/kg$_{fuel}$ and where the reference has given the LHV for the fuel in the same units, the gasification thermal efficiency has been determined. The comparison of the modelled thermal efficiencies of IGCC plants given later shows similar efficiencies for air and oxygen gasification. This shows that the energy content of the gas compared with that in the fuel is similar for the two processes, as shown in Table 4.1 in terms of MJ/kg$_{fuel}$. This is not apparent when the LCV is expressed in MJ/kg$_{syngas}$.

The presence of CH$_4$ in the syngas is a measure of inefficient gasification. If the gasifier is operated at the optimum rich equivalence ratio for coal, there should be no CH$_4$ as it is not an equilibrium product of rich combustion, which should only have H$_2$, CO, CO$_2$ and N$_2$ if an air gasifier is used. Water added to the gasifier in the slurry increases the H$_2$ yield and decreases the CO (Aponte and Gordillo, 2013), but the mixture with the greatest energy in the syngas is richer as water increases and has an equivalence ratio of 5 for a water-to-fuel ratio of 1 compared with an equivalence ratio of 3 for the maximum energy in the syngas with a dry feed and no water or steam added (Aponte and Gordillo, 2013). Thus, the gasifier operating conditions have to be matched to that of the fuel, as the stoichiometric A/F depends on the fuel composition.

A disadvantage of air gasification is that the volumetric flow is greater than that for oxygen gasification. This increases flow pressure losses in the gasifier and the combustor has to be designed to accommodate the higher syngas flow rate and cannot be the same combustor as for oxygen gasification. However, this is not a problem in combustor and fuel system design. The stoichiometric flame temperature is lower for air gasification syngas compared with oxygen gasification, but this merely means that a different air flow split in the combustor has to be used. The lower peak flame temperature also means that NO$_x$ emissions will be lower for air gasification.

The literature results for biomass gasification in Table 4.2 show a range of CO of 11.5–27% for air gasification and a range of H$_2$ on a dry basis from 8.7% to 25%. The

Table 4.2 Biomass gasification syngas composition.

CV feed (MJ/kg)	Soft wood chips 19.6	Oak Dry 20.5	Sorghum Dry 14.3 17.4 daf (dry ash free)	Wood	Wood	Wood	Biomass 15.5	Dairy Biomass 21.5 daf
Gasifier	Air	Air	Air	Air fixed-bed updraft	Air fixed-bed downdraft	Air fluidised bed	Air fixed bed updraft	Fixed bed updraft Ø = 3 Air/steam Steam/fuel 0.8
CO (%)	16.0	18	14.6	24	21	14	27	11.5
H$_2$ (%)	7.7 8.7 dry	16	12.9 14.2 dry	11	17	9	17.3	25
CO$_2$ (%)	15.3	13	15.2	9	13	20	9.0	26
CH$_4$ (%)	7.6	1.5	5.4	3	1	7	4.0	1.5
H$_2$O (%)	11.0	–	9.4	–	–	–	4.1	–
N$_2$ (%)	41.5	48	42.3	53	48	50	38.6	36
Ar (%)	–	–	–	–	–	–	–	–
MW	–	24.8	25.4				24.05	
LHV (MJ/m^3)	5.59	4.8	5.35 6.52 dry	5.5	5.7	5.4		4.9
LHV (MJ/kg$_{gas}$)	4.72	4.35	4.72 6.16 dry				5.38	
LHV (MJ/kg$_{fuel}$)	14.2	13.0	14.1 wet 18.5 dry				15.04	
Efficiency (%)	72%	63.6%	99% wet 106 dry				97%	62%
Reference	Lepszy and Chmielniak (2010)	Pinta and Vergnet (1994)	Baldacci et al. (1994)	Gordillo et al. (2009)	Gordillo et al. (2009)	Gordillo et al. (2009)	Porta et al. (2006)	Gordillo et al. (2009)

CO range is below that for air gasification of coal but the range for H_2 spans that for coal. For CO_2 biomass, atmospheric pressure gasification gives 13–15.3%, which is much higher than for coal with a range of 1.7–4.4% and most plants at 3.2%. If a WGS reactor is added to the biomass gasification plant, the total hydrogen production would be less than for coal as the CO content of the syngas is lower. The three biomass-based syngases in Table 4.1 have high CH_4, 1.5–7.6%, and, in addition to poor rich combustion in the gasifier, the gasifier could be operating too rich where equilibrium CH_4 starts to occur (Aponte and Gordillo, 2013). If a coal gasifier is operated on biomass with no change in the air or oxygen flows, the stoichiometry of the gasification zone will be richer as the stoichiometric A/F for woody biomass is typically 6/1 compared with 12/1 for coal. It is considered that the biomass gasifiers in Table 4.1 may not be optimally designed and may be operating at conditions more appropriate for coal gasification.

The LHV range for biomass air gasification in Table 4.2 is 4.35–5.38 MJ/kg$_{syngas}$ and this is similar to that of coal and the maximum value is higher compared to any coal gasifier. Thus, the air gasification of biomass generates a syngas having energy content and composition similar to that of coal air gasification. The thermal efficiency of biomass gasification varies from 62% to 100%. These estimated thermal efficiencies indicate that the thermal efficiency of biomass gasification is similar to that of coal gasification.

Most of the gasifiers in Table 4.1 use a dry solid fuel feed. Some gasifiers are designed around a coal/water slurry fuel feed, as shown in Table 4.1. This results in a high water content of the gasified gas, which can be removed by condensation, but this is energy inefficient and the water is usually kept in the gas, which results in a lower CV, as shown in Table 4.1. Dry-based systems are more efficient and for biomass more practical.

Biomass has very poor grindability in comparison to coal due to its fibrous nature, which leads to higher operational energy costs. Higher wear on the milling surfaces is an issue for fuels that contain silica as trace elements. The fuel density is also substantially different from coal, which can affect the ability to produce a uniform slurry mixture. The particle sizes and distribution of sizes will also affect mixing. The pretreatment of the fuel prior to grinding could be considered to improve this step, for instance, by torrefaction to improve grindability; however, an energy penalty would be incurred. A dry-based fuel feed into the gasifier is preferred for biomass.

One of the most challenging aspects of biomass gasification is the avoidance of tar and inorganic deposits building up within the system, leading to efficiency loss and blockages. Slagging and fouling deposits are typically associated with elements such as sodium and potassium that are present within the biomass fuel, whereas tar is formed from partially pyrolysed fuel that has converted into heavy hydrocarbons, which condense in cooler sections of the plant. In addition, certain trace elements within the biomass can potentially lead to corrosion at high temperature conditions.

Table 4.1 also shows some syngas compositions with pre-combustion CO_2 removal (indicated as CCS), using the WGS reaction and solvent extraction of the CO_2. Table 4.1 shows that the carbon removal process is not perfect and that there is residual CO_2 in the syngas. However, the reduction in the gas turbine exhaust CO_2 is significant and typically 90% of carbon is captured. For oxygen-blown gasification, the syngas has 3–8.7% N_2 depending on the gasifier, and this is due to inefficient separation of oxygen from air in the ASU. Ninety five percentage efficient separation will yield ~4% N_2 in the syngas. The argon results in Table 4.1 indicate that some ASUs are inefficient at removing argon, which is about 1% in air; the table shows argon values between 0% and 1% for oxygen-based gasifiers.

The WGS reactor and the gasifier together are a means of producing a hydrogen-rich gas, as shown in Table 4.1. For air gasification of coal, the hydrogen content in Table 4.1 for the two systems with pre-combustion CCS are 37.3% and 40.5% hydrogen, with the rest being mainly nitrogen. For oxygen gasification with pre-combustion CCS, the two studies in Table 4.1 show hydrogen content on a dry gas basis of 86% and 90%. Majoumerd et al. (2015) assessed the cost of hydrogen from coal, for a 250-MW electric IGCC plant with pre-combustion carbon capture, as 16.5 €/GJ, compared with 8 €/GJ for natural gas gasification or reforming to produce hydrogen, which is double the cost of hydrogen. Producing hydrogen from natural gas with carbon capture is the lowest-cost way of producing low carbon electricity for natural gas as the fuel.

Table 4.1 has only two references for oxygen gasification of coal with pre-combustion CCS, and on a dry basis the hydrogen contents of the gas at 85.7% and 90% are in good agreement. The thermal efficiency of gasification with a WGS reactor is very high due to the conversion of hydrogen in water into gaseous hydrogen. In some cases, this gives more energy in the syngas than in the original fuel, as shown in Table 4.1. We have not found any oxygen gasification results for biomass.

4.4 Carbon Dioxide Separation Technologies

4.4.1 Physical Absorption

Absorption techniques use liquid solvents and usually consist of two separate reactor stages, which are in series:

Absorption stage. The solvent stream is passed counter current to the gas stream so that CO_2 and other acidic gases, such as hydrogen sulphide, are absorbed into the solvent.
Stripping stage. The solvent is regenerated, usually by heating, in order to release CO_2, which can be subsequently captured. This is the part of the system that is energy intensive as the solvent is boiled using steam taken from the steam turbine part of the CCGT and thus, as a consequence, less electricity is generated.

Amine gas treatment to remove hydrogen sulphide and carbon monoxide from gas streams is already an established process in petroleum refineries and natural gas processing. The process is also known as 'scrubbing' or 'sweetening'. Typical amine-based solvents include monoethanolamine (MEA), diethanolamine (DEA) and methyldiethanolamine (MDEA). Solvents used for IGCC are generally hydrophilic and include proprietary mixtures such as Selexol and Purisol. Other solvents include Rectisol, which is essentially methanol. Each system has relative benefits and costs in terms of solvent costs versus the process energy requirements. A number of authors have used the levelised cost of electricity (LCOE) to compare performance (e.g. Siefert et al., 2016; Herzog, 1999; Li et al., 2012). New solvent developments have focused on

CO_2-reactive aspects for CCS applications and include production of ionic liquids. Siefert et al. (2016) compared a hydrophobic solvent with hydrophilic Selexol as a baseline and found that the costs of CO_2 capture could potentially be reduced subject to the commercial-scale synthesis cost of their ionic solvent. This was largely because it was not necessary to cool the syngas for water removal; hence, an energy penalty was avoided.

A key problem with all solvent-based CO_2 removal techniques is that CO_2 has to be released from the solvent by heating or boiling it before recycling. The heating of the amine solvent to release CO_2 uses steam generated from the GT exhaust, which is then not available for steam turbine electric power generation and so the steam-cycle export power production is reduced. The heat required to release CO_2 from MEA was estimated to be $1\,MJ/kg_{CO_2}$ by Consoni and Vigano (2005) and $1.32\,MJ/kg_{CO_2}$ by Hoffmann et al. (2008), which illustrates the uncertainty in CCS process modelling. This is about $3\,MJ/kg_{fuel}$ and for NG with a LCV of $50\,MJ/kg_{fuel}$ this is a 6% reduction in the available output energy of the IGCC process, which for 50% thermal efficiency is 12% of the plant input energy or a thermal efficiency loss of about 12%.

4.4.2 Adsorption Processes

A review of recent research publications by Theo et al. (2016) has indicated that over the last 3 years most research has focused on adsorption technologies. Adsorption processes generally use solid adsorbents and involve the formation of chemical bonds between CO_2 and the material.

Typical adsorbents include activated carbon, zeolites, mesoporous silica, calcium oxide, metal–organic frameworks (MOFs) and porous polymers and alkali-metal-based materials. Others include carbon-based materials such as activated carbon and graphene. There is ongoing research on a wide range of novel materials (e.g. Zhao et al., 2014; Seggiani et al., 2013; Narasimharao and Ali, 2016). A review of promising materials by Lee and Park (2015) concluded that more development is needed in terms of the working adsorption capacity, cycle lifetime and multicycle durability.

4.4.3 Clathrate Hydrates

Another method that is being investigated is the capture of CO_2 within clathrate hydrate crystallisation (Linga et al., 2007), whereby CO_2 is held in a 'cage' of hydrogen-bonded water molecules (Englezos, 1993; Sloan, 1998). The formation of the gas hydrate is often promoted using tetrahydrofuran (THF) (Babu et al., 2016; Lee et al., 2010).

4.4.4 Membrane Technologies

Gas separation can be achieved via diffusion across a selective membrane, which can be either H_2 selective or CO_2 selective. A review of membrane types was conducted by Scholes et al. (2010), who looked at the advantages and disadvantages of metallic, porous inorganic and polymeric membranes. It was concluded that a significant issue in terms of assessing technologies is the lack of pilot plant trials to test the feasibility. Another conclusion was that more focus should be on membranes designed for CO_2 selectivity rather than H_2 (Scholes et al., 2010). Cobot et al. (2014) showed that membrane separation of CO_2 from flue gases results in between 16% and 30% lower energy costs compared to MEA solvent extraction. Hoffmann et al. (2008) showed that potentially CO_2 selective membranes could offer reduced energy consumption relative to amine solvent extraction. For a 493-MW F-class IGCC pre-combustion, membrane separation of CO_2 after the WGS reactor could increase the cycle thermal efficiency based on MEA absorption from 47.5% to 50.7% for an 80% CO_2 capture. However, no specific membrane was involved in this study, which was based on an assumed performance of a membrane separator.

4.4.5 Cryogenic Separation

Cryogenic separation is a refrigeration-based technique that is used for gas streams with high levels of CO_2. It has the advantage of being a simple process, but the disadvantages of being energy intensive and requiring the pre-separation of water from the gas stream in order to prevent ice formation.

4.4.6 Post-combustion Chilled Ammonia

Amines are formally derivatives of ammonia. Similar to other amines, ammonia can absorb CO_2 at atmospheric pressure, but at a slower rate than that of MEA. It is designed to operate with slurry. The process requires the flue gas to be chilled to 1.6 °C before it enters the CO_2 removal system. The cooled flue flows upwards in a spray tower with a slurry of ammonium carbonate and bicarbonate sprayed downwards. The CO_2 is absorbed and subsequently released under pressure. The advantage over MEA is the lower energy required for regeneration as the heat of absorption and reaction is low. The regeneration can be carried out using low-grade waste heat instead of high-grade steam heat used for CO_2 release from MEA. The cost of the process has been estimated to be half of the cost of MEA (Zachary, 2009).

4.5 Chemical Looping Processes

Chemical looping combustion is a method of using redox reactions, whereby the bed material is first oxidised in the air reactor bed, then the carbon-based fuel is oxidised by the bed material to produce CO_2 in the fuel reactor bed. The process is combustion without direct contact of the fuel and air, and so produces an almost pure CO_2 gas stream.

The bed material is an oxygen carrier, which is typically a metal oxide. An example of nickel reacting with pure carbon is shown below:

$$2Ni(s) + O_2(g) = 2NiO(s) \tag{4.3}$$

$$C + 2NiO(s) = CO_2 + 2Ni(s) \tag{4.4}$$

The reduced metal is then re-oxidised in the air reactor before being reintroduced back into the fuel reactor, completing the loop. The resultant gas consists of carbon dioxide and water vapour as well as trace pollutants. In this case, the overall reaction is oxidation (combustion) of the carbon to produce heat energy with nickel acting as a catalyst. However, high oxygen content and low carbon content of biomass may be an issue in successful operation.

The processes are typically conducted using two interconnected fluidised beds, but other systems including packed beds or moving beds have also been used. A review of chemical looping systems has been conducted by Nandy et al. (2016), including analysis of reactor types, fuel types and oxygen carrier types (Nandy et al., 2016). The process was found to have good potential for gaseous, liquid and solid fuels.

4.6 Existing Schemes

The Global CCS Institute (Global CCS Institute, 2017) has reviewed large projects that are already in operation or planned and have typically been associated with natural gas processing or coal gasification. The Institute lists 10 CCS schemes applied to NG CCGT, 7 of which have enhanced oil recovery (EOR) use for CO_2 and three with dedicated geological storage (DGS). A total of 27 Mt/year of CO_2 are recovered in these plants. It also details seven IGCC plants with CCS (six with EOR) although four of the plants are only at the planning stage and are for operation post 2020. Three CCS plants are in China, two in the United States, one in Canada and one in the United Kingdom, which is due to open with DGS in 2022. The reason for the dominance of EOR in the use of recovered CO_2 in existing plants is that this enables an income to be generated from the CO_2, whereas DGS will always involve a cost. Although there are a number of commercially operating IGCC plants, they are rarely used for coal, with petroleum coke as the main fuel for refinery use. Few IGCCs are equipped with CCS technologies; this is not for any technical reason but is a purely economic decision. Until there is a legal obligation to build the plants and put the costs onto the price of electricity, as is currently done for nuclear and offshore wind, no CCS plants will be built for grid-connected electricity.

Buggenum IGCC power station in Eemshaven, The Netherlands, is a multi-fuel power plant owned by the utility company Vattenfall. The plant was designed with the option to retrofit a pre-combustion unit. Most of the process engineering experience relevant to the reactor design and syngas properties was within the chemical industries rather than power production, so a small demonstration plant was built for testing ahead of the full commercial facility (van Dijk et al., 2014). The CCS pilot plant consists of syngas conditioning, WGS, physical absorption and CO_2 compression. The physical solvent is DEPEG (dimethyl-ether of polyethylene glycol) and the overall capture efficiency is 80–85% (van Dijk et al., 2014). Tests are ongoing at the facility, which has a fully instrumented system to investigate transient operations and can be used to validate models (Trapp et al., 2015).

The Puertollano IGCC power plant in Spain began commercial operation using syngas in 1998 and has a capacity of 335 MW (Casero et al., 2013). It has a 14-MW (thermal) CCS pilot plant and hydrogen production plant that has been in operation since 2010. The pilot plant consists of a WGS reactor, CO_2 separation by amine-based absorption and PSA hydrogen purification. The CO_2 capture rate achieved is 91.7% with hydrogen purity of 99.995% (Davidson, 2011). The plant is operated by ELCOGAS who have research aims of supporting sustainable zero-emissions technology based on coal, wastes and biomass (Casero, 2013).

4.7 Modelling of IGCC Plant Thermal Efficiency With and Without Pre-combustion CCS

Most of the published work on IGCC with and without CCS has been modelling work, as none of the CCS options have been built due to the lack of government policy on payment for the higher cost of electricity. The models of IGCC and CCS plants use process-modelling software, with actual gas turbine compressor and turbine efficiencies for the size of plant modelled. The key energy losses in IGCC are energy costs of operating the gasifier and gas clean-up system, which are increased by the cost of oxygen production if an oxygen-based gasifier is used.

For CCS there are two additional costs: the boiling of the amine solvent to release the CO_2, which uses steam generated from the GT exhaust, as discussed earlier, and the cost of compressing CO_2 to 60–100 bar for injection into the CO_2 pipeline. The compression energy consumption was estimated as 12% of the net electrical energy output by Baldwin (2009), and, hence, there is a 12% reduction in the thermal efficiency. However, Majoumerd et al. (2013) estimated the CO_2 compression energy as 5.7% of the 402 MW electrical power for a CCS IGCC system and in 2015 the same authors estimated it to be 5.2%. Sander and Span (2008) calculated the compression energy as 3.3% of 1201 MW IGCC system, and Lindfeldt and Westermark (2006) had it at 1.8%. Koopman and Bahr (2010) showed that CCS roughly doubles the cost of electricity and about 1/3 of this is the cost of CO_2 compression. This also indicates the uncertainty in process modelling of IGCC with CCS, as the energy cost of CO_2 compression should be known with some consistency. The additional cost of the pipeline flow pressure losses and the further compression prior to injection into underground storage are additional costs not covered in this review.

The costs for IGCC and CCS are compared in Table 4.3 in three ways:

1) Reduction in plant thermal efficiency compared with a CCGT with NG fuel
2) Cost per tonne of CO_2 removed or the carbon tax necessary to break even
3) Impact on the price of electricity relative to coal-fired steam plants or to IGCC coal or NG plants.

Some literature examples of ICGG and CCS process modelling are reviewed in Table 4.3. First, the thermal efficiency reduction is compared with NG CCGT when an IGCC plant is used to produce syngas for combustion in the CCGT. Several thermodynamic analyses of the IGCC plant are summarised in Table 4.3 for F-class baseline NG CCGT plant with a thermal efficiency of 58% without CCS and 47.5–48.8% with pre-combustion oxygen-based CCS. The thermal efficiency penalty with CCS was 9.2–10.5% for NG CCGT with pre-combustion CCS.

Table 4.3 has three examples of membrane separators of H_2/CO_2 and the thermal efficiencies with pre-combustion CCS for oxygen gasification ranged from 32.5% to 48.8% and the efficiency loss due to CCS was 6.5–8.7%. The best result here is as good as the NG CCS thermal efficiency. Manzolini et al. (2006) also modelled post-combustion MEA solvent CCS with a thermal efficiency of 47.4% compared with 48.8% for pre-combustion hydrogen-based CCS. This 1.4% gain in thermal efficiency indicates the advantage of pre-combustion CCS.

Table 4.3 shows that for air gasification of coal the IGCC thermal efficiencies ranged from 44.3% to 48.9% and with MEA solvent CCS these were reduced to 32.9% to 38.1%

Table 4.3 Comparison of the thermal efficiency penalty for the IGCC plant relative to the NG CCGT plant and the thermal efficiency penalty for pre-combustion CO$_2$ removal (with one comparison with post-combustion CCS). Pre-combustion CO$_2$ separation by MEA solvent absorption for all except where membrane separators are indicated.

Reference	Net electrical power (MW$_e$) No CCS	Thermal eff. elec. No CCS	Thermal eff. elec. with H$_2$ CCS	Loss in thermal efficiency due to CCS	Net elec. power (MW$_e$) CCS	Gas turbine class	Turbine entry temperature (K)	Cost of carbon removal per tonne CO$_2$	Cost of electricity increase with CO$_2$ capture for IGCC baseline	Cost of electricity increase relative to NG CCGT
Schultz (2016) Air gasifier 1 bar Duke Edwardsport	618 Real	38.5% IGCC Coal				GE F 58% NG	~1600			
Hoffmann et al. (2008) NG O$_2$ gasifier For H$_2$ CCGT	410	58% NG	47.5% NG	10.5%	455	GE 9FB		$46		
Majoumerd et al. (2015) NG oxygen gasifier H$_2$ CCGT	430	58% NG	48.7% NG	9.3%	360	F	~1600			
Manzolini et al. (2006) NG O$_2$ gasifier Membrane separator for H$_2$/CO$_2$ H$_2$ CCGT	770	56.8% NG CCGT	48.8% (47.4 for MEA post-combustion)	8.0%	770			€31.2 (€48.6 for MEA post-combust.)	56.7 €/MWh	40.3% NG with CCS

(Continued)

Table 4.3 (Continued)

Reference	Net electrical power (MWe) No CCS	Thermal eff. elec. No CCS	Thermal eff. elec. with H$_2$ CCS	Loss in thermal efficiency due to CCS	Net elec. power (MWe) CCS	Gas turbine class	Turbine entry temperature (K)	Cost of carbon removal per tonne CO$_2$	Cost of electricity increase with CO$_2$ capture for IGCC baseline	Cost of electricity increase relative to NG CCGT
Lindfeldt and Westermark (2006) Coal slurry O$_2$ gasifier High P Membrane separator for H/CO$_2$ AZEP Alstom H$_2$ CCGT Membrane O$_2$ separation	39.0% (Low gasifier efficiency)	32.5%		6.5%	371					
Sander and Span (2008) Coal O$_2$ gasifier High P Membrane separator H$_2$/CO$_2$	43.0	34.3		8.7%	412					
Li and Wang (2012) Air gasifier, 24 bar, dry feed	245 Model IGCC Coal	44.3% IGCC Coal	37.9% IGCC Coal Excludes CO$_2$ compressor	6.4%	227	GE 7FA 58% NG	~1600			
Giuffrida et al. (2012) Air gasifier MHI 27.6 bar Coal A	970	48.9%	38.1%	10.8%	1000	Siemens SGT5-4000F 58% Eff.	1608			

Reference												
Giuffrida (2014)	Air gasifier MHI Coals B&C	970	48.1–47.7			1000						
Narula and Wen (2010)	Coal O_2 IGCC	630 640	38.5% 38.2%	32.0% 32.5%	6.5% 5.7%	560					103 $/MWh +32%	
Kreutz et al. (2010)	Oxygen gasifier 1 bar	272	47.7%									
Vascellari et al. (2006)	Oxygen gasifier 1 bar	452	46.7%	39.8%	6.9%	393	F 58%	~1600	€29.5	44%		
Pronske et al. (2006)	Coal Cryo Oxygen gasifier Pressure		45	39	6.0				$19.7		$69 MWh +30.2%	
Sipocz et al. (2011)	Coal O_2	309	47.2%	37.4%	9.8%	324	Siemens 94.3A					
Li and Wang (2012)	Oxygen gasifier 24 bar dry	238	46.0%	36.5% Excludes CO_2 comp.	9.5%	255	GE 7FA					
Majoumerd et al. (2013)	Oxygen Gasifier 45 bar Coal	470		35.7%		402	F-class Siemens SGT5-4000F 58% efficiency	~1600 K				45% IGCC w/o CCS
Majoumerd et al. (2015)	Oxygen gasifier 45 bar Coal	470	47.0%	35.7%	11.3%	395	Buggenum IGCC F-class 58% NG	~1600		53%		50%

(Continued)

Table 4.3 (Continued)

Reference	Net electrical power (MWe) No CCS	Thermal eff. elec. No CCS	Thermal eff. elec. with H₂ CCS	Loss in thermal efficiency due to CCS	Net elec. power (MWe) CCS	Gas turbine class	Turbine entry temperature (K)	Cost of carbon removal per tonne CO₂	Cost of electricity increase with CO₂ capture for IGCC baseline	Cost of electricity increase relative to NG CCGT
Kawabata et al. (2012)	Oxygen gasifier 25bar 380	51.0%	39.9% Post-combustion	11.1%	306	MHI J-class 63% NG	1973 K			
Galanti et al. (2009)	Coal O₂ gasifier Steam power 660	41.2% Brindisi ENEL	43.0		250				€53.MWh 28% Excludes CO₂ compressor	
Stępczyńska et al. (2012)	Supercritical steam Double reheat 1012	48.6%			702					
Majoumerd et al. (2015)	Supercritical steam 760	45.5	33.4% Post-comb. MEA	12.1%	550					
Hustad (2010)	Supercritical steam	47	27 Post-comb. MEA	20%						
Hustad (2010)	Supercritical steam	46	35 Post-comb. MEA	11%						

with CCS penalties of 6.4–10.8%, which are similar to the range for NG CCS and the membrane separator IGCC CCS. Table 4.3 also shows a practical F-class IGCC with CCS at the GE IGCC Duke Edwardsport plant and the thermal efficiency with CCS was 38.5% (Schultz, 2016). This is at the higher end of the above range of thermal efficiencies. Duke Edwardsport is an atmospheric pressure air gasification system and all the predictions in Table 4.3 are for pressurised gasification systems. This indicates that the potential improved thermal efficiencies for locating the gasifier downstream of the compressor have not been demonstrated in the process models.

Table 4.3 shows many results for oxygen-based coal gasification with a thermal efficiency range for F-class IGCC with no CCS of 38.2% to 47.7%. There is also one prediction for the future technology J-class at 51% thermal efficiency. With pre-combustion CCS, the thermal efficiency range falls to 32–39.8% and the CCS thermal efficiency penalty ranges from 5.7% to 11.3%. This range is very similar to that for air gasification and the Duke Edwardsport thermal efficiency is towards the top end of this range. Thus, it can be concluded that there is no advantage of using oxygen gasification compared with air gasification, as was concluded above in relation to the quality of the gasified gas. However, there is a significant cost saving in not using an ASU plant.

Also compared in Table 4.3 are the thermal efficiencies for modern supercritical steam coal plant, as this is the alternative clean coal technology. Table 4.3 shows that the modern steam plant has a thermal efficiency range from 45.5% to 48.6% without CCS and 27–35% with post-combustion CCS. This shows that IGCC with pre-combustion CCS had a higher thermal efficiency by about 5% for the best system, with the potential for greater benefits with J-class IGCC.

No open literature studies of biomass-fuelled IGCC plants have been found, although in the United Kingdom one small 5 MW plant was built at Arbro, but it was unsuccessful as there were gasifier problems. This review of biomass gasification shows similar properties to coal syngas and so it should be capable of operating with pre-combustion separation of CO_2, with similar thermal efficiencies to coal-based systems. However, new plants have to be built to demonstrate this as pre-combustion carbon capture is not a retrofit option. Without financial incentives, such as subsidies equal to those for the nuclear industry, this is unlikely to happen. Thus, the potential for negative CO_2 emissions from biomass IGCC will not be realised.

4.8   Summary and Research Challenges

Pre-combustion CCS technologies as part of an IGCC process are a technically feasible method for biomass carbon capture with negative CO_2 emissions as a consequence. All components of such plants already exist for NG and coal-based systems with commercial gasifiers, WGS reactors and syngas CO_2 removal are already existing. However, to build such plants, fiscal incentives will be required, similar to those for nuclear and offshore wind-energy generators.

Gasification of coal and petroleum coke is an established and reliable technique, and this review has shown that biomass gasification can produce a similar quality syngas to coal gasification, but there are potential issues with new feedstocks such as biomass.

The use of air for gasification of coal results in a similar proportion of the energy in the fuel in the syngas as for oxygen gasification. There is also no advantage to oxygen gasification of coal in an IGCC pre-combustion CCS system. Thus, there is no advantage to oxygen gasification and thus no requirement for ASU plants, and this enables the capital cost of the plants to be reduced. In many cases, the literature assumes that oxygen gasification must be the best, without a parallel study of air gasification. The resultant syngas is of lower calorific value, but this just requires a different gas supply system and combustor for the gas turbine, and there is no major problem in burning the air-gasified syngas. It also has the advantage that the stoichiometric flame temperature is lower and this gives lower NO_x emissions than for oxygen-based gasification. This is especially the case for pre-combustion CCS as the syngas is a mixture of H_2 and N_2 and this will burn with low NO_x.

Most of the models for IGCC and IGCC with pre-combustion CCS assume that the gasifier will operate at pressure fed by the high-pressure compressor exit gases. In practice, as shown in this review, there is no proven advantage to the high-pressure route and it is more expensive in capital costs. The commercial 618 MW Duke Edwardsport IGCC plant has an atmospheric pressure oxygen-blown gasifier. The gas requires compression to inject into the IGCC combustion system and this method gives small changes to the gas turbine plant. For air gasification this leads to a disadvantage that the mass flow of the syngas with N_2 is greater than for oxygen-blown gasifiers, and this can create problems as the turbine mass flow is then greater than the compressor mass flow, which gives matching problems. So for air gasification a pressure system might be preferable.

There have been few trials with biomass gasification and IGCC plants. Significant difficulties that could arise when using biomass are due to inconsistent fuel quality, presence of trace elements associated with deposit formation and relatively low calorific value. Tar and inorganic deposits are potentially problematic if fluidised beds are used. Corrosion can also occur in higher-temperature reactors.

There is substantial research activity ongoing regarding the H_2/CO_2 separation stage of the process, especially in the areas of identifying novel or modified materials as adsorbents, solvents or separation membranes. There is evidence that membrane separation offers the potential for reduced energy costs in the CO_2 removal process. However, models of complete IGCC with pre-combustion CCS show the same range of CCS thermal efficiency penalties as for MEA CO_2 extraction. More work is required to make membrane separation a practical technique that would supplant MEA solvent extraction of CO_2. Essentially, the energy to overcome the pressure loss of the membrane separation is not sufficiently lower than the heat energy to release CO_2 from the solvent, for the membrane technology to have a decisive advantage.

Most of the studies of IGCC with pre-combustion CCS have been based on F-class CCGT, which is gas turbine technology in the late 1990s. An F-class CCGT on NG has a thermal efficiency of 58% and if pre-combustion carbon removal was used for NG then the thermal efficiency would be reduced to typically 48%. If coal is gasified, the IGCC F-class plant has a modelled best thermal efficiency of 48% and at worst 38%. The commercial Duke Edwardsport IGCC plant on coal has a thermal efficiency of 38.5%, which may indicate that some of the models are optimistic. The addition of pre-combustion CCS to IGCC reduces the efficiency further to at best 40% and at worst 32%.

The change in thermal efficiency essentially increases the cost of electricity produced and this would be 21% for natural gas pre-combustion CCS and for coal compared with the baseline NG CCGT it would be an increased cost in the range of 45–81% and a biomass system is likely to be in this range. Specific models of the economics of IGCC pre-combustion CCS indicate an increase in cost for NG due to pre-combustion CCS of 40% and for coal-based IGCC with pre-combustion CCS of 45–50% relative to NG. The additional cost of adding pre-combustion CCS to IGCC plant varies between 30% and 53%, if those studies that excluded the cost of CO_2 compression are omitted. These increased costs of electricity exclude the cost of CO_2 pipeline charges and the further recompression of the gas at the underground storage site, so the actual increase in the cost of electricity will be higher than the above. However, as new nuclear power plants in the United Kingdom is subsidised at 100% above the 2015 cost of electricity, increasing with inflation, it is clear that pre-combustion CCS with NG CCGT or coal/biomass IGCC could provide low CO_2 electricity at a lower cost than nuclear.

In terms of the cost of electricity, this ranges from 69 to 103 €/MWh (only two studies) for coal and 57 €/MWh for NG. In terms of the cost per tonne of CO_2 removed, the NG cost is 47 and for IGCC with CCS it has been estimated at 20–30 per tonne of CO_2 in two studies, which are probably too low as the NG figures are more reliable. This is the minimum carbon tax that is required on the price of electricity to make this method of CO_2 removal economically acceptable. Currently, the UK government has a carbon tax on coal-based electricity of £0.44 per GJ, which, using 25 GJ per tonne as the gross CV for coal, converts to £11 per $tonne_{coal}$ and with about 3 kg of CO_2 per kg of coal this converts to £3.7 per $tonne_{CO_2}$. Clearly, at this carbon price no IGCC CCS plant will be built and this carbon policy has simply accelerated the closure of existing coal-based electricity production. The advantage for biomass is that it does not have a carbon tax applied to it and this is why there is interest in biomass-based IGCC with pre-combustion CCS.

References

Aponte, J. and Gordillo, G. (2013). Wild cane potential to produce gaseous fuels via air-steam thermal gasification. ASME Turbo Expo 2013: Turbine Technical Conference and Exposition.

Babu, P., Ong, H.W.N., and Linga, P. (2016). A systematic kinetic study to evaluate the effect of tetrahydrofuran on the clathrate process for pre-combustion capture of carbon dioxide. *Energy* **94**: 431–442.

Baldacci, A., Gradassi, A.T., Zeppi, C. et al. (1994). Syngas production from sorghum at Greve in Chianti gasification plant. *Proceedings of the 8th European Biomass Conference*, Vol. 3, Vienna (3–5 October 1994). Biomass for Energy Environment Agriculture and Industry, pp. 1807–1813.

Baldwin, P. (2009). Capturing CO_2 – gas compression vs liquefaction. *Power Magazine* (May).

Bonzani, F. (2006). Syngas burner optimization for fuelling a heavy duty gas turbine with various syngas blends. ASME Turbo Expo 2006: Power for Land, Sea, and Air, Barcelona. ASME Paper GT2006-90761.

Carbo, M.C., Jansen, D., Dijkstra, J.W., et al. (2007). Precombustion decarbonisation in IGCC: abatement of both steam requirement and CO_2 emissions. 6th Annual Conference on Carbon Capture and Sequestration Combined Cycle (IGCC) Technology, Pittsburgh, USA.

Casero, P., Coca, P., García-Peña, F., and Hervás, N. (2013). CO_2 emissions reduction technologies in IGCC: ELCOGAS's experiences in the field. *Greenhouse Gases: Science and Technology* **3**: 253–265. doi: 10.1002/ghg.1351.

Casero, P., Garcia-Pena, F., and Coca, P. (2013). ELCOGAS pre-combustion carbon capture pilot. Real experience of commercial technology. *Energy Procedia* **37**: 6374–6382.

Cobot, G., Calbry, M., Xavier, P., et al. (2014). Effect of CO_2 capture on CCGT efficiency using membrane separation, EGR and OEA effects on combustion characteristics. Proceedings of the ASME Turbo Expo 2014. ASME paper GT2014–25781.

Consoni, S. and Vigano, F. (2005). Decarbonized hydrogen and electricity from natural gas. *International Journal of Hydrogen Energy* **30**: 701–718.

Corradetti, A. and Desideri, U. (2006). Analysis of biomass integrated gasification fuel cell plants in industrial CHP applications. ASME 2006 4th International Conference on Fuel Cell Science, Engineering and Technology 2006.

Davidson, R. (2011). Pre-combustion Capture of CO_2 in IGCC Plants. IEA clean energy report. CCC/191, p. 98, 978-92-9029-511-2, December 2011.

Englezos, P. (1993). Clathrate hydrates. *Industrial and Engineering Chemistry Research* **32**: 1251–1274.

Galanti, L., Franzoni, A., Traverso, A., and Massardo, A.F. (2009). Electricity and hydrogen co-production from coal and biomass. Proceedings of ASME Turbo Expo 2009: Power for Land, Sea and Air. GT2009, Orlando, FL, USA (8–12 June 2009).

Giuffrida, A. (2014). Impact of low-rank coal on air-blown IGCC performance. ASME Turbo Expo 2014: Turbine Technical Conference and Exposition.

Giuffrida, A., Romano, M.C., and Lozza, G. (2012). CO_2 capture from air-blown gasification-based combined cycles. Proceedings of ASME Turbo Expo 3. ASME Paper GT2012-69787, pp. 395–404.

Global CCS Institute (2017). Large CCS facilitieshttp://www.globalccsinstitute.com/projects/large-scale-ccs-projects (accessed 20 April 2017).

Gordillo, G., Annamalai, K., and Carlin, N. (2009). Adiabatic fixed-bed gasification of coal, dairy biomass, and feedlot biomass using an air–steam mixture as an oxidizing agent. *Renewable Energy* **34** (12): 2789–2797.

Herzog, H.J. (1999). The economics of CO_2 capture. In: *Greenhouse Gas Control Technologies* (ed. P. Reimer, B. Eliasson and A. Wokaum). Oxford: Elsevier.

Hoffmann, S., Bartlett, M., Finkenrath, M., et al. (2008). Performance and cost analysis of advanced gas turbine cycles with pre-combustion CO_2 capture. ASME Turbo Expo 2008: Power for Land, Sea, and Air. Volume 2: Controls, Diagnostics and Instrumentation; Cycle Innovations; Electric Power, Berlin, Germany (9–13 June 2008). Paper No. GT2008-51027, pp. 663–671, 9 pages.

Hustad CW. (2010). Deployment of Low and Zero Emission Fossil Fuel Power Generation in Emerging Niche Markets. ASME. Turbo Expo: Power for Land, Sea, and Air, Volume 2: Controls, Diagnostics and Instrumentation; Cycle Innovations; Electric Power: 397–409. doi: 10.1115/GT2008-50106.

Iki, N., Tsutsumi, A., Matsuzawa, Y., and Furutani, H. (2009). Parametric study of advanced IGCC(Conference Paper). Proceedings of the ASME Turbo Expo. Volume 4: 2009 ASME Turbo Expo, Orlando, FL, USA (8 June 2009 through 12 June 2009), Code 80492. ASME Paper GT2009-59984, pp. 405–412.

Kawabata, M., Kurata, O., Iki, N. et al. (2012). Advanced integrated gasification combined cycle (A-IGCC) by exergy recuperation-technical challenges for future generations. *Journal of Power Technologies* **92** (2): 90–100.

Koopman, A.A. and Bahr, D.A. (2010). The impact of CO_2 compressor characteristics and integration in post combustion carbon sequestration comparative economic analysis. ASME Turbo Expo 2010: Power for Land, Sea, and Air.

Kreutz, T., Martelli, E., Carbo, M., Consonni, S., Jansen, D. (2010). Shell gasifier-based coal IGCC with CO_2 capture: partial water quench vs. novel water-gas shift Proc. ASME. 43963; Volume 1: Aircraft Engine; Ceramics; Coal, Biomass and Alternative Fuels; Education; Electric Power; Manufacturing Materials and Metallurgy: 583–592. (October 10, 2010), GT2010-22859. doi: 10.1115/GT2010-22859.

Lee, H.J., Lee, J.D., Linga, P. et al. (2010). Gas hydrate formation process for pre-combustion capture of carbon dioxide. *Energy* **35** (2010): 2729–2733.

Lee, S.-Y. and Park, S.-J. (2015). A review on solid adsorbents for carbon dioxide capture. *Journal of Industrial and Engineering Chemistry* **23**: 1–11.

Lepszy, S. and Chmielniak, T. (2010). Technical and economic analysis of biomass integrated gasification combined cycle(Conference Paper). Proceedings of the ASME Turbo Expo Volume 1: ASME Turbo Expo 2010: Power for Land, Sea, and Air, GT 2010, Glasgow, UK (14 June 2010 through 18 June 2010), pp. 631–637.

Li, M., Rao, A.D., and Scott Samuelsen, G. (2012). Performance and costs of advanced sustainable central power plants with CCS and H_2 co-production. *Applied Energy* **91**: 43–50. doi: 10.1016/j.apenergy.2011.09.009.

Li, X. and Wang, T. (2012). A parametric investigation of integrated gasification combined cycles with carbon capture. ASME Turbo Expo 2012: Turbine Technical Conference and Exposition, Copenhagen, Denmark. ASME Paper GT2012-69519.

Lindfeldt, E.G. and Westermark, M.O. (2006). An integrated gasification zero emission plant using oxygen produced in a mixed conducting membrane reactor. ASME Turbo Expo 2006: Power for Land, Sea, and Air.

Linga, P., Kumar, R., and Englezos, P. (2007). The clathrate hydrate process for post and pre-combustion capture of carbon dioxide. *Journal of Hazardous Materials* **149**: 625–629.

Majoumerd, M.M., Assadi, M., and Breuhaus, P. (2013). Techno-economic evaluation of an IGCC power plant with carbon capture(Conference Paper). Proceedings of the ASME Turbo Expo Volume 2: ASME Turbo Expo 2013: Turbine Technical Conference and Exposition, GT 2013, San Antonio, TX, USA (3 June 2013 through 7 June 2013). Code 101331.

Majoumerd, M.M., Assadi, M., Breuhaus, P., and Arild, Ø. (2015). System integration and techno-economy analysis of the IGCC plant with CO_2 capture – results of the EU H2-IGCC project(Conference Paper). Proceedings of the ASME Turbo Expo. Volume 3: ASME Turbo Expo 2015: Turbine Technical Conference and Exposition, GT 2015, Montreal, Canada (15 June 2015 through 19 June 2015). Code 113665.

Manzolini, G., Macchi, E., Dijkstra, J.W., and Jansen, D. (2006). Technical economic evaluation of a system for electricity production with CO_2 capture using a membrane reformer with permeate side combustion(Conference Paper). Proceedings of the ASME Turbo Expo. Volume 4: 2006 ASME 51st Turbo Expo, Barcelona, Spain (6 May 2006 through 11 May 2006). Code 68506. pp. 89–99.

Nandy, A., Loha, C., Sai, G. et al. (2016). Present status and overview of chemical looping combustion technology. *Renewable and Sustainable Energy Reviews* **59**: 597–619.

Narasimharao, K. and Ali, T.T. (2016). Effect of preparation conditions on structural and catalytic properties of lithium zirconate. *Ceramics International* **42**: 1318–1331.

Narula, R.G. and Wen, H. (2010). The battle of CO_2 capture technologies. ASME Turbo Expo 2010: Power for Land, Sea, and Air. Volume 1: Aircraft Engine; Ceramics; Coal,

Biomass and Alternative Fuels; Education; Electric Power; Manufacturing Materials and Metallurgy, Glasgow, UK (14–18 June 2010).

Parulekar, P.S. (2011). Comparison between oxygen-blown and air-blown IGCC power plants: a gas turbine perspective. ASME 2011 Turbo Expo: Turbine Technical Conference and Exposition.

Pinta, F. and Vergnet, L.F. (1994). Testing of wood gasification pilot plant for industrial heat generation. Proceedings of the 8th European Biomass Conference, Volume 3, Vienna (3–5 October 1994), pp. 1791–1800.

Porta, M., Traverso, A., and Marigo, L. (2006). Thermoeconomic analysis of a small-size biomass gasification plant for combined heat and distributed power generation. Paper No. GT2006-90918. ASME Turbo Expo 2006: Power for Land, Sea, and Air. Volume 4: Cycle Innovations; Electric Power; Industrial and Cogeneration; Manufacturing Materials and Metallurgy, Barcelona, Spain, (8–11 May 2006), pp. 357–365; 9 pages.

Pronske, K., Trowsdale, L., Macadam, S., Viteri, V., Bevc, F., Horazak, D. (2006). An overview of turbine and combustor development for coal-based oxy-syngas systems Proc. ASME. 42398; Volume 4: Cycle Innovations; Electric Power; Industrial and Cogeneration; Manufacturing Materials and Metallurgy: 817–828. (January 01, 2006), GT2006-90816. doi: 10.1115/GT2006-90816.

Sander, F. and Span, R. (2008). Model of a coal fired IGCC process with hydrogen membrane reactor and capture of CO_2. ASME Turbo Expo 2008: Power for Land, Sea, and Air.

Scholes, C.A., Smith, K.H., Kentish, S.E., and Stevens, G.W. (2010). CO_2 capture from pre-combustion processes – strategies for membrane gas separation. *International Journal of Greenhouse Gas Control* **4**: 739–755.

Schultz, M. (2016). GE energy gas turbines for syngas and hydrogen combustion. Presented at the Leeds University Training Course on Ultra Low NO_x Gas Turbines (January).

Seggiani, M., Puccini, M., and Vitolo, S. (2013). Alkali promoted lithium orthosilicate for CO_2 capture at high temperature and low concentration. *International Journal of Greenhouse Gas Control* **17**: 25–31.

Siefert, N.S., Agarwala, S., Shia, F. et al. (2016). Hydrophobic physical solvents for pre-combustion CO_2 capture: experiments, computational simulations, and techno-economic analysis. *International Journal of Greenhouse Gas Control* **49**: 364–371.

Sipocz, N., Mansouri, M., Breuhaus, P., and Assadi, M. (2011). Development of H2 rich syngas fuelled GT for future IGCC power plants- establishment of a baseline. Proceedings of the ASME Turbo Expo 2011, Vancouver. ASME Paper GT2011-45701.

Sloan, E.D. Jr. (1998). *Clathrate Hydrates of Natural Gases*, 2e. Marcel Dekker.

Stępczyńska, K., Łukowicz, H., Dykas, S., and Rulik, S. (2012). Thermo-economic analysis of the ultra-supercritical 900 MW power unit with the various configurations of the auxiliary steam turbine. Proceedings of ASME Turbo Expo 2012. GT2012, Copenhagen, Denmark (11–15 June 2012). ASME Paper GT2012-68159.

Theo, W.L., Lim, J.S., Hashim, H. et al. (2016). Review of pre-combustion capture and ionic liquid in carbon capture and storage. *Applied Energy* **183**: 1633–1663.

Trapp, C., de Servi, C., Casella, F. et al. (2015). Dynamic modelling and validation of pre-combustion CO_2 absorption based on a pilot plant at the Buggenum IGCC power station. *International Journal of Greenhouse Gas Control* **36**: 13–26.

van Dijk, H.A.J., Damen, K., Makkee, M., and Trapp, C. (2014). Water energy procedia water–gas shift (WGS) operation of pre-combustion CO_2 capture pilot plant at the Buggenum IGCC. *Energy Procedia* **63**: 2008–2015.

Wang, Y. and Weng, Y. (2014). Analysis on integrated gasification humid air turbine system with air blown gasifier. ASME Turbo Expo 2014: Turbine Technical Conference and Exposition, Volume 3A: Coal, Biomass and Alternative Fuels; Cycle Innovations; Electric Power; Industrial and Cogeneration, Düsseldorf, Germany (16–20 June 2014). ASME Paper GT2014-26770.

Zachary, J. (2009). Design challenges for combined cycles with post-combustion CO_2 capture. ASME Turbo Expo 2009: Power for Land, Sea, and Air. Volume 1: Aircraft Engine; Ceramics; Coal, Biomass and Alternative Fuels; Controls, Diagnostics and Instrumentation; Education; Electric Power; Awards and Honors, Orlando, FL, USA (8–12 June 2009). ASME Paper GT2009-59381.

Zhao, Y., Cao, Y., and Zhong, Q. (2014). CO_2 capture on metal-organic framework and graphene oxide composite using a high-pressure static adsorption apparatus. *Journal of Clean Energy Technologies* **2**: 34–37.

5

Techno-economics of Biomass-based Power Generation with CCS Technologies for Deployment in 2050[*]

Amit Bhave[1], Paul Fennell[3], Niall Mac Dowell[3], Nilay Shah[3] and Richard H.S. Taylor[2]

[1] *CMCL Innovations, Castle Park, Castle Street, Cambridge, UK*
[2] *E4tech, Westminster, UK*
[3] *Centre for Environmental Policy, Imperial College, London, UK*

All costs reported were 'next-of-a-kind' costs and were for plants constructed in 2011, though cost reductions as a function of time were also estimated. Significant increases in capital costs (between 45% and 130%) were noted for electricity plants, though it was found that plant scale (MW_e) was the most important parameter for capital cost per MW_e, rather than CO_2 capture technology choice. The levelised cost of electricity (LCOE) was lower for co-firing than it was for dedicated plants, where direct comparison was possible. It was also found that there were only marginal differences in the cost of CO_2 avoided, in comparison to a reference coal plant (at the 50 MW_e scale) between amine scrubbing, oxy-fuel and IGCC for co-fired plants (£78 ± 20 per tonne). Here, the cost of CO_2 *captured* $(£ / tCO_2)$ is defined as the LCOE $(£ / MWh_e)$ × annual electricity output (MWh_e/yr) divided by the CO_2 emissions captured (tCO_2/yr). The *cost of* CO_2 *avoided* $(£ / tCO_2)$ is the difference between the LCOE for the technology being studied and a reference unabated coal plant from the International Energy Agency (IEA),

---

[*]On behalf of the entire TESBiC team:
Amit Bhave[1], Richard H.S. Taylor[2], Paul Fennell[3], William R. Livingston[4],
Nilay Shah[3], Niall Mac Dowell[3], John Dennis[5], Markus Kraft[5,6],
Mohammed Pourkashanian[7], Mathieu Insa[8], Jenny Jones[7], Nigel Burdett[9],
Ausilio Bauen[2], Corinne Beal[10], Andrew Smallbone[1], Jethro Akroyd[1]

[1]CMCL Innovations, UK
[2]E4tech, UK
[3]Imperial College London, UK
[4]Doosan Power Systems, UK
[5]Department of Chemical Engineering and Biotechnology, University of Cambridge, UK
[6]Nanyang Technological University, Singapore
[7]Faculty of Engineering, University of Leeds, UK
[8]EDF Energy, France
[9]Drax Power, UK
[10]Alstom Boiler France, France

Biomass Energy with Carbon Capture and Storage (BECCS): Unlocking Negative Emissions, First Edition.
Edited by Clair Gough, Patricia Thornley, Sarah Mander, Naomi Vaughan and Amanda Lea-Langton.
© 2018 John Wiley & Sons Ltd. Published 2018 by John Wiley & Sons Ltd.

divided by the difference between the CO_2 emissions from the technology under consideration and those from the reference unabated coal plant.

Co-fired carbonate looping, reflecting the lower technology readiness level (TRL) of the technology, had a slightly higher central cost estimate, but much greater uncertainty in the final cost (£88 ± 48 per tonne). Again at the 50 MW$_e$ scale, for dedicated biomass-fired plants, chemical looping had the lowest cost of CO_2 avoided, but the largest relative uncertainty (£62 ± 33 per tonne), with oxy-fuel combustion and amine scrubbing being around (£84 ± 32 per tonne) and IGCC marginally lower (£76 ± 28 per tonne).

5.1 Introduction

The TESBiC (Techno-Economic Study of Biomass to Power with CCS) project took place in 2011–2012. The purpose of the project was to gain insight into whether the use of biomass for power generation, with integrated carbon capture, was likely to be a viable technology – and to elucidate the most viable technologies. As defined within this project, biomass power with carbon capture and storage (CCS) has three main components:

- A biomass feedstock supply chain
- A power plant conversion system
- A carbon capture technology

There are numerous combinations of the above that could be used to produce a viable biomass CCS system. The processes involve the combustion or gasification of biomass (potentially co-fired with fossil fuels), with post-combustion, oxy-combustion or pre-combustion CO_2 capture.

The use of CCS with fossil-based fuels is a bridging technology, allowing transition to a longer-term solution. The particular utility of power with CCS is that it allows power to be despatched regardless of the time of day or weather conditions, thus compensating for intermittency of supply in many classes of renewable electricity production. On the other hand, critics have suggested that CCS results in perpetuation of fossil fuel use and may retard the transition to a fully renewable power generation system. Biomass CCS does not suffer from these issues and is one of the few technologies available currently that offers the realistic potential for economic large-scale net CO_2 removal from the air (Mac Dowell and Fajardy, 2016).

The UK ETI conducted extensive modelling using its Energy Systems Model (ESME) toolkit, which evaluated a large number of potential methods to keep to the United Kingdom's emissions reduction targets to 2050, at least cost. A significant fraction of the model runs rely on the use of Biomass-Enhanced CCS (BECCS) for lowest overall cost decarbonisation, owing to the very high assumed costs of mitigation in certain hard-to-decarbonise areas. In essence, some industries (for example, aviation and shipping) may be so challenging to decarbonise at source that it could be cheaper to allow them to emit at source and to decarbonise via the purchase of negative emissions. Furthermore, the development of negative emissions technologies could be important if global emissions significantly overshoot the aimed-for trajectories.

The development of BECCS has been slower than that for coal- or gas-fired CCS, in particular, because initial demonstration costs for fossil-based CCS are lower, partly because fossil fuels are cheaper and partly because existing infrastructure burning fossil fuels could be used to enable demonstrations. There are significant gaps in the understanding of BECCS: what the key barriers to decarbonisation are, what the technology is likely to cost and which technologies could be feasibly deployed by 2050 (in the absence of a dedicated Manhattan Project equivalent). As of 2016, there is still no commercial incentive to apply CO_2 capture technologies to dedicated biomass power plants (or to power plants or industry in general), and specific incentives associated with the removal of CO_2 from the atmosphere (negative emissions credits) will need to be introduced to support these activities. This is not as yet being discussed seriously at government level.

The main objective of the TESBiC project was to provide information and enhanced understanding of the potential technology development and demonstration acceleration opportunities for the ETI. Initial discussions with the ETI led to the exclusion of waste fuels from the project, due to other ETI projects covering the scope of these initial steps of this value chain.

It should be noted that the power industry is historically very conservative in character, and the current electricity generation, distribution and trading systems have become accustomed over many decades to the degree of flexibility of operation provided by the pulverised coal boiler and the combined-cycle gas turbine (in terms of ability to respond to short-term variations in electricity demand or to provide associated ancillary grid services). It may also be the case that in the future, when a higher proportion of the United Kingdom's power generation will be from intermittent sources, the flexibility of operation provided by thermal power plants will be of even more importance.

The technical options for the capture of CO_2 from thermal power plants fall into the following three categories:

Post-combustion technologies. Post-combustion technologies involve the addition of a chemical process plant to the flue gas outlets, downstream of the current environmental control equipment. These technologies are the most familiar to the electricity supply industry, since they have been applying the liquid or slurry scrubbing of flue gases for the control of gaseous emissions for several decades. In general terms, these will not have a significant impact on the flexibility of operation of the power plant, except in cases where there is a requirement for an oxygen supply, e.g. for the carbonate-looping system, where there is oxygen-blown combustion of a secondary fuel in the calciner. The addition of an air separation unit (ASU) may have a significant impact on the flexibility of operation of the overall plant. This can be reduced by the provision of oxygen storage facilities on-site, although it is clearly a process- and site-specific issue.

Oxy-fuel technologies. Oxy-fuel technologies conventionally involve a fairly major reconfiguration of the fuel firing systems, the draft plant and the boiler and the installation of an ASU for oxygen production. In the case of chemical-looping combustion technologies, the reconfiguration of the firing and heat transfer systems is fairly radical. For all oxy-fuel technologies, the ASU will have a significant impact on the flexibility of operation of the plant, although the impact of this can be reduced by the provision of oxygen storage facilities on-site.

Pre-combustion technologies. Pre-combustion technologies involve the gasification of the solid fuel followed by CO_2 capture by physical absorption, with combustion of the hydrogen-rich product gas in a gas turbine in a combined-cycle configuration. The co-firing of biomass in an IGCC system is relatively well understood technology and the capture of CO_2 from the syngas is fairly conventional. The dedicated biomass IGCC system, on the other hand, is more complex and less well understood, and the uncertainties become particularly large due to lack of experience at smaller scales ($\sim 30-50$ MW$_e$).

As part of the initial paper-based screening, the TESBiC consortium gathered the information shown in Table 5.1 for each biomass power conversion technology and carbon capture technology considered in the scope of the project.

For some of the technologies or combinations investigated, there was not sufficient information available in the literature to allow conclusive answers to the questions posed earlier. In this case, the collective judgement of the consortium was used to form an opinion, or else the lack of objective evidence was noted, meaning that the technology would be less likely to be progressed for further study.

The following 11 co-firing and dedicated biomass conversion technologies were initially reviewed in depth, leading to the production of a longlist for combustion technology:

- *Pulverised coal combustion:* with direct co-firing of biomass, or conversion to 100% biomass
- *IGCC coal gasification:* with direct co-firing of biomass, or conversion to 100% biomass

Table 5.1 Information gathered on each component technology.

Category	Information gathered
Introduction	Description
Development aspects and prospects	Key drivers for development
	Key development issues/barriers, potential show-stoppers
	Main players internationally
	Pilot/demonstration/commercial plants and R&D activities
	UK activities and capabilities
	Technology Readiness Level (TRL), and likely TRL in 2020
Technical and economic characteristics	Equipment scales, suitability for small-scale applications ($10-30$ MW$_e$)
	Efficiency
	CO_2 capture rate
	Ability to load follow flexibility
	CAPEX, OPEX
Feasibility	Maximum % co-firing allowable/dedicated biomass
	Ease of changing to high co-firing/complete conversion
	Implications of retrofitting capture versus new build

- *Dedicated biomass combustion*: bubbling or circulating fluidised bed (CFB) or grate
- *Dedicated biomass gasification*: bubbling, circulating or dual fluidised bed, or entrained flow.

14 carbon capture technologies were also reviewed and a similar longlist produced:

- *Post-combustion*: (solvent scrubbing; low-temperature solid sorbents; ionic liquids; enzymes; membrane CO_2 separation and high-temperature solid sorbents)
- *Oxy-combustion*: (cryogenic O_2 separation; membrane O_2 separation and chemical-looping combustion using solid oxygen carriers)
- *Pre-combustion*: (integrated gasification combined cycle (IGCC) with physical absorption; membrane H_2 separation; membrane syngas generation; sorbent-enhanced reforming using carbonate looping; and Zero-Emission Coal Alliance (ZECA) concept).

Using the information collected for each of the different power and capture technologies, 28 feasible combinations of component technologies were proposed. The TESBiC project partners then assessed each combination against a set of criteria, which were agreed with the ETI, with the key benefits and risks highlighted. The assessment criteria covered a range of different development, techno-economic, feedstock, feasibility and UK aspects. The most important criteria were identified as: the likely technology readiness level (TRL) (GCCSI, 2016) in 2020; key technical issues; plant efficiency with capture; capital costs with capture and potential for UK deployment. For the TRL criterion, it was assumed that only technologies which had the potential to reach TRL 5 by 2020 could be deployable at a mass-production scale by 2050.

Based on the key advantages and disadvantages for each combination of technologies, the TESBiC consortium collectively came to an opinion as to which technologies to progress and reject and provided evidence for these rejections. This led to only 8 of the 28 technology combinations being recommended for progression (see Table 5.2).

- Low-temperature solid sorbents, ionic liquids, enzymes and membrane CO_2 separation combinations (3, 4, 5, 6, 5a, 6a, 7, 8) potentially have reduced capital costs compared to amine scrubbing, but they generally only have marginal efficiency benefits, and there are uncertainties regarding operating costs, as well as several major technical issues yet to be resolved.
- Membrane O_2 separation, membrane H_2 separation, membrane production of syngas, sorbent-enhanced reforming and the ZECA concept combinations (11a, 12a, 17, 18, 19, 20, 21, 22, 23, 24) potentially have high plant energy efficiency, but there are numerous technical issues in addition to uncertain capital costs and paucity of available data for the earliest stage concepts.
- Dedicated biomass with carbonate looping (10) was not progressed, as it was not fully apparent that the calciner could be biomass-fired – i.e. co-firing percentages might be limited to <70%. For this reason, the consortium began by exploring only the co-firing option (9).
- Co-firing chemical-looping combustion (13) was not progressed, since coal gasification rates are slower than those for biomass, and unreacted char leads to carryover and loss of CO_2. Also, chemical looping cannot be retrofitted to a pulverised coal plant – a CFB boiler is needed. Hence, the dedicated biomass option (14) was preferred for progression.

Table 5.2 Power-capture technology combinations proposed for progression/rejection.

Progress / Reject	Post-combustion						Oxy-combustion				Pre-combustion			
	Solvent scrubbing, e.g. monoethanolamine, chilled ammonia	Low-temperature solid sorbents, e.g. supported amines	Ionic liquids	Enzymes	Membrane separation of CO$_2$ from flue gas	High-temperature solid sorbents, e.g. carbonate looping	Oxy-fuel boiler with cryogenic O$_2$ separation	Oxy-fuel boiler with membrane O$_2$ separation	Chemical-looping combustion using solid oxygen carriers	IGCC with physical absorption e.g. Rectisol, Selexol	Membrane separation of H$_2$ from synthesis gases	Membrane production of syngas	Sorbent-enhanced reforming using carbonate looping	ZECA concept
Coal IGCC gasification — Direct co-firing	Not feasible						Not feasible			15	17	19	21	23
Coal IGCC gasification — Conversion to 100% biomass														
Pulverised coal combustion — Direct co-firing	1	3	5	5a	7	9	11	11a	13		Not feasible			
Pulverised coal combustion — Conversion to 100% biomass														
Dedicated biomass combustion — Fixed grate	2	4	6	6a	8	10	12	12a	14					
Dedicated biomass combustion — Bubbling fluidised bed	Not feasible						Not feasible							
Dedicated biomass combustion — Circulating fluidised bed														
Dedicated biomass gasification — Bubbling fluidised bed										16	18	20	22	24
Dedicated biomass gasification — Circulating fluidised bed														
Dedicated biomass gasification — Dual fluidised bed														
Dedicated biomass gasification — Entrained flow														

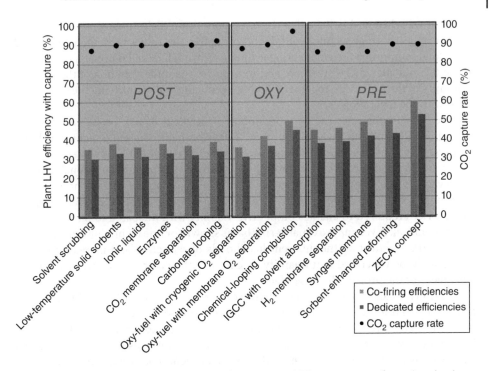

Figure 5.1 Estimated plant LHV efficiencies with capture, and CO_2 capture rates, for each technology combination (error bars are quantified in the detailed analysis).

With feedback from the ETI and a group of independent technical reviewers, this selection process left eight technology combinations recommended for progression to further consideration:

(1)   Co-firing combustion, with post-combustion amine scrubbing
(2)   Dedicated biomass combustion, with post-combustion amine scrubbing
(9)   Co-firing combustion, with post-combustion carbonate looping
(11) Co-firing oxy-combustion, with cryogenic O_2 separation
(12) Dedicated biomass oxy-combustion, with cryogenic O_2 separation
(14) Dedicated biomass chemical-looping combustion using solid oxygen carriers
(15) Co-firing IGCC, with physical absorption
(16) Dedicated biomass IGCC, with physical absorption

An overall view of the combinations recommended for progression or rejection is given in Table 5.2. The recommendations covered all three main capture categories and also gave an equal split between large-scale co-firing combinations and small-scale dedicated biomass combinations.

Further quantitative data are provided below for each technology combination, comparing factors such as the plant efficiency with capture, the CO_2 capture rate (Figure 5.1) and the estimated cost of avoided CO_2 (based on costs available from the literature, and occasionally with significant assumptions made based on analogies with similar systems). These were the best estimates available at a relatively early stage of the TESBiC project. It was clear at this stage that the error bounds on the estimates were especially large for the early-stage technologies. Eight in-depth case studies were then conducted (see below) to improve the engineering estimates (through detailed process modelling) for the technologies shortlisted here and to improve the understanding of the techno-economics of the processes, together with both reducing and quantifying the risks associated with the technologies.

Figure 5.2 demonstrates the very preliminary initial sift of risk versus reward. The higher the TRL, and the fewer the number of development issues and technical showstoppers, then the lower the 'risk' was judged to be. The cost of CO_2 avoided (in comparison to a reference coal power station) was deemed to represent a reasonable proxy for 'rewards', since it includes a variety of factors such as capture rate, plant efficiency and capital costs with capture in its calculation, and is also a useful indication of the carbon prices required to enable competitive viability with unabated fossil fuel or biomass generation. A very approximate error bar is shown for one technology in Figure 5.2 (+40% – 30% – reflecting the different likelihoods of reduction and increase in cost), but it should be reiterated that quantifying carefully the error bounds was part of the later stages of the project, discussed later.

A clear justification was thus demonstrated for why the shortlist of eight technologies was chosen for progression. These eight technologies were assessed to have the lowest risk, i.e. were further left on the *x*-axis, and hence were considered to be most likely to be developed in time for 2050 mass deployment. At this stage, it was felt that there was insufficient evidence to exclude any major technology on the grounds of cost

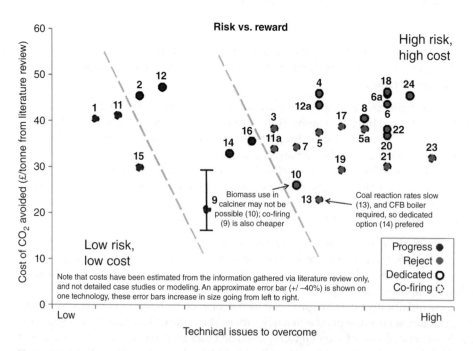

Figure 5.2 Initial rough cut of risk versus reward in the first stage of the project. (*See colour plate section for the colour representation of this figure.*)

(such information being developed more fully later in the project). This means that whilst attractive in terms of potential deployment, all eight technologies still covered a broad range of avoided CO_2 costs (and all were potentially deployable at scale by 2050):

- The 'benchmark' near-term cases of co-firing with amine scrubbing (**1**) and oxy-fuel with cryogenic O_2 separation (**11**) had average costs of avoided CO_2.
- The corresponding dedicated biomass systems (combinations **2** and **12**) were more expensive, and at a slightly earlier stage of development, but there were not expected to be major technical differences to the co-firing cases. However, these abate more CO_2 per MWh of power produced than a co-fired system.
- Both co-firing (**15**) and dedicated biomass (**16**) IGCC with physical absorption were considered to have the potential to be cheaper than the previous options, mainly due to their higher efficiencies. However, (**16**) had only been considered theoretically, and there was not a clear development pathway. There were, however, no major technical showstoppers, and knowledge spill-over from (**15**) and biofuels applications could accelerate (**16**)'s development. Of the dedicated biomass gasification combinations, (**16**) was judged to be more promising overall at this stage than (**18**), (**20**), (**22**) or (**24**), both in terms of risk and reward (in particular because of their low TRL). There may also be interesting options for small-scale integration with future syngas infrastructure, or H_2 storage.
- The more technically risky options of dedicated biomass chemical-looping combustion (**14**) and co-firing with post-combustion carbonate looping (**9**) show low costs of avoided CO_2. (**9**) also has the potential benefit of cement industry decarbonisation at low cost. (**14**) could have even higher efficiencies (above 50%) via process integration options with gas turbines or H_2 production, and was therefore judged to be the technology most suited to small-scale power applications.

It is important to reiterate that this was a high-level classification and that the purpose of retaining a wide range of technologies was to avoid incorrectly excluding potential technologies at this early stage of the project. At this stage, all technologies considered to have the potential to reach TRL 5 by 2020 were retained.

5.2 Case Study Analysis

The eight shortlisted technologies chosen were each subjected to a more detailed techno-economic assessment involving the list/steps below, in order to assess opportunities for development and acceleration of deployment. The techno-economic assessment included a modelling exercise and benchmarked technologies against a number of key performance indicators. The following items of engineering and other work were prepared:

- An overview of the total process for the combined technologies and the relevant engineering standards
- A preliminary process flow diagram with energy, material and carbon balances
- A list of the major items of equipment with short performance specifications covering the key process factors
- A high-level process control philosophy

- An environmental performance summary, including the process inputs and estimates of the emissions to air, land and water, including the major prescribed pollutant species and carbon in all forms
- An estimate of the project development and capital costs for new build and retrofit projects, as appropriate
- An estimate of the fixed and variable operating and maintenance costs
- An overview evaluation of the system performance and a critical assessment of the knowledge gaps, technical risks and subject areas that will require further development work.

The selected technologies and their component parts spanned a very wide range of TRLs, from as low as TRL 4 (i.e. components tested in the laboratory) up to 7 (pilot plant >5% of commercial scale). For this reason, it proved to be challenging, and in some cases just not possible to obtain sufficiently detailed information on the key criteria across all the technologies considered. To reiterate, the following technology combinations were selected for further study:

- The co-firing of biomass materials in a pulverised coal-fired power plant, with post-combustion carbon dioxide capture by amine scrubbing (co-fire amine)
- The combustion of biomass materials in a dedicated power plant with post-combustion carbon dioxide capture by amine scrubbing (bio amine)
- The co-firing of biomass materials in a pulverised coal power plant, with carbon dioxide capture by using oxy-fuel firing technology, with cryogenic O_2 separation (co-fire oxy)
- The combustion of biomass materials in a dedicated power plant with carbon dioxide capture by oxy-fuel firing technology, with cryogenic O_2 separation (bio oxy)
- Biomass co-firing combustion, with CO_2 capture by post-combustion carbonate looping (co-fire carb loop)
- Dedicated biomass chemical-looping combustion using solid oxygen carriers (bio chem loop)
- Co-firing IGCC, with CO_2 capture by physical absorption (co-fire IGCC)
- Dedicated biomass, i.e. Biomass Integrated Gasification Combined Cycle (BIGCC) (for small scale, i.e. <50 MW_e), with CO_2 capture by physical absorption (bio IGCC).

Thus, the key carbon capture technologies from all three categories (pre-combustion, post-combustion and oxy-fuel) were covered. An overview of the key features of the selected technology combinations is presented in Table 5.3. Outline process flow diagrams with listings of the key flows and process conditions were prepared, as well as calculations of the overall process efficiencies (LHV (lower heating value) basis). High-level assessments of the process control and environmental issues have also been performed, together with sensitivity analyses. The level of detail of the studies varied, depending largely on the status of development of the technologies involved.

Overall, the vast majority of the industrial expenditure up to the time of the project (2011) had been on the development of carbon dioxide capture technology for conventional large coal-fired power plants, i.e.:

- The application of CO_2 capture by solvent-scrubbing technologies to the flue gas streams from large pulverised coal-fired boilers and fluidised bed boilers firing coal and the compression and drying of the CO_2 prior to transportation offsite.

Table 5.3 An overview of the eight selected technology combinations, studied in detail.

Criteria	Co-firing amine scrubbing	Dedicated biomass with amine scrubbing	Co-firing oxy-fuel	Dedicated biomass oxy-fuel	Co-firing carbonate looping	Dedicated biomass chemical looping	Co-firing IGCC	Dedicated biomass BIGCC
Current TRL	6–7	4	6	5	4–5	4	5–6	4
Key technical issues	Scale-up, amine degradation	Scale-up, amine degradation	Corrosion, O_2 energy costs, slow response	Corrosion, O_2 energy costs, slow response	Calciner firing, solid degradation, large purge of CaO	Loss in activity, reaction rates, dual-bed operation	Complex operation, slow response, tar cleaning, retrofit impractical	Complex operation, slow response, tar cleaning, retrofit impractical
Suitability for small scale	Low	High	Low	High	Low	High	Low	High
Plant efficiency with capture	OK	Low	OK	Low	Good	Good	High	Good
Capital costs with capture	OK	Expensive	OK	High ASU costs	OK	Low cost	OK	Expensive
UK deployment potential	Immediate capture retrofit opportunities	Retrofit opportunities, high long-term potential	Retrofit opportunities, long-term doubtful	Retrofit opportunities, high high long-term potential	Capture retrofit opportunities, cement integration	Likely first demos in Europe, UK in ~2020. High long-term potential	No current UK plants, several demos by 2020. Long-term doubt	No current UK plants, demo unlikely by 2020. High long-term potential

- The application of oxy-fuel firing technologies to large coal-fired boilers, and the purification, compression and drying of the product CO_2 prior to transportation offsite.
- The application of physical CO_2 absorption techniques to the syngas produced by large coal-fired IGCC power plants.

The development of all three technologies has been pursued in parallel, with oxy-fuel and IGCC at the pilot- and small-demonstration scale, with post-combustion capture via alkanolamine scrubbing being the most highly developed in 2016 – in particular owing to the Boundary Dam scheme (SaskPower, 2016). At the time of the TESBiC project, it was challenging for the electricity supply industry and their equipment suppliers to come to a clear decision as to which technology was preferred for future application, and indeed it remains an open topic for discussion which technologies would be best applied in the context of biomass utilisation, as opposed to fossil fuels.

It is also clear that even in 2016, there has been relatively little activity on the application of the carbon capture technologies to dedicated biomass power plants or to coal power plants which are co-firing biomass. For most applications, the biomass fuel is significantly more expensive than coal, generally by a factor of more than two on a £/GJ basis, and the application of CO_2 capture and storage technology is also relatively expensive.

As discussed earlier, one of the key challenges encountered was associated with the wide variation in the scales of operation and TRLs of the technology combinations studied. These varied from TRL 4, i.e. laboratory component testing, to values as high as TRL 7, i.e. pilot plant >5% of commercial scale. One of the consequences of the variability in the TRL values was that the quality of the technical and economic data available for the technology combinations varied accordingly.

The quality of the technical and economic data for the technology combinations 1–4 (amine and oxy) and 7 (co-fire IGCC), i.e. the technology combinations with relatively high TRL values, was in general very good. These data were largely from published literature, supplemented with in-house information and using validated open-source simulation tools. Overall, the gaps in knowledge of the more advanced technologies are modest and these will be filled by proceeding to the large-scale demonstration stage, with the associated engineering and R&D activities required to address any emergent technical difficulties. This will be done in the first instance for fossil fuel-fired plants, possibly with biomass co-firing. It is likely that the application of the carbon dioxide capture technologies to dedicated biomass plants can then be achieved without a further specific demonstration phase on a biomass plant. It was anticipated at the end of the project in 2012 that the technologies studied would have been ready for market by 2015–2020.

The equivalent technical and economic data for the technology combinations 5, 6 and 8 (carbonate/chemical looping, bio IGCC) were not available in anything like a comparable level of detail, and the process flow sheets and other data have had to be built up using commercial or bespoke academic software. Similarly, the economic data were based on very limited engineering knowledge or experience of the proposed plant configuration, and hence are subject to relatively large bounds of uncertainty. At this stage, plant scale was not harmonised; later work produced results for plant scales of 50, 250 and 600 MW_e.

The gaps in the technical and economic data on the less developed technology combinations were at the time of the TESBiC project substantial. It is important to note that the results of the TESBiC project were a snapshot of progress in 2012 and that the gaps may well be different in 2016, owing to work progressing since, including work conducted by the ETI.

In 2016 (and certainly in 2012), to fill these gaps still requires significant further work at laboratory and pilot scale, extensive programmes of technical work at large pilot scale, a large amount of engineering work and finally component testing and demonstration projects to provide information similar to that available for technology combinations 1–4. In some cases, there are ongoing collaborative projects in these areas with financial support from the British government and the European Commission.

The following approach was adopted within the TESBiC project to help take account of this issue:

- As far as possible common approaches were adopted for both the technical and economic assessment work.
- Where possible, the technical performance and economic parameters for items of equipment, e.g. steam raising, ASU, fuel reception storage and handling, etc., which were common to a number of the technology combinations, were harmonised.
- The basic assumptions about some items of the capital costs (operation and utilities, civil works and land costs, project development costs, etc.), of the fixed plant operating costs (maintenance and labour, insurance, etc.), of the feedstock prices and of the plant capacity factors, etc. were harmonised as far as was possible. These were employed in calculating the overall capital expenditure (CAPEX) and operating and maintenance costs (OPEX) figures for the eight technology combinations. At this stage, errors were not fully quantified, nor were the base-case scenarios harmonised to a common size basis. However, the error associated with e.g. net efficiency would be 2–3%, with potential errors in costings of up to $(-20 + 40)$% (lower for higher TRL technologies, higher for lower TRL technologies). Tables 5.4 and 5.5 present the figures used in the base-case analysis.

The key performance and economic parameters for the eight technology combinations are listed in Tables 5.4 and 5.5. All the eight combinations are also represented in Figure 5.3 in terms of their total LCOE and TRL, at the various range of scales presented in Tables 5.4 and 5.5. Looking in the first instance at the gross and net electricity generation efficiency values obtained for all eight technology combinations, the negative impacts of the carbon dioxide capture technologies were evident. In the cases studied, the operation of the CO_2 capture and compression/cleaning systems (wherever directly comparable with unabated coal plant) decreased the net electrical efficiency of the power plant in the range of 6–13 percentage points. This had also the effect of increasing the specific investment costs significantly. In cases where direct comparison was possible, the increase in the specific CAPEX was in the range of 45–130%, as would be expected. The impact of the operation of the CO_2 capture plant on the total annual O&M costs of the plants was more modest, i.e. an increase in the range 4–58%. The specific O&M cost per MWh of power was clearly much higher for the cases with CO_2 capture because of the significant reduction in the net power output of the plant.

As discussed earlier, it should be also noted that an additional challenge encountered was the lack of a single common engineering software platform that can evaluate all

Table 5.4 Summary of economic parameters for technology combinations 1–4 at base-case capacities.

	1 Biomass co-firing in pulverised coal boiler		2 Dedicated biomass combustion		3 Biomass co-firing in pulverised coal boiler		4 Dedicated biomass combustion	
	Without CCS	With solvent scrubbing	Without CCS	With solvent scrubbing	Without CCS	With oxy-fuel firing	Without CCS	With oxy-fuel firing
Gross power (MW_e)	545.2	474.1	83.5	68.5	545.2	545.2	83.4	84.6
Net power (MW_e)	518.9	398.9	75.8	48.9	518.9	388.7	75.79	48.85
Gross efficiency %	47.1	40.9	39.6	32.5	47.1	47.1	39.5	40.1
Net efficiency %	44.8	34.4	35.9	23.2	44.8	33.6	35.9	23.2
Total installed cost(£M)	509.3	638.2	143.5	198.1	509.3	709.1	143.5	213.2
Total investment cost (£M)	662	830	187	258	662	922	187	277
Specific CAPEX (£M / MW_e)	1.28	2.08	2.46	5.27	1.28	2.37	2.46	5.65
Total O&M costs (£M/yr)	146	158.5	33.3	37.1	175	173.4	33.3	34.3
Energy generated (MWh/yr) × 10^6	3.86	2.97	0.56	0.36	3.86	2.89	0.56	0.36
OPEX and fuel (£/MWh)	37.78	53.36	59	101.9	45.64	58.52	59	94.5

Table 5.5 Summary of economic parameters for technologies 5–8 at base-case capacities.

	5 Carbonate looping		6 Chemical looping		7 IGCC co-firing with CO$_2$ absorption		8 BIGCC	
	Without CCS	With carbonate looping	Without CCS	With chemical looping	Without CCS	With CO$_2$ capture	Without CCS	With CCS
Gross power (MW$_e$)	N.A.	326	N.A.	325	549	549	64.7	73
Net power (MW$_e$)		300		300	484	461	47.5	40
Gross efficiency %		43.9		44.5	45.8	41.7	56.0	56.0
Net efficiency %		40.4		41.0	40.4	35.0	41.0	34.0
Total installed cost (£M)		644		509	650	848	109	167
Total investment cost (£M)		837		662	845	1102	142	217
Specific CAPEX (£M / MW$_e$)		2.79		2.21	1.74	2.40	2.98	5.43
Total O&M costs (£M/yr)		174.70		192.36	128.47	148.45	22.55	35.25
Energy generated (MWh/yr) × 10^6		2.23		2.23	3.61	3.43	0.35	0.29
OPEX and fuel (£/MWh)		78.20		86.60	35.60	43.20	63.80	118.40

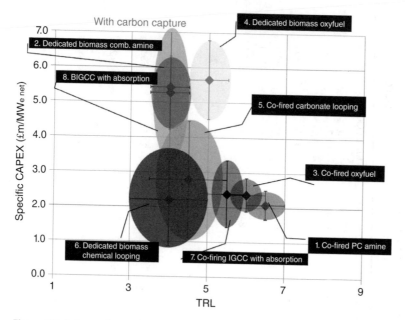

Figure 5.3 Estimated capital costs versus TRL for all eight technology combinations at their base-case scale (40–461 MW$_e$), accounting for uncertainties. (*See colour plate section for the colour representation of this figure.*)

eight biomass CCS technology combinations. Consequently, four software platforms ranging from commercial flow-sheeting software, validated open-source power plant and carbon capture simulation tools to in-house bespoke academic computer programmes were used to assess the eight technologies in detail. The harmonisation explained earlier was beneficial in systematically specifying the boundary conditions for the techno-economic simulations.

The data presented here also illustrate the difficulties the industry faced (and still faces in 2016) in identification of the preferred technology between solvent-scrubbing and oxy-fuel firing. The estimated capital costs of these technologies and the impacts on the net plant efficiencies are generally of the same order. The oxy-fuel technology involves the installation and operation of an ASU, with the associated cost penalties and impacts on power plant flexibility of operation. The solvent-scrubbing process does not require an ASU but has a similar impact on the net power plant efficiency, and there are additional environmental and other issues associated with the on-site storage handling and use of large quantities of organic chemicals.

For the carbonate looping, chemical-looping combustion and the BIGCC systems, the quoted net electrical efficiency of the systems were in the order of 40–41%. These values are significantly higher than those for the more conventional combustion-based solvent-scrubbing and oxy-fuel firing systems studied here.

The specific OPEX + fuel cost estimates for the technology combinations indicated that the specific operating costs for the large coal power plants co-firing biomass and IGCC plants were significantly lower (£/MWh) than those for the smaller dedicated biomass combustion and biomass gasification power plants. The differences between the specific fuel + OPEX costs for the solvent-scrubbing and the oxy-fuel-firing processes were fairly modest.

As stated earlier, the study involved technology combinations (and the associated individual components) spanning a wide range of TRLs, from as low as TRL 4 up to TRL7. It was necessary to account for the uncertainties associated with most of the process and economic factors. For instance, the estimation of the overall electrical generation efficiency values for the technologies is also clearly affected by the TRL. For case studies 1–4 and 7, where good-quality data for current technology at state-of-the-art steam conditions are available, these have been employed in the assessment. For case studies 5, 6 and 8 where good-quality data are not available and the level of engineering detail of the configuration of the process plant and heat exchangers are not available, significant speculation about future operating conditions was required, and this followed through into the wider error bars shown.

The estimated capital costs of the eight technology combinations for the base case (not harmonised to a common scale) are plotted against the TRLs in Figure 5.3. It is evident that the uncertainties associated with economic parameters for the better known and understood technology combinations, for example, co-firing and dedicated biomass combustion with amine scrubbing and oxy-fuel, are relatively low, i.e. of the order of 20%, whereas for relatively new technologies, such as dedicated biomass combustion and gasification as well as chemical looping (TRL around 4), the uncertainties were considered to be in the range of 35–60%.

It was clear that given the very significant influence of plant nameplate capacity (MW$_e$) on the specific investment costs, it was necessary to also compare all eight biomass CCS technologies when built at the same common plant scale. This benchmarking process was carried out to give results at 50 MW$_e$, as well as at 250 MW$_e$, to show the impact of building each technology at small scale, as well as at intermediate scales. An even larger scale (e.g. 500 MW$_e$) was not chosen, since dedicated biomass combustion power plants are typically limited to the <300 MW$_e$ range, i.e. some of the technologies (for example, CFBs) were thought to be reaching their physical limits and could not realistically be compared at very large scales. However, it is clear in 2016 that some assumptions were a little conservative – the 460 MW$_e$ Lagisza power plant in Poland, commissioned in 2009 and operating since, is an example of a very large-scale CFB.

For each technology, using the known base-case total investment costs given at a certain nameplate capacity, we derived new total investment costs at the new benchmark scales by applying a generic engineering scaling exponent (0.7) to the ratio of plant scales. The specific investment cost was then derived at the new benchmark scale (i.e. the scaling factor applies to total costs, not to the specific costs). For the fixed operating costs, the same assumption was made that they are 5%/yr of total installed costs at the new scale, with the variable non-fuel operating costs assumed to scale linearly in proportion to nameplate capacity. This was clearly a significant approximation.

One of the key results of the TESBiC project was therefore a set of estimates for each of the eight technologies, yielding the cost of CO$_2$ avoided and cost of CO$_2$, with agreed error bars based on the errors associated with the individual components of the cost estimates and taking into account the TRL of each technology. It is important to note that for a BECCS plant compared to a standard coal-fired power station, the cost of CO$_2$ avoided is lower than the cost of CO$_2$ captured (this is the opposite to applying CCS to a coal-fired power station, where the cost of CO_2 captured is lower than the cost of CO_2 avoided).

Figures 5.4–5.7 show the results for the different technologies studied at the benchmark scales. Comparing Figures 5.4–5.7, it is clear that the overriding cost effect is actually the plant scale, with differences in cost between the different technologies being significantly lower than the differences in cost between different plant scales. The overall message of the project was that though some technologies do have significantly better central cost estimates, there was no technology that was a clear winner over all others, when technological maturity was taken into account.

The following *short summary statements* on the activities and findings of the first two parts of the TESBiC project can be made:

- This was the first project that tried to look objectively across and compare different BECCS technologies. The results made it clear that it was not possible to objectively pick a 'winning' form of BECCS. Each technology had its benefits and its risks, with technologies that had higher potential efficiency generally having the disadvantage of lower TRL and consequent greater difficulty in assessing costs.
- At the time of the project (2012), there was relatively little activity at industrial scale on the application of the carbon capture technologies to dedicated biomass power plants or to coal power plants which are co-firing biomass.
- The industry was moving to large-scale demonstration projects on fossil-fuel-based power technologies but because these are relatively expensive and require significant government subsidies, progression to the demonstration phase is relatively slow. This point is probably even more true in 2017 than it was in 2011, after the cancellation of the UK CCS demonstration programmes, which aimed to demonstrate full-chain CCS at the ~300 MW_e scale, with the UK government committing around £1 billion to the demonstrations.
- The alternative technologies with lower TRL values continue to be developed at laboratory and, in some cases, at pilot plant scales. More up-to-date information regarding progress in chemical and carbonate looping can be found in Fennell (2015).
- The TRLs for the eight technology combinations considered here varied over a wide range from TRL 4 to TRL 7.
- Within the TESBiC project, a number of software platforms were employed, as appropriate, for process assessment, and a harmonised set of boundary conditions and assumptions were implemented across the eight technology combinations, in order to facilitate the comparison of the relevant techno-economic parameters.
- The process engineering calculations for mass and energy flows, equipment lists, high-level control philosophy, environmental performance, basic process economics and existing gaps in understanding as well as future development requirements were evaluated for the eight technology combinations.
- Wherever direct comparison was feasible from a similar unabated plant, it was observed that the net efficiency decrease due to carbon capture varied in the range of 6–13 percentage points, the specific investment costs increased significantly in the range of 45–130% and the annualised operating and maintenance costs were increased by between 4% and 58%.

It was clear at the end of the project that a number of parameters still had significant potential error ranges and there were a number of areas (such as heat exchanger design

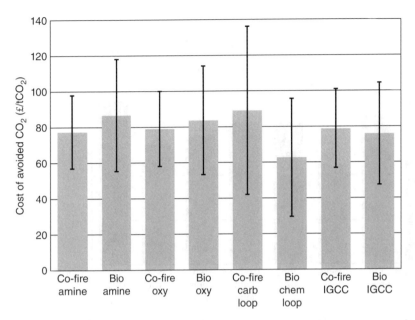

Figure 5.4 Costs of CO_2 avoided for the eight technologies assessed, harmonised to 50 MW$_e$ scale. All biomass as pellets. *Source:* From Bhave et al. (2017). Reproduced with the permission of Applied Energy.

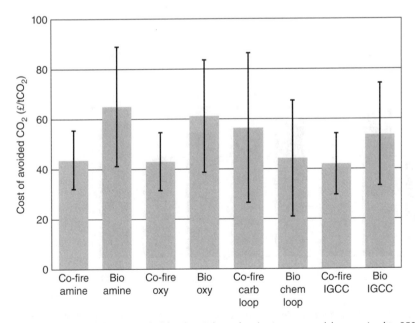

Figure 5.5 Costs of CO_2 avoided for the eight technologies assessed, harmonised to 250 MW$_e$ scale. All biomass as pellets.

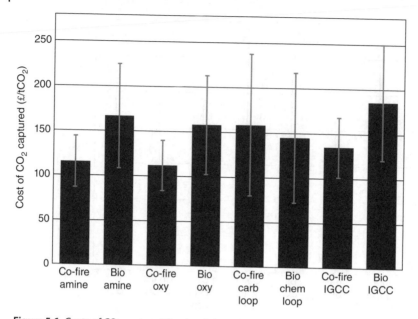

Figure 5.6 Costs of CO_2 captured for the eight technologies assessed, harmonised to 50 MW$_e$ scale. All biomass as pellets.

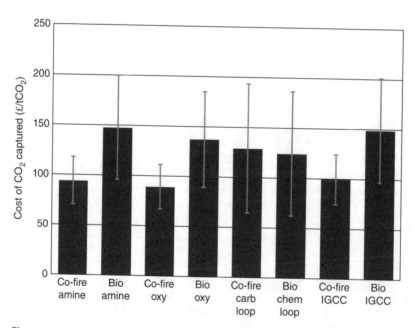

Figure 5.7 Costs of CO_2 captured for the eight technologies assessed, harmonised to 250 MW$_e$ scale. All biomass as pellets.

for the chemical and calcium-looping technologies) where questions still remained unanswered. For this reason, the ETI funded a follow-up study (TESBiC 2), which, through structured interviews with pilot plant operators and industrial process developers, succeeded in further quantifying a number of the key parameters and thereby reduced the uncertainties surrounding the technologies. Further information on the project can be found in a recent publication (Bhave et al., 2017).

Acknowledgements

The TESBIC project was commissioned and funded by the ETI, a public-private partnership between the UK government and industry. The authors are grateful to the ETI as a whole for allowing publication of the results and to Dr. Geraldine Newton-Cross of the ETI in particular for helpful comments on early drafts of this chapter.

References

Bhave, A., Taylor, R.H.S., Fennell, P. et al. (2017). Screening and techno-economic assessment of biomass-based power generation with CCS technologies to meet 2050 CO_2 targets. *Applied Energy* **190**: 481–489.

Fennell, P. (2015). 1 - calcium and chemical looping technology: an introduction. In: *Calcium and Chemical Looping Technology for Power Generation and Carbon Dioxide (CO₂) Capture (ed. P. Fennell and B. Anthony). Woodhead Publishing.*

GCCSI (2016). *CO₂ capture technologies: technology options for CO₂ capture.* https://hub.globalccsinstitute.com/publications/technology-options-co2-capture/technology-readiness-level-trl (accessed 2 November 2016).

Mac Dowell, N. and Fajardy, M. (2016). On the potential for BECCS efficiency improvement through heat recovery from both post-combustion and oxy-combustion facilities. *Faraday Discussions* **192**: 241–250.

SaskPower (2016). *Boundary dam capture project.* http://www.saskpower.com/our-power-future/carbon-capture-and-storage/boundary-dam-carbon-capture-project/ (accessed 13 December 2017).

Part II

BECCS System Assessments

6

Life Cycle Assessment

Temitope Falano and Patricia Thornley

Tyndall Centre for Climate Change Research, School of Mechanical Aerospace and Civil Engineering, University of Manchester, England

6.1 Introduction

This chapter discusses the rationale for and role of life-cycle assessment (LCA) in relation to bioenergy with carbon capture and storage (BECCS). There have been many published LCAs of bioenergy systems but few of these have also covered carbon capture and storage. Obtaining a plausible LCA of the system is the key if we are to be confident that BECCS systems deliver real carbon reductions. This chapter explains the origin and background of LCA, what it can be used for and why it is so important for bioenergy systems. Since the attraction of BECCS is the 'net negative' greenhouse gas (GHG) footprint, it is critical to ensure that this 'net negative' figure has been generated from robust and consistent analysis of both the supply chain and the engineering conversion system, with comparable depth of study for each component of the bioenergy system.

This chapter therefore presents the key issues likely to be encountered in generating an appropriately comprehensive LCA of a BECCS system and discusses the parameters for which it is important to establish sensitivity results. The limited existing publications detailing GHG balances for BECCS systems are reviewed and the need for more comprehensive and robust analyses highlighted.

6.2 Rationale for Supply-Chain Life-Cycle Assessment

Provision of energy from a fossil fuel source generally results in release of GHGs. This is commonly attributable to the combustion of fossil carbon, which was previously sequestered in the ground and is being released to the atmosphere during the combustion process. However, it is also important to take into account other sources of GHG emissions resulting from fossil fuel energy use that are often overlooked. These include energy embodied in materials and infrastructure, e.g. the GHG emissions associated with construction of a power station and the GHG emissions associated with processing, transport and provision of the fuel consumed during the plant operation. It may

Biomass Energy with Carbon Capture and Storage (BECCS): Unlocking Negative Emissions, First Edition.
Edited by Clair Gough, Patricia Thornley, Sarah Mander, Naomi Vaughan and Amanda Lea-Langton.
© 2018 John Wiley & Sons Ltd. Published 2018 by John Wiley & Sons Ltd.

also be necessary to include emissions associated with maintenance and perhaps even treatment and disposal of wastes and end-of-life plant components. Furthermore, accounting for the emissions beyond the direct releases to include embodied carbon enables comparisons to be made between the relative performance of other renewable energy sources, for example, by including the manufacturing of wind turbines or photovoltaic cells. An approach is therefore needed which takes all these factors into account if we are to make a truly representative assessment of the GHG burden associated with a particular method of supplying energy.

One method which is commonly used is LCA. This technique was developed in the 1960s to evaluate the net environmental impact of manufacturing products in different ways. The key feature of LCA is that it looks beyond the production process that is evident to consider the 'upstream' processes that supply feedstock, reagents and material to the process and it also considers final disposal of the product and any associated wastes. This 'cradle to grave' approach allows a much more holistic approach to assessing the environmental impact of a product or service. It also allows us to consider not only the material but also the energy inputs into the system and their full impact.

Bioenergy systems are low carbon by virtue of the fact that they sequester carbon from the atmosphere during growth, and when the biomass is converted to energy an equivalent amount of carbon dioxide is returned to the atmosphere. Therefore, there is no net increase in the atmospheric carbon dioxide burden: the biomass acts as a 'carrier' in which carbon dioxide is stored until it is re-released when energy is extracted.

However, this does mean that there are dynamic variations based on whether the carbon dioxide is present in the atmosphere or biosphere at any particular point in time. For fast-growing crops, this is not a significant issue, as the timing of sequestration during crop growth is close to the timing of release and so the two can be considered to cancel each other out within a reasonable lifetime or analysis period. Taking account of the temporal aspects becomes practically and conceptually more difficult when dealing with plants that take longer to grow. For example, forests have their fastest carbon sequestration periods during the early periods of growth but the carbon can then stay locked up for many years, even after the tree is felled. However, as soon as the wood is used for energy, the carbon is immediately released to the atmosphere. So, the carbon dioxide in the atmosphere is being absorbed during a relatively long period of time and then being released at a single point in time. In the long term, averaging over all the trees growing globally, this is not a problem as it simply represents a cycling of carbon dioxide. However, it could be problematic if that release occurred at a time that exacerbated the atmospheric emission levels, i.e. when there was insufficient 'carbon budget' left for further emissions. Therefore, when dealing with biomass that locks up carbon for a longer period of time, it may also be important to take into account the temporal dimensions of the GHG balance. Traditional LCA is unable to do this, but methods for addressing it are emerging (Röder and Thornley, 2016).

However, it has to be acknowledged that establishing the bioenergy system itself may entail the use of fossil fuels and the consequent carbon impact needs to be taken into account in the calculations. This can range from the GHG impact of the manufacture of steel to build the power plant through to fuel used in delivery and processing of fuel. A comprehensive assessment would also consider the consequences of bioenergy implementation on other adjacent systems. An obvious example is that their fossil fuel emissions would be avoided, but if biomass demand directly led to earlier forest harvests, the amount of carbon locked up in forests might change. Decisions about whether and how to include these different impacts depend very much on the question that the LCA study is designed to address. LCA is a technique used to address a variety of different environmental impact questions and can provide a whole range of different results, depending on the scope of the system that is included, the emission factors used, whether consideration is given to displaced products and alternative uses, etc. Therefore, in accordance with the established ISO guidelines (Adams et al., 2013, ISO, 2006), a very important first step in any LCA study is to pause to consider what question is actually being asked, as this will inevitably influence its goal and scope. It has been shown (Thornley et al., 2015) that subtly different framings of research questions, for example, whether seeking to minimise system GHGs or maximise reductions in absolute or relative terms, can lead to different rank orderings of different bioenergy system alternatives. Therefore, for BECCS systems, it might be important, for example, to consider whether the objective is to identify the largest GHG reductions per unit of biomass used or the smallest level of GHGs emitted per unit of CO_2 biologically sequestered. However, it must be clearly understood that these decisions need to be made at the outset of an LCA study as they affect scope, assumptions and method. Extreme care must therefore be taken when factors are taken from energy-system and other models designed for a different purpose and containing assumptions that may not be consistent with the BECCS study objectives.

Studies of BECCS usually include the sequestration of carbon dioxide during plant growth and the storage of carbon in geological formations at the end of the process. This means that BECCS systems often result in a net transfer of carbon dioxide from the atmosphere to the ecosphere, whereas a fossil-fuel-based energy system usually entails a transfer of carbon from the ecosphere to the atmosphere. The BECCS system effectively reverses the conventional flow of emissions, hence the concept of 'negative emissions', that is, providing energy from a system that removes GHGs from the atmosphere rather than emitting them.

At the 2015 Conference of Parties meeting in Paris, the global agreement on future GHG emissions resulted in challenging reduction targets for many countries. Balancing these with trends in population, economic growth and energy consumption is extremely difficult. Even if a country's electricity sector were completely decarbonised and transport burdens shifted primarily to this low-carbon electricity sector, substantial emissions would remain related to land use, food production and other 'difficult-to-decarbonise' sectors such as industrial manufacture, waste management and heating. Therefore, any energy technology options which deliver 'negative' emissions tend to be particularly prioritised in energy-system models, as this eases the burden on other sectors. Few technologies can achieve net GHG removal. From an energy-system modeller's perspective, BECCS achieves this while simultaneously delivering energy to the system. This is very valuable, particularly at a time of higher carbon prices, and so many future energy-system scenarios that meet emissions targets rely on BECCS as part of the energy mix. This context should be taken into account when developing an LCA question. For example, the primary 'currency' of most of these models is energy, so if the objective is to verify the correct emission factor for BECCS to input into an energy-system model, then the functional unit for the LCA is most likely to focus on energy. Furthermore, the extent to which embodied emission and related impacts are taken into account should be consistent with the assumptions in the energy-system model. Thus, consideration should be given to whether the energy-system model is including energy for equipment manufacture elsewhere in its calculations, which might ordinarily be included in an LCA model, and the choice of emissions factors should be consistent. In the LCA, we might know that a particular material is likely to originate from a particular geopolitical region and use a corresponding emission factor for the GHG consequences of electricity in that region, whereas the energy-system model might use more restrictive geographical emission factors. Similarly, LCA studies can take account of the time of day/year that energy is supplied and whether that produces a different carbon footprint, but energy-system models often use annual averages for projected emissions factors. This can be particularly important when considering least cost dispatch and the associated carbon footprint of different technologies.

6.3 Variability in Life-Cycle Assessment of Bioenergy Systems

As outlined in Section 6.2, LCA can calculate the net impact of a BECCS system over a full life cycle, combining the carbon sequestration during plant growth with the emissions associated with biomass harvesting, processing and delivery, as well as those associated with the energy infrastructure and conversion process. It is therefore ideal for evaluating the net climate impact of a BECCS system. However, LCAs of bioenergy systems published in the literature often show significant degrees of variability and are sometimes disputed and contested.

It is important to understand that LCA enables (but does not ensure) a systematic analysis of net impact. It is a flexible tool which can be adapted to answer a host of different questions, with a correspondingly wide variety of 'answers'. Even when following established standards and guidelines, there is significant room for different interpretations which would use different methods and assumptions. This drives variability in results and some of the key contributory factors are discussed later. More details of all of these can be found in Adams et al. (2013) and Gohin (2016).

6.3.1 Variability Related to Scope of System

The goal and scope of an LCA must be consistent with a detailed evaluation of the actual LCA 'question' being asked. In the context of BECCS, clarity around the inclusion or exclusion of the following items is the key:

6.3.1.1 Land-Use Emissions

Land-use emissions relate to releases of GHGs from soil during biomass growth. They are highly variable and depend on ecological zone, agronomy, timing/weather, etc., making them challenging to measure and control. It is therefore important that the treatment of and assumptions about land-use emissions are explicit in LCA calculations for BECCS systems.

This includes emissions such as N_2O from soil during crop cultivation caused by soil microbial activity, which can be significant for some bioenergy feedstocks, particularly where there are high levels of N fertiliser application. Key considerations in determining an appropriate scope for the LCA are as follows:

- Whether the land-use emissions are causally linked to the biomass production, e.g. for agri-wastes and residues it could be argued that they are a by-product of the main agricultural activity.
- Whether an appropriate counterfactual[1] is being considered that characterises the land-use emissions if there is no bioenergy system.

It should be noted that land-use emissions are often highly variable and can be a key driver of uncertainty in the overall system calculations.

6.3.1.2 Land-Use Change Emissions

In bioenergy LCA calculations, it is possible to include direct GHG emissions related to land-use change. However, it should be noted that in the Intergovernmental Panel on Climate Change (IPCC) framework, land-use change emissions are reported and managed separately from energy-system emissions. It may be appropriate for LCA calculations to take account of land-use change emissions if there is an explicit land-use change that can be linked to the bioenergy system, such as if waste land is planted with energy crops or scrub land is converted to agricultural cultivation. This would include emissions associated with clearing the previous crop, appropriate tillage and restoration sufficient to plant the bioenergy crop. However, great care must be taken to ensure that it is reasonable to include these emissions. This involves consideration of issues such as:

- Has the bioenergy system directly caused the land-use change?
- What land-use pattern might otherwise have been implemented and is it appropriate to consider emissions associated with that in the LCA?

6.3.1.3 Indirect Land-Use Change Emissions

If direct land-use change is attributed to a particular bioenergy system, then it may be appropriate to include the GHG consequences of that direct land-use change in an LCA evaluation. However, using a piece of land for bioenergy production means that it is not available for production of food or other agricultural products. If the land would not have been in productive use, then it is reasonable to ignore the displaced product, but if the land would otherwise have been used to produce a product that would have satisfied an identified demand, then it may also be legitimate to consider the indirect effect of the land-use change, i.e. it will now be necessary to produce that product somewhere else and that will incur another (indirect) land-use change to ensure that the net amount of the original product available to end users remains the same (Achten et al., 2013, Gohin, 2016, Plevin and Kammen, 2013, Schmidt et al., 2011). The potential impact of this can only really be assessed at a larger scale and depends very much on context and land-use drivers. Careful thought is required to establish whether it is appropriate to include indirect land-use change in a particular calculation. A higher-level assessment of competing land-use drivers is also needed to establish the magnitude of any such impact and the extent to which it can be related back to the bioenergy system or other causes/drivers.

---

1 The counterfactual is the energy or product provision that would be used if the bioenergy (or other system being studied) is not being used.

6.3.2 Variability Related to Methodology

LCA methodologies conventionally fall into two categories: attributional LCA and consequential LCA. The former, which focuses on process, supply chain and disposal, seeks to attribute an environmental impact to an activity or process. It is appropriate to use attributional LCA where there is a direct causal link between the process and the impacts being measured and the process can be considered to be reasonably independent of other impacts, systems or side-effects. Consequential LCA, by contrast, is used when there is an explicit need to take into account other factors apart from those directly related to the process. These could include, for example, looking at the impact of land-use changes, product substitution and markets, and hence interfaces with the food, energy, water and land-use systems. In this way, the practitioner can consider the wider 'consequences' of the process, system or product being examined. Attributional LCAs usually involve a defined production system, but more than one product. Ideally, this would be addressed by system expansion, but where this is not possible, allocation approaches are used instead. The most common allocation options are to distribute the environmental burdens of the system across the different products in accordance with their mass, energy content or cost. This choice should be made based on what is most appropriate given the goal and scope definition of the system. For example, if the goal is to assess the carbon footprint of producing a unit of a chemical in a factory, an attributional approach would be appropriate. If, however, the aim is to assess the lowest GHG impact from a range of energy technologies, a consequential approach might be more appropriate to enable consideration of the interplay between the different technologies, what might be displaced and the consequences of those interactions and displacements for the overall carbon footprint. Overall, consequential analysis is normally best suited to policy- or system-level questions. Nevertheless, there may be cases where the choice is not clear-cut and/or this may give rise to significant discrepancies between different LCA analyses of what appear to be the same product/system.

6.3.3 Variability Related to System Definition

Defining the goal and scope of the LCA will result in certain parts of the system being defined as either within or outside the scope of consideration, and variations in this scope will give rise to variations in results. Common areas where this tends to be an issue include emissions from land/soil; emissions associated with land-use change; emissions associated with counterfactual land use, waste disposal or material/product manufacture. For example, if the goal is to assess how forest products can be used to produce energy, it may be legitimate to include afforestation and management, but if the goal is to assess different options for management of waste forest residues, including energy production, it has to be assumed that there is a primary market for the main forest product and so it may be legitimate to exclude establishment and management.

6.3.4 Variability Related to Assumptions

Assembling the appropriate data and assumptions is a very significant part of the LCA study. Specialist software is available that can accurately capture the impacts of even relatively small components of a process, e.g. the toxicity impacts of metals present in tyres depositing on land during use. However, there are also disadvantages and one is that such software usually comes with a pre-programmed database of assumptions ranging from electricity mix to the carbon intensity of fertiliser or steel production. These figures can vary significantly depending on the actual process employed and the geographic location: for example, the carbon intensities of electricity in the US and France are very different; fertiliser carbon intensity is sensitive to production method, transport, etc., which will be different in a factory using coal from one using natural gas; steel production carbon intensity may vary significantly with age of plant and assumptions about transport emissions are sensitive to the type of vehicle and mode of operation assumed.

So, even when the high-level system description appears the same, there may be significant differences in the assumptions that mean that different analyses are actually comparing different things, giving rise to different results.

All these variations mean that there is significant scope for variations in the calculated impact of what is ostensibly the same biomass or bioenergy product. Muench (Saikkonen et al., 2014) carried out a comprehensive review of bioenergy LCAs and noted the impact of assumptions and methodological choices on results, concluding that there was a need for a higher level of transparency to improve the comparability of evaluations.

6.4 Published LCAs of BECCS

A comparative evaluation was performed by Spath & Mann (Thompson et al., 2011), which showed the clear GHG benefits of biomass systems over fossil fuel systems, particularly when combined with carbon capture and storage (CCS) approaches. Since then the GHG balance of bioenergy systems has been a topic of significant interest. There are hundreds of peer-reviewed LCA studies of bioenergy systems based on high-level process models, which made significant assumptions about plant performance. Practical considerations about how such systems would be engineered were not investigated fully and there was only limited exposition of key assumptions, parameters and scale approaches (Scovronick and Wilkinson, 2013).

There are many high-level assessments of BECCS indicating positive GHG benefits, which have informed energy-system models and projections. Typical of these is Kemper (2015), which reviews BECCS from a climate change perspective and concludes that BECCS shows significant potential to achieve net CO_2 removal, but focuses on gaps in policy and enabling frameworks, concluding that deployment will require links with the food-water-energy-climate nexus to be addressed.

Other studies (for example, Laude et al. (2011), Strogen et al. (2013)) have looked at the carbon balance and economics of BECCS systems but have not given detailed technical assessment of the LCA. Many studies (e.g. Luckow et al. (2010), Djuric Ilic et al. (2014)), have evaluated the impact and potential of BECCS by comparing the utilisation of biomass in different conversion options and concluded that BECCS forms a very attractive part of a long-term energy solution under stringent GHG reduction quotas. There has been consideration of the optimal scale and other major process options on the overall balance (Achten et al., 2013), but little detailed analysis of the GHG balance itself.

Corti and Lombardi (Soimakallio and Koponen, 2011) evaluated the more advanced option of biomass gasification and CCS by removing CO_2 from the syngas prior to the energy conversion stage at a 20 MW scale, and at a larger scale with Carpentieri (Lechon et al., 2011). They concluded that net GHG balances of −410 to −594 g/kWh were achievable, with the greater negativity achieved by the larger-scale facility. However, these were based on very high-level models of the actual plant performance and limited LCA assumption transparency and practical validation.

Two studies that have considered the LCA aspects of BECCS in detail focused on co-firing in a coal-fired power plant as the option closest to market and so with more readily available technical detail (Falano and Thornley, 2015, Plevin and Kammen, 2013). The key conclusions of both studies are that carbon neutrality and net negative emissions are achievable with BECCS systems. Schakel finds that at 30% co-firing of biomass net negative emissions of the order of 67–85 g/kWh are achieved, closely consistent with Falano's findings that net negative emissions were achieved at 30% co-firing ratio; increasing this to 60% increased the GHG savings very significantly. Both studies concluded that while global warming potential (GWP) decreased, other environmental impacts increased as a result of introducing biomass co-firing. This is consistent with other works on bioenergy LCA (Thornley and Gilbert, 2013), which focused on trade-offs associated with bioenergy systems, showing that maximising reductions in GHGs may increase other impacts such as eutrophication and acidification. These wider environmental impacts are legitimate components of the analysis; however, they also reinforce the importance of the system's scope. When we compare a bioenergy system with a conventional energy system, the functional unit on which the comparison is based is usually an energy unit. However, most bioenergy systems will involve some degree of land occupation and this practically always involves environmental exchanges, which give rise to the eutrophication and acidification impacts. If a scrupulously fair comparison were being carried out, consideration could legitimately be given to an alternative use for the land as well as the energy. Then the work could reach quite different conclusions.

6.5 Sensitivity Analysis of Reported Carbon Savings to Key System Parameters

When carrying out LCAs, practitioners will attempt to use the best data to which they have access. However, data acquisition is never perfect. There is natural variability associated with the nature of a real, organic system using land, where yields and other parameters may vary. In addition, there is variability related to differences in assumptions or contexts (Adams et al., 2013). Where significant uncertainty exists, it is therefore extremely important to carry out a thorough sensitivity analysis that tests the impact of changing key parameters on a wide variety of results.

6.5.1 Impact of CO_2 Capture Efficiency

One of the most novel aspects of most BECCS systems is the actual CO_2 removal and capture. There is limited practical/operational experience of carbon capture and storage facilities and practically no operating experience with a biomass-derived fuel gas. If considering a physical sorbent, such as an amine scrubber, assumptions can be made in relation to the efficiency of removal, but these are untested and so should be subject to sensitivity. Capture rates might typically vary from 85% to 99% depending on system design, which is influenced by economic and other considerations. This will directly impact the GHG balance and so could have a significant impact on the system performance characterisation. Calculations of this impact on a life-cycle basis indicate that these feasible variations in capture efficiency could result in a variation in CO_2 intensity of the electricity produced of over 100 kg CO_2/MWh (Falano and Thornley, 2015). This variation is of a similar order of magnitude to the actual carbon intensity of potential low-carbon energy supply options and so is very significant: the variability effectively equates to the difference between one BECCS plant being required to offset every unit of low-carbon generation or two BECCS plants being required per unit to be offset.

6.5.2 Variation of Energy Requirement Associated with CO_2 Capture

Energy penalties associated with CO_2 capture include the use of steam for the CO_2 stripping column, pumping electricity for the amine sorbent, power for the combustion air compressors and electricity for the air separation unit (see Chapters 5 and 7). This energy use effectively reduces the net efficiency and electrical output of the BECCS system. In other words, the CO_2 capture/reduction in CO_2 intensity is at the expense of higher internal energy consumption.

This internal energy consumption is accounted for in most high-level BECCS models. However, it is difficult to evaluate accurately at this stage of development. Therefore, it would be wise to evaluate the potential impact of variation in this energy requirement on the overall CO_2 balance. Preliminary assessments have shown that implementing BECCS at a 15% co-firing rate with a post-combustion amine scrubber increases the internal energy demand of the plant by 13% (Falano and Thornley, 2015), which means that very significant carbon savings are achieved at the expense of a small but notable increase in energy consumption. This and other impacts on efficiency should always be taken into account in LCA and energy-system models.

6.5.3 Variation of Biomass Yield

Life-cycle analysis, by definition, includes consideration of the conversion plant and the upstream biomass production; thus waste and residual feedstocks may have fairly low environmental impacts (since they do not represent a dedicated land use or activity), but energy crops developed specifically for the purpose of providing biomass fuel represent a dedicated activity that must be taken into account. Crop yield makes a huge difference to economic viability and so it makes sense to test the response of the carbon intensity of the BECCS system to crop yield. However, variations in crop yield of up to 20% have only a marginal impact on carbon intensity, with increases in yield of this magnitude reducing carbon intensity by only of the order of 1 kg CO_2/MWh.

6.6 Conclusions

LCA is a tool that can be used to assess the GHG impact of BECCS systems, since it quantitatively balances upstream carbon sequestration in biomass with releases associated with processing of the feedstock, energy production and energy consumption associated with storage. LCAs have shown that BECCS systems can achieve significant negative net emissions. However, in order to ensure that the calculations are appropriate to the overall objective, it is important that the assumptions and scope of the system analysed are clear and transparent. While many LCA studies of bioenergy systems have been carried out, only a limited number to date have focused on BECCS, and these have typically used models and high-level assumptions, which have not always been transparent. If there is to be confidence around the ability of BECCS to really deliver negative emissions, it is critical that more detailed engineering studies are developed that reflect the reality of likely BECCS systems performance based on experience accumulated with other bioenergy technologies appropriately developed and extrapolated.

References

Achten, W.M.J., Trabucco, A., Maes, W.H. et al. (2013). Global greenhouse gas implications of land conversion to biofuel crop cultivation in arid and semi-arid lands – lessons learned from Jatropha. *Journal of Arid Environments* **98**: 135–145.

Adams, P., Bows, A., Gilbert, P., et al. (2013). Understanding greenhouse gas balances of bioenergy systems. *SUPERGEN Bioenergy Hub Expert Workshop*, Manchester, UK: Greenhouse Gas Balances of Bioenergy.

Carpentieri, M., Corti, A., and Lombardi, L. (2005). Life Cycle Assessment (LCA) of an integrated biomass gasification combined cycle (IBGCC) with CO_2 removal. *Energy Conversion and Management* **46**: 1790–1808.

Corti, A. and Lombardi, L. (2004). Biomass integrated gasification combined cycle with reduced CO_2 emissions: Performance analysis and life cycle assessment (LCA). *Energy* **29**: 1209–2124.

Djuric Ilic, D., Dotzauer, E., Trygg, L., and Broman, G. (2014). Introduction of large-scale biofuel production in a district heating system – an opportunity for reduction of global greenhouse gas emissions. *Journal of Cleaner Production* **64**: 552–561.

Falano, T. and Thornley, P. (2015). Air/oxy biomass conversion with carbon capture and storage. European Biomass Conference, ETA Florence 2015. Vienna.

Gohin, A. (2016). Understanding the revised CARB estimates of the land use changes and greenhouse gas emissions induced by biofuels. *Renewable and Sustainable Energy Reviews* **56**: 402–412.

Hennecke, A.M., Faist, M., Reinhardt, J. et al. (2013). Biofuel greenhouse gas calculations under the european renewable energy directive – a comparison of the BioGrace tool vs. the tool of the roundtable on sustainable biofuels. *Applied Energy* **102**: 55–62.

ISO (2006). *Environmental management – Life cycle assessment – Principles and framework*. ISO.

Kemper, J. (2015). Biomass and carbon dioxide capture and storage: A review. *International Journal of Greenhouse Gas Control* **40**: 401–430.

Laude, A., Ricci, O., Bureau, G. et al. (2011). CO_2 capture and storage from a bioethanol plant: carbon and energy footprint and economic assessment. *Int. J. Greenhouse Gas Control* **5**: 1220–1231.

Lechon, Y., Cabal, H., and Sáez, R. (2011). Life cycle greenhouse gas emissions impacts of the adoption of the EU directive on biofuels in Spain. Effect of the import of raw materials and land use changes. *Biomass and Bioenergy* **35**: 2374–2384.

Luckow P., Wise M. A., Dooley J. J., and Kim S. H. (2010), Biomass Energy for Transport and Electricity: Large Scale Utilization under Low CO_2 Concentration Scenarios, Pacific Northwest National Laboratory, prepared for the U.S. Department of Energy.

Plevin, R.J. and Kammen, D.M. (2013). Indirect land use and greenhouse gas impacts of biofuels. In: *Encyclopedia of Biodiversity*, 2ee. Waltham: Academic Press.

Röder, M. and Thornley, P. (2016). Bioenergy as climate change mitigation option within a 2 °C target – uncertainties and temporal challenges of bioenergy systems. *Energy, Sustainability and Society* **6**.

Saikkonen, L., Ollikainen, M., and Lankoski, J. (2014). Imported palm oil for biofuels in the EU: profitability, greenhouse gas emissions and social welfare effects. *Biomass and Bioenergy* **68**: 7–23.

Schmidt, J., Gass, V., and Schmid, E. (2011). Land use changes, greenhouse gas emissions and fossil fuel substitution of biofuels compared to bioelectricity production for electric cars in Austria. *Biomass and Bioenergy* **35**: 4060–4074.

Scovronick, N. and Wilkinson, P. (2013). The impact of biofuel-induced food-price inflation on dietary energy demand and dietary greenhouse gas emissions. *Global Environmental Change* **23**: 1587–1593.

Soimakallio, S. and Koponen, K. (2011). How to ensure greenhouse gas emission reductions by increasing the use of biofuels? – suitability of the European Union sustainability criteria. *Biomass and Bioenergy* **35**: 3504–3513.

Strogen, B., Horvath, A., and Zilberman, D. (2013). Energy intensity, life-cycle greenhouse gas emissions, and economic assessment of liquid biofuel pipelines. *Bioresource Technology* **150**: 476–485.

Thompson, W., Whistance, J., and Meyer, S. (2011). Effects of US biofuel policies on US and world petroleum product markets with consequences for greenhouse gas emissions. *Energy Policy* **39**: 5509–5518.

Thornley, P. and Gilbert, P. (2013). Biofuels: balancing risks and rewards. *Interface Focus* **3**.

Thornley, P., Gilbert, P., Shackley, S., and Hammond, J. (2015). Maximizing the greenhouse gas reductions from biomass: the role of life cycle assessment. *Biomass and Bioenergy* **81**: 35–43.

7

System Characterisation of Carbon Capture and Storage (CCS) Systems

Geoffrey P. Hammond

Department of Mechanical Engineering, Institute for Sustainable Energy and the Environment (ISEE), University of Bath, UK

7.1 Introduction

7.1.1 Background

Carbon dioxide (CO_2) capture and storage (CCS) may constitute an important transitional component of a wider low-carbon strategy for the future in the industrial world (Helm, 2012; Watson, 2012; Hammond and Pearson, 2013). CCS facilities coupled to fossil-fuelled, biomass or co-fired power plants or industrial sites provide a climate-change mitigation strategy that potentially permits the continued use of fossil fuel resources, whilst reducing (or burying) CO_2 emissions below ground. A collaborative study between the Energy Technologies Institute (a public–private partnership of key industrial companies and United Kingdom (UK) funders of energy RD&D) and the Ecofin Research Foundation recently (ETI/ERF, 2012) examined the conditions required for mobilising private-sector financing of CCS in the United Kingdom. They argue that this technology would be a 'huge prize' that could cut the annual costs of meeting the UK government's 2050 target of 80% carbon reduction by up to 1% of gross domestic product (GDP). Nevertheless, they noted that the prevailing financial market conditions were demanding. In order to meet this challenge, they suggest that the United Kingdom needs to build confidence in long-term policy, develop attractive pricing for CCS contracts with suitable risk sharing, put in place an appropriate regulatory and market framework and devise new ways to offset North Sea storage liability risks. It currently exhibits a significant cost premium over its competitors and will rely on cost reduction to become commercially viable.

The UK Coalition Government launched a 2012 CCS roadmap (Hammond and Spargo, 2014) that included the European Union (EU)-stimulated requirement on any new fossil-fuelled power station to demonstrate that it was 'captureready'. It simultaneously announced its latest competition (known as the CCS Commercialisation Programme) for £1 bn capital funding to build a commercial-scale, coal- or natural-gas-fuelled power plant and capture facility in Great Britain to be operational by

Biomass Energy with Carbon Capture and Storage (BECCS): Unlocking Negative Emissions, First Edition.
Edited by Clair Gough, Patricia Thornley, Sarah Mander, Naomi Vaughan and Amanda Lea-Langton.
© 2018 John Wiley & Sons Ltd. Published 2018 by John Wiley & Sons Ltd.

2016–2020 with an appropriate storage site offshore. The UK government also established a CCS Cost Reduction Task Force as an industry-led joint venture to assist in making the technology a commercially viable operation by the early 2020s. But recently (25 November 2015), it cancelled its £1 bn CCS public capital investment funding competition, which suggests that this technology may have an uncertain future in Britain. This took place only weeks before the final competition submissions by the two short-listed schemes – the Peterhead Project in Aberdeenshire, Scotland, and the White Rose Project at the Drax site in North Yorkshire (Hammond and Spargo, 2014) – were due to be lodged. The UK Prime Minister, David Cameron, subsequently stated in Parliament that CCS is a 'technology that isn't working' (Prime Minister's Questions, House of Commons, 16 December 2015). However, the House of Commons' Energy and Climate Change Committee (ECCC, 2016) in their report from a subsequent examination of the future of CCS argued that meeting the United Kingdom's climate-change targets of an 80% reduction in carbon emissions by 2050 against a 1990 baseline would be more challenging and costly without CCS deployment. They advocated the development of a new gas-focused CCS strategy with an emphasis on the infrastructure requirements for CO_2 transport and storage. The Committee felt that without such a strategy the United Kingdom would lose CCS investment and assets and ultimately expertise and knowledge in this critical area. The closure of the CCS Commercialisation Programme resulted in a severe lack of confidence by industrial partners who view the decision as putting future investment and applications for EU funding at risk.

Large-scale CCS in the power sector (especially emerging biomass energy and carbon capture and storage (BECCS)) is at an early stage of worldwide research, development and demonstration (RD&D). The Boundary Dam Power Station (near Estevan, Saskatchewan, Canada) only came online as the world's first integrated, full-scale coal-fired power station – (post-combustion) CCS plant – in the autumn of 2014. Commissioning activities for the next large-scale plant at the Kemper County Energy Facility in Mississippi, the United States, were not expected to start until the middle of 2016, according to the Global CCS Institute (2015). Nevertheless, Hammond et al. (2011) and Hammond and Spargo (2014) evaluated fossil-fuelled (coal and natural gas) plants with CCS infrastructure in terms of various energy, environmental (particularly climate change) and economic criteria. They found that CO_2 capture can reduce emissions by over 90% or about 70% when upstream emissions were accounted for Hammond et al. (2013) and Hammond and O'Grady (2014). However, this will reduce the efficiency of the power plants concerned, incurring energy penalties of 14–30% compared to reference plants without capture. Hammond and Spargo (2014) examined potential design routes for the capture, transport and storage of CO_2 from UK power plants. Energy and carbon analyses were performed on power stations with and without CCS, based on both currently available and novel CCS technologies. Due to lower operating efficiencies, the CCS plants show a longer energy payback period (EPP) and a lower energy gain ratio (EGR) compared to conventional plants (Hammond and Spargo, 2014). Costs of capture, transport and storage were concatenated by Hammond et al. (2011), indicating that the whole CCS chain cost of electricity (COE) increased by 30–140% depending on the option adopted. More recent cost estimates have been obtained by UK industry-led attempts to determine opportunities for cost reductions across the whole CCS chain (CCRTF, 2013).

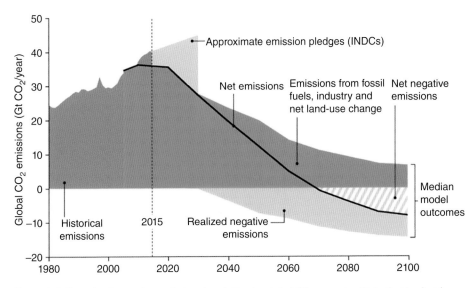

Figure 1.2 The role of negative emissions in relation to global CO$_2$ scenarios. Note that 'realised negative emissions' correspond to 'negative emissions described in Figure 1.1b and that 'net negative emissions in this figure correspond to 'global net negative emissions' described in Figure 1.1d. *Source:* Anderson and Peters (2016).

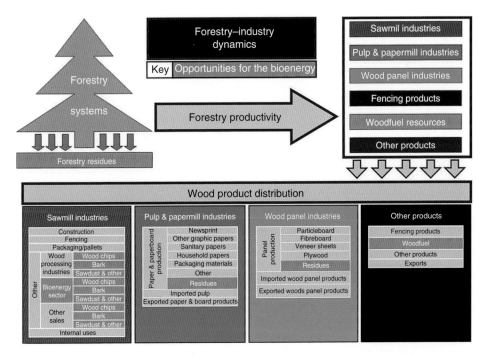

Figure 2.3 Example of the analysis boundaries and developed approach for evaluating forestry system and industry dynamics, as modelled for biomass resource assessment research (Welfle, 2014).

Biomass Energy with Carbon Capture and Storage (BECCS): Unlocking Negative Emissions, First Edition.
Edited by Clair Gough, Patricia Thornley, Sarah Mander, Naomi Vaughan and Amanda Lea-Langton.
© 2018 John Wiley & Sons Ltd. Published 2018 by John Wiley & Sons Ltd.

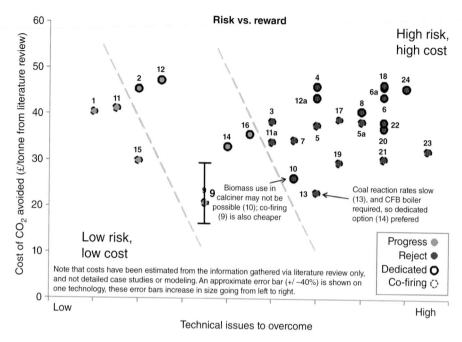

Figure 5.2 Initial rough cut of risk versus reward in the first stage of the project.

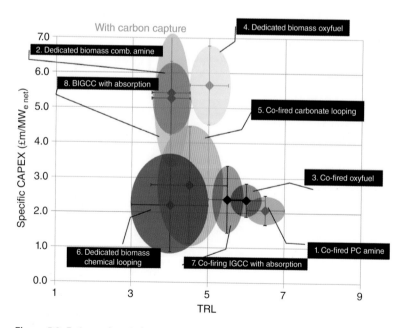

Figure 5.3 Estimated capital costs versus TRL for all eight technology combinations at their base-case scale (40–461 MW$_e$), accounting for uncertainties.

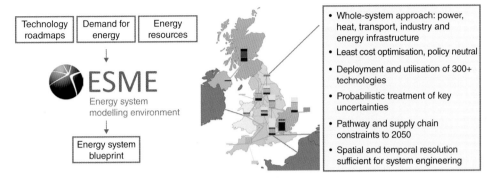

Figure 8.1 Overview of the Energy System Modelling Environment (ESME).

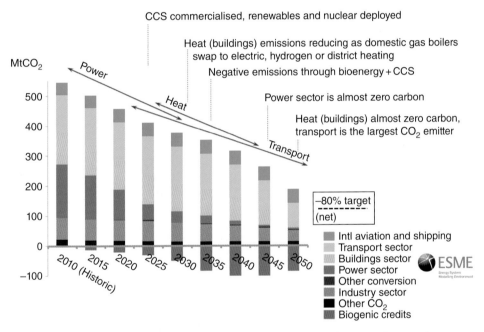

Figure 8.2 Typical decarbonisation pathway modelled in ESME, needed to meet the United Kingdom's 2050 GHG emission reduction targets. *Source:* Data from v4.2.

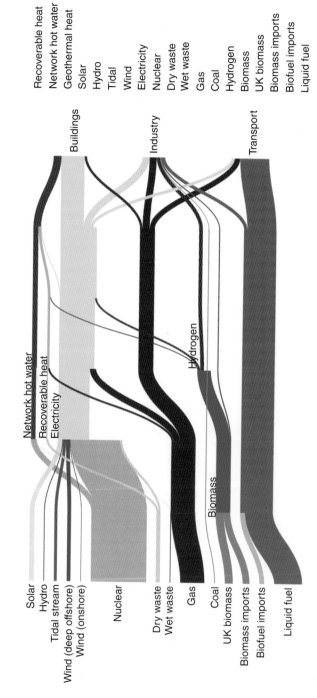

Figure 8.3 Typical (average) energy flows in 2050: Sankey diagram from ESME. *Source:* Data from v4.2.

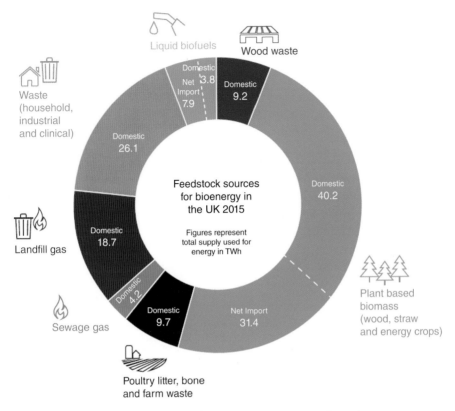

Figure 8.4 2015 biomass and waste feedstock sources. *Source:* BEIS, (2016a).

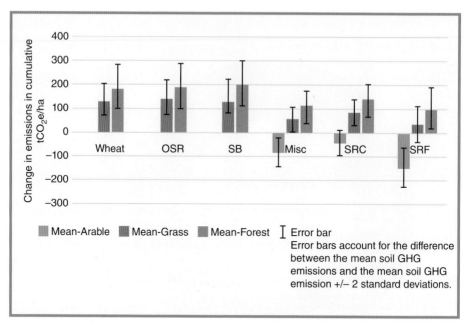

Figure 8.6 Estimates derived from the ELUM model on mean soil GHG emissions over 40 years (relative to counterfactual land use), expressed as net GHG emissions per hectare across the United Kingdom. The model was validated using empirical data collected during the ELUM project (Smith et al., 2010).

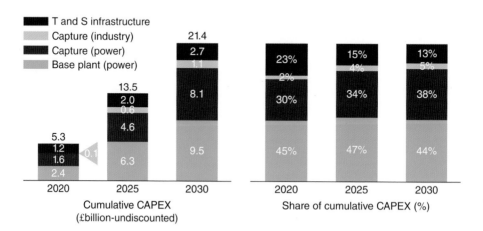

Figure 8.7 Capital costs of building a 50 Mt/year CO_2 (~10 GWe) CCS network (£bn 2014 undiscounted). *Source:* FromETI, (2016d). Reproduced with permission of ETI.

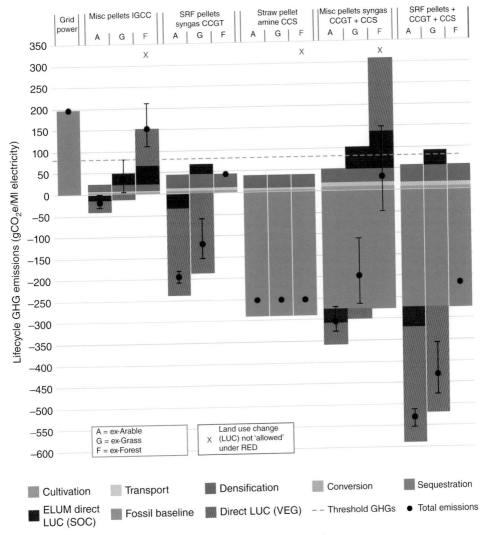

Figure 8.8 Quantifying the impact of dLUC emissions and CCS on UK bio-electricity value chains (life-cycle GHG emissions: gCO₂e/MJ).

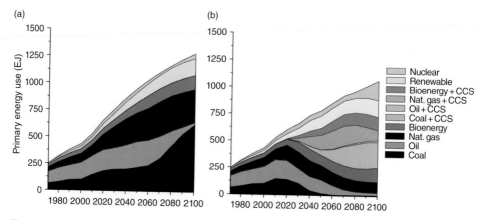

Figure 9.1 Trends in global energy use for (a) baseline scenario and (b) RCP2.6 scenario. Note in (b) BECCS use (bright green) from 2020, non-CCS bioenergy use (dark green) and fossil fuel with CCS use (blue, pale orange and grey). *Source:* van Vuuren et al., 2011, Figure 2 (p102).

7.1.2 The Issues Considered

BECCS technologies are aimed at providing a net negative emission option within a low-carbon transition pathway to 2050 and beyond. In order to prepare for the development of a new phase in the deployment of emerging BECCS in the power sector and industry, the performance of CCS systems will need to be characterised and evaluated according to a number of criteria. However, given the present state of play with CCS RD&D, many of the system performance characteristics need to be determined on a first-of-a-kind (FOAK) basis (Hammond et al., 2011; Hammond and Spargo, 2014). Nevertheless, they are based on well-established, often quantifiable, economic, energy-related and environmental (including climate-change) criteria. Carbon-capture facilities will hinder the performance of power plants and give rise to an energy penalty, which, in turn, lowers the system (thermodynamic) efficiency. The LCOE can then be used as an indicator of the impact of adding capture equipment on plant economics. All such cost figures should be viewed as indicative or suggestive (Hammond and Spargo, 2014). They nonetheless help various CCS stakeholder groups (such as those in industry, policymakers (civil servants and the staff of various government agencies) and civil society and environmental non-governmental organisations (NGOs)) to assess the role of this technology in national energy strategies and its impact on local communities. Finally, the environmental performance of CCS developments can be assessed in terms of climate-change impacts (including parameters such as *carbon* intensity and the cost of carbon avoided or captured), as well as effects on biodiversity, land use and water resources. Values for various CCS performance criteria are illustrated by reference to data for contemporary fossil-fuelled systems (Hammond et al., 2011). They represent the challenges against which BECCS will need to be evaluated. Carbon dioxide geological storage will, in many cases, have potential consequences for the marine environment. Such impacts vary as to whether they have consequences on a global, regional or local scale and to which stage of the CCS life cycle they relate.

7.2 CCS Process Characterisation, Innovation and Deployment

7.2.1 CCS Process Characterisation

The maturity of the CCS processes is characterised in Table 7.1 (Hammond et al., 2011). The current knowledge of CCS elements is ranked according to their technological maturity, following the characterisation adopted by the Intergovernmental Panel on Climate Change (IPCC) in their Special Report on CCS (IPCC, 2005). 'Research' indicates a process that is currently undergoing simulation trials, but is not yet reached significant developments or demonstrations. 'Demonstration' processes are those that have undergone post-research development *via*, for example, pilot-scheme projects. In the United Kingdom, the government originally aimed to encourage the development of four CCS demonstrators for operation by 2014–2015 (Gibbins and Chalmers, 2008; Chalmers et al., 2009; DECC, 2009; Gough et al., 2010), with commercial retrofits being introduced after 2025 on a pre-commercial basis. Similarly, the EU wanted to select 12 CCS demonstration projects for operation across Europe by 2015 (Gibbins and Chalmers, 2008; Gough et al., 2010). 'Commercialisation' reflects the introduction of

Table 7.1 Global state-of-the-art of CCS technologies.

CCS element	CCS process	Research	Demonstration	Commercial	Mature
Capture	Post-combustion			■	
	Pre-combustion			■	
	Oxy-fuel combustion		■		
Transport	Pipeline				■
	Shipping			■	
Geological storage	Depleted oil/gas fields			■	
	EOR				■
	Saline formations			■	
	ECBM		■		
Oceanic storage	Dissolution type	■			
	Lake type	■			
Other storage	Mineral carbonation	■			
	Industrial usage				■

NB: Filled cells indicate the greatest maturity level for specific elements, although there are less mature technologies associated with the other elements. EOR, enhanced oil recovery; ECBM, enhanced coal-bed methane.

Source: Hammond et al. (2011); adapted from the IPCC Special Report on CCS (IPCC, 2005).

processes into a fully competitive market. Finally, 'mature' processes are those that have displayed operation in the market over a reasonable period. The UK Parliamentary Office of Science and Technology (POST) recognises that current commercial CCS operations are all at a much smaller scale than is required for, say, a 500 MW power station (POST, 2005). Even the largest plant in the United States (Trona, CA) is less than 10% of the capacity needed for such a large-scale power station (POST, 2005).

But all has not gone well with proposed CCS developments (Hammond, 2014). It was the social considerations, not principally the technical issues, which led to community rejection of some of these projects. Indeed, public protest can be a showstopper for CCS (see also Markusson et al., 2012). Thus, the highly publicised failure of Shell's Barendrecht project in the Netherlands was caused by community activism and protest (Hammond, 2014). The company had wanted to store CO_2 from its nearby Pernis Refinery beneath the town in two depleted gas fields with a combined capacity of over 10 million tonnes. However, the Municipality of Barendrecht voted against CO_2 storage, following persistent community objections over safety concerns, and then in 2010 the Dutch government felt obliged to cancel the project. Similarly, Vattenfall abandoned its Beeskow project in Brandenburg in December 2011, due to well-organised local opposition (together with uncertainty over Germany's proposed CCS law). It had intended to demonstrate large-scale CO_2 capture *via* oxy-fuel and post-combustion processes (coupled to 'new-build' power plants), transported *via* a steel pipeline, and then to permanently store it in the Birkholz-Beeskow sub-structure at a depth of approximately 1300 m, covered by two layers of 'cap rock'. Local protests, led here by

NGOs, focused on possible CO_2 leakages along the transport route. For these and related reasons, an understanding of the potential economic and social constraints on CCS is critical for its successful implementation (Hammond, 2014).

7.2.2 CCS Innovation and Deployment

There is a large body of literature concerning innovation and innovation theory (Allen et al., 2008). The UK Department for Transport (DfT, 2007) has presented a useful, but simplified, representation of the process of innovation that they attribute to the Carbon Trust; see Figure 7.1 (Allen et al., 2008). This incorporates various actors and institutions, along with the relationships between them. It implies a linear process (from basic R&D to the diffusion of a commercial technology), although it is important to emphasise that innovation is a dynamic, non-linear process, as acknowledged by the DfT (2007). Thus, the full picture is more complex, as feedback loops exist between the different stages, and there are important links between technological and institutional changes that must be considered (Hammond et al., 2011). A whole-system perspective on the innovation process is therefore appropriate (as opposed to considering each stage in isolation), and it is from such a perspective that policy guidance should be drawn. The market penetration of a (successful) new technology typically varies in the manner of the hypothetical S-shape, or 'logistic', curve (Foxon et al., 2005; Midttun and Gautesen, 2007) shown in Figure 7.2 (Allen et al., 2008). Take-up of the technology begins slowly, then as commercial viability is reached, production

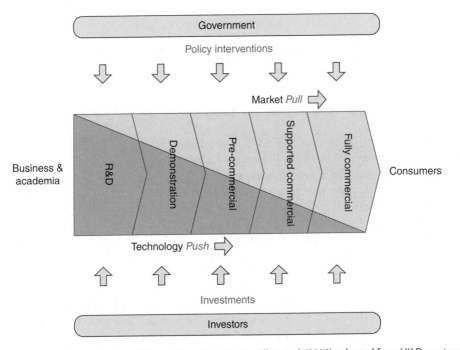

Figure 7.1 The innovation chain and its actors. *Source:* Allen et al. (2008); adapted from UK Department for Transport (DfT, 2007).

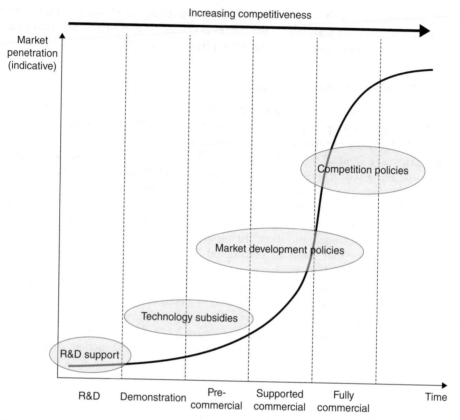

Figure 7.2 S-curve of technology development and policy categories. *Source:* Allen et al. (2008); adapted from Foxon et al. (2005) and Midttun and Gautesen (2007).

takes off and finally the technology rapidly diffuses before gradually slowing down as the market saturates.

A roadmap for the deployment of CCS in the United Kingdom has recently been devised by Gough et al. (2010) on the basis of a two-stage process involving a CCS landscape review and a high-level (i.e. expert) stakeholder workshop. They originally envisaged that the development phase would extend over the period to 2015, followed by commercialisation out to 2050. The cost of production of a technology tends to reduce as production volumes increase, a phenomenon reflected by the so-called technology learning curves or experience curves (IEA, 2000; Riahi et al., 2004; Mukora et al., 2009). The causes of cost reduction vary, but can include learning-by-doing improvements and economies of scale. It is therefore clear that higher costs of new technologies present a barrier to entry when competing with established technologies. This contributes to the lock-in of incumbent technologies and highlights the path dependence of development, both of which can discourage innovation (Allen et al., 2008; Hammond et al., 2011). In order to promote innovation and create a market for diverse technology options, these processes must be considered in the context of policymaking.

7.3 CCS Options for the United Kingdom

In order to estimate the current potential of CCS in the United Kingdom, a set of the most likely technological options was identified by Hammond et al. (2011), along with assumptions and appropriate data requirements. Pulverised coal-fired steam-cycle (PC) and natural gas combined-cycle (NGCC) plants are the main types of electricity generation systems in the United Kingdom (POST, 2005). However, integrated coal gasification combined-cycle (IGCC) plants were also examined in order to represent a potential advanced power technology for the United Kingdom. The most common forms of capture currently being developed were included in the assessment (POST, 2005). The PC and NGCC systems utilise globally used, post-combustion monoethanolamine (MEA) absorption techniques (Boyle et al., 2003; POST, 2005; Hammond and Ondo Akwe, 2007), whereas the IGCC plant was assumed to use the globally predominant physical, pre-combustion capture solvent known as Selexol.

The potential CCS options for the United Kingdom are summarised in Table 7.2. The United Kingdom is an island nation, and the majority of the opportunities for CO_2 storage lie offshore. Transport of CO_2 will vary depending on the distance between the power station, where capture takes place, and the offshore storage facility. To provide a comparative assessment of each CCS system in the UK context, the transport requirements were assessed by Hammond et al. (2011) from the largest current power station, the coal-fired Drax station in Yorkshire in the north-east of England. The transport of CO_2 from this location would require both onshore and offshore methods to potential storage sites. PC, NGCC and IGCC plants were therefore all assumed here to be geographically located at the Drax site for comparison purposes.

Hammond et al. (2011) observed that, in previous economic studies, enhanced oil recovery (EOR) and storage in depleted oil and gas wells were shown to be the most financially beneficial options and provide the highest degree of storage permanence. Estimates of safe geological storage beneath the Norwegian sector of the North Sea suggest about 600 years (Boyle et al., 2003), although the gas leakage rate over such very

Table 7.2 Potential UK CCS options for contemporary power plants. *Source:* Hammond et al. (2011).

Plant	Capture	Transport	Storage
PC	No capture: reference plant		
PC	Amine capture	Teesside/North Sea	EOR
PC		Humberside/North Sea	Depleted gas fields
NGCC	No capture: reference plant		
NGCC	Amine capture	Teesside/North Sea	EOR
NGCC		Humberside/North Sea	Depleted gas fields
IGCC	No capture: reference plant		
IGCC	Selexol capture	Teesside/North Sea	EOR
IGCC		Humberside/North Sea	Depleted gas fields

NB: PC, pulverised coal-fired steam cycle; NGCC, natural gas combined cycle; IGCC, integrated coal gasification combined cycle; EOR, enhanced oil recovery.

long timescales has to be monitored and verified (IPCC, 2005; Gibbins and Chalmers, 2008; Orr Jr, 2009; Gough et al., 2010). Hammond et al. (2011) examined two storage options that appear feasible from the Drax location and were therefore employed as a benchmark for UK CO_2 transport requirements (see Table 7.2): EOR storage in the North Sea, which could exploit existing pipelines from Teesside; and storage in depleted gas fields off the coast of East Anglia, which could exploit existing pipelines from Humberside.

7.4 The Sustainability Assessment Context

Sustainable development implies the balancing of economic and social development with environmental protection: the 'Three Pillars' model (Hammond and Jones, 2011; Hammond, 2016). The interconnections between these pillars can be illustrated by the sustainability Venn diagram shown in Figure 7.3, where the three types of constraints overlap. This is a simplified model, and the UK government later added two additional principles of sustainable development to the original three pillars (Defra, 2005; Hammond, 2016): (i) promoting good governance and (ii) using sound science responsibly, i.e. adopting 'evidence-based' approaches (Hammond and Jones, 2011). In the long term, Planet Earth will impose its own constraints on the use of its physical resources and on the absorption of contaminants, whilst the laws of the natural sciences, including, for example, those of thermodynamics (Hammond, 2004), and human creativity will limit the potential for new technological developments. The three pillars of sustainability imply that differing professional disciplines and insights are required in order to address each dimension (Hammond and Jones, 2011).

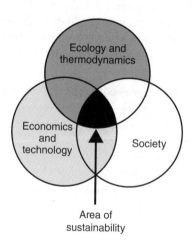

Figure 7.3 Sustainability assessment Venn diagram. *Source:* Hammond (2004); after Clift (1995) and Parkin (2000).

7.4.1.1 The Environmental Pillar
This can be tackled in *quantitative* terms *via* energy and environmental performance appraisal (see, for example, the interdisciplinary study by Hammond and Winnett

(2006)), typically on a life-cycle or 'full fuel-cycle' basis. These can be undertaken using the techniques of thermodynamic (energy and exergy) analysis and environmental *life-cycle assessment* (LCA) (Hammond et al., 2015). Typically, the uncertainty band in the resulting estimates of energy system performance parameters is of the order of perhaps ±20% (Hammond and Jones, 2011).

7.4.1.2 The Economic Pillar

This pillar can also be addressed in quantitative terms using methods such as cost–benefit analysis (CBA). However, Hammond and Winnett (2006) found that estimates of costs and benefits associated with energy technologies exhibited a wide variation. These were found to reflect variations of several orders of magnitude, i.e. factors of ten. They consequently argued that this demonstrated the frailty of the present generation of monetary valuation methods.

7.4.1.3 The Social Pillar

Here the approaches that can be applied are typically qualitative, although fully or semi-quantitative social science methods are available. They include (Hammond and Jones, 2011) analytic and deliberative processes (e.g. stakeholder engagement), mapping of socio-technical systems, customer surveys in response to new technologies (such as smart meters) and business models and the ethical reflection on energy system impacts and futures. Clift (2007) observes that this pillar should encompass inter- and intra-generational equity concerns.

Attempts have been made to bring the above perspectives together using a variety of sustainability assessment approaches, including a simple sustainability checklist; ecological or environmental footprinting (see, e.g. Hammond, 2016); multi-criteria decision analysis (MCDA) (Elghali et al., 2007); sustainability maps or 'tortilla' diagrams and a sustainability appraisal framework, as advocated by the UK sustainability NGO Forum for the Future, founded by the environmentalists Sara Parkin and Jonathan Porritt (Hammond and Jones, 2011). The participatory multi-criteria mapping and decision-conferencing approach developed by Elghali et al. (2007) for the sustainability assessment of bioenergy systems is perhaps the most comprehensive thus far devised. However, MCDA typically aggregates various distinct impacts arising from alternative technological options (Hammond and Jones, 2011). Thus, Allen et al. (2008) suggested that there is a number of reasons for discouraging such aggregate methods including both CBA and MCDA. Decision-makers are presented with a single, aggregate decision criterion, which actually hides many disparate environmental impacts (Hammond et al., 2015). Allen et al. (2008) therefore argued that it is vitally important that implications of these impacts are faced, particularly by politicians, rather than obscured by the methodology.

Developing a means of system characterisation for CCS systems, including emerging BECCS, will inevitably involve some or all of the three pillars of sustainability. Thus, it is common practice (IPCC, 2005; Hammond et al., 2011; Hammond and Spargo, 2014) to base the characteristics of such systems on economic and (energy and) environmental pillars, sometimes referred to as two-dimensional or two-pillar model of sustainability. In fact, Hammond (2016) argued that environmental footprinting is a tool for coupling considerations of economy and ecology and represents thereby a partial measure of the extent to which an activity is sustainable. The social implications of CCS in terms of, for example, employment and inequality are more difficult to embrace as part of a 'systems integration framework' for sustainability assessment (Liu et al., 2015; Hammond, 2016). The focus of this chapter is therefore on CCS system characteristics that emanate from the economic and environmental pillars of sustainability.

7.5 CCS Performance Metrics

7.5.1 Energy Analysis and Metrics

In order to determine the primary energy inputs needed to produce a given amount of product or service, it is necessary to trace the flow of energy through the relevant industrial system (Slesser, 1978; White and Kulcinski, 2000; Hammond and Winnett, 2006; Allen et al., 2008; Hammond and Jones, 2008). This idea is based on the First Law of Thermodynamics, that is, the principle of conservation of energy, or the notion of an energy balance applied to the system. Thus, the First-Law or energy analysis (EA) (Hammond and Winnett, 2006; Allen et al., 2008; Hammond and Jones, 2008; Hammond et al., 2015) can be employed to estimate the energy requirements of building new electricity generation plant. Analysis is performed over the entire life cycle of the power cycle: 'from birth to death' (White and Kulcinski, 2000) or 'from cradle to grave' (Hammond and Winnett, 2006; Allen et al., 2008; Hammond and Jones, 2008). Energy analysis (EA) implies the identification of feedback loops (Slesser, 1978; Hammond and Jones, 2008), such as the indirect or embodied, energy requirements for materials and capital inputs. Several differing methods of energy analysis have been developed, the most significant being statistical analysis, input–output table (I-O) analysis, process analysis (or energy 'flow charting') and hybrid analysis (Hammond and Winnett, 2006; Allen et al., 2008; Hammond and Jones, 2008; Hammond et al., 2015). It yields, among others, the metrics described in the following text.

The EPP is the time taken for a power system to repay the energy that has been invested in its construction. The values calculated are typically obtained on the basis of a static EA approach (Slesser, 1978). The number of years in which the electricity generated by the power plant equals the primary energy invested in the system is therefore known as the EPP. It is defined (Allen et al., 2008) by.

$$EPP = E_{construction}/E_{output/year} \qquad (7.1)$$

In contrast, the EGR is the ratio of the net energy output over the life of a power system divided by the total energy consumed (Slesser, 1978) during construction. The net energy produced is the total net electricity generation – converted from watt-hours to joules for consistency (White and Kulcinski, 2000; Allen et al., 2008; Cheng and Hammond, 2017):

$$EGR = E_{output}/E_{input} \qquad (7.2)$$

The energy input to the system is that used for materials production, construction, operation and decommissioning (White and Kulcinski, 2000). The boundaries of the system studied coincided with inputs of raw materials, such as coal, gas or biomass fuel resources, as well as steel and concrete used in the construction of power stations. Energy is required for processing and manufacturing of raw materials used during construction (predominantly through the burning fossil fuels in the transport of materials and the operation of construction machinery). Obviously, the major energy requirement is associated with power plant operation. An allowance has also to be made for the energy needed to decommission and demolish the power station at the end of its life.

Another indicator of CCS performance that is often employed to highlight the importance of the energy requirements of capture is the energy penalty in percentage terms (% variation from MW_{ref}). This is the most commonly used metric that can be determined using the following expression adapted from Rubin et al. (2007):

$$\text{Energy penalty} = (1 - (\text{Efficiency}_{cap}/\text{Efficiency}_{ref})). \tag{7.3}$$

It takes into account the net plant efficiencies with and without capture. Plants with capture have lower efficiencies, due to increased energy requirements inherent in the capture process. Efficiency values for power stations can be used in Eq. (7.3) to determine the associated energy penalties with capture.

7.5.2 Carbon Accounting and Related Parameters

Carbon emissions are the currency of debate in a climate-constrained world. In order to estimate the embodied energy and embodied carbon of a product or service, the technique of environmental LCA is used (Hammond et al., 2015). This involves identifying all of the processes that are needed to create the product or service, whilst assigning energy requirements and carbon emissions to upstream processes. All of these individual processes are summated, and the embodied energy and carbon within the prescribed system boundary can be calculated. A full LCA requires a detailed investigation that is often time consuming. Therefore, wherever possible, the embodied energy and carbon data are taken from existing studies, such as the inventory of carbon and energy database (ICE) developed at the University of Bath (Hammond and Jones, 2008; Hammond and Jones, 2011).

Carbon emission factors (CO_2 emissions per kWh) reflect the amount of carbon that is released as a result of using one unit of energy and have units of typically of magnitude $kg\,CO_2/kWh$.

CCS plants require larger energy input and higher energy to operate due to processes such as air separation and CO_2 compression. But without capture, this would result in a higher level of CO_2 emitted to the atmosphere. The more advanced novel technologies, such as pressurised oxy-fuel combustion, have the potential to capture over 98% of the operational CO_2 emissions emitted from a power station chimney or 'stack' that would otherwise be exhausted to the atmosphere. However, these plants and their CO_2 emissions must be assessed based on their entire life cycle, not just the CO_2 emitted during operation (Hammond and Spargo, 2014). Global warming results from the release of a variety of greenhouse gas (GHG) emissions into the atmosphere. Each of the GHGs has a different level of potency and can be normalised relative to the impacts of one unit of CO_2. Thus, a unit of methane is regarded as being 25 times more harmful than a single unit of carbon dioxide (on a 100-year timescale). It is therefore considered to have a global warming potential of 25 (Hammond and Jones, 2011). The units of the basket of six GHGs incorporated in the Kyoto Protocol (1997), an extension to the United Nations Framework Convention on Climate Change (UNFCCC) the second instalment of which ends in 2020, are measured in terms of carbon dioxide equivalent (CO_{2e}), i.e. typically, $kg\,CO_{2e}$.

7.5.3   Economic Appraisal and Indicators

Economic appraisal evaluates the costs and benefits of any project, programme or technology in terms of outlays and receipts accrued by a private entity (household, firm, etc.) as measured through market prices (Brent, 1996). Financial appraisal is used by the private sector and omits so-called environmental externalities. In contrast, economic CBA is applied to take a society-wide perspective, with a whole-system view of the costs and benefits (Hammond and Winnett, 2006; Allen et al., 2008). It accounts for private and social, direct and indirect, tangible and intangible elements, regardless of to whom they accrue and whether they are accounted for in purely financial terms (Brent, 1996). A further distinction between financial appraisal and CBA is in the use of the discount rate to value benefits and costs occurring in the future (Hammond and Winnett, 2006; Allen et al., 2008; Hammond et al., 2012). Financial appraisal uses the market rate of interest (net of inflation) as a lower bound and therefore indicates the real return that would be earned on a private-sector investment.

This approach takes account of the time value of money, and discounting, in order to obtain the appropriate investment appraisal criteria (Hammond and Winnett, 2006; Kohyama, 2006; Allen et al., 2008). The LCOE is typically employed to compare the economic performance of different power generators. This is the price at which electricity must be sold in order to recover all costs incurred during generation. The net present value (NPV) of the sum of the capital cost, maintenance and operational costs and, potentially, decommissioning is calculated over the life of the project, along with the NPV of the total electricity generated. This yields the LCOE in *pence per kilowatt hour* (p/kWh_e) for the systems evaluated, which can then be compared to that for alternatives. This method can be used to effectively compare different energy options with a various life spans, capital costs and efficiencies so that the most cost-effective option can be determined. The discounted cash flow (DCF) over the life of each project – here assumed to be 120 years – is calculated as shown in Eq. 7.4:

$$\text{Discounted cash flow} = \sum_{t=1}^{t=120} \frac{R_t}{\left(1+\text{TDR}\right)^t} \tag{7.4}$$

where R_t is the net receipts (income less cost), t is the time in years for the total foreseen life of the project and r is the discount rate. In the case of public-sector investments, a test discount rate (TDR) is utilised. It is typically derived from a comparison with private-sector discount rates or weighted average cost of capital (WACC). In the United Kingdom, HM Treasury (HMT, 2003) recommends that the TDR for projects with durations of less than 30 years should be taken as 3.5%, then falling in line with the profile indicated in Table 7.3.

Table 7.3 The declining long-term UK Test Discount
Rate. *Source:* HMT (2003).

Period of years	0–30	31–75	76–125
Discount rate	3.5%	3.0%	2.5%

7.6 CCS System Characterisation

7.6.1 CO$_2$ Capture

7.6.1.1 Technical Exemplars

Three generic systems may be used to capture CO$_2$ from these three types of power stations: PC and NGCC plants are currently operational in the United Kingdom and globally; IGCC plants are being introduced into the global market.

Post-combustion capture separates CO$_2$ from the exhaust (flue) gas after combustion. This system typically exploits chemical solvents such as amines (IPCC, 2005; Orr Jr, 2009; Hammond et al., 2011; Hammond and Spargo, 2014) such as MEA, to absorb CO$_2$. This is the most common method of capture, and, therefore, has the most operational experience. However, the low concentration of CO$_2$ in the flue gas inhibits the capture process. It therefore requires powerful chemical solvents and large-scale processing equipment to handle the emissions. This is both costly and energy intensive. Nevertheless, it offers significant potential for the retrofitting of capture systems to current PC systems and, for that reason, has been favoured by the UK government.

Pre-combustion capture (IPCC, 2005; Orr Jr, 2009; Hammond et al., 2011; Hammond and Spargo, 2014) separates CO$_2$ from the gas stream before combustion, where the concentration of CO$_2$ in the gas stream is high. This aids the capture process and enables less selective capture techniques, such as physical absorption using Selexol. The quantity of gas involved is lower, reducing the need for large equipment, and this can reduce the energy requirements. But the process involves more drastic changes to the power station.

Oxy-fuel combustion capture (IPCC, 2005; Orr Jr, 2009; Hammond et al., 2011; Hammond and Spargo, 2014) involves combustion of fuel in oxygen instead of air. This produces a gas rich in CO$_2$ that aids the capture process significantly. The process is nonetheless expensive and is presently only at the demonstration phase. Research is currently examining more effective chemical and physical absorbents, as well as the development of novel capture techniques. The latter include new adsorbents, membranes and cryogenics that may lower the costs and energy penalties associated with carbon capture (IPCC, 2005; Orr Jr, 2009; Hammond et al., 2011; Hammond and Spargo, 2014). Technical and cost data associated with these routes have been described in the IPCC Special Report on CCS (IPCC, 2005); although the IPCC SRCCS (2005) provided technical and cost data for the types of power generators listed in Table 7.1, Hammond et al. (2011) employed data from Parsons et al. (2002).

7.6.1.2 Energy Metrics

The effect of capture on plant efficiency is illustrated in Figure 7.4 on a lower heating value (LHV), or net calorific value, basis. NGCC system is the most efficient with and

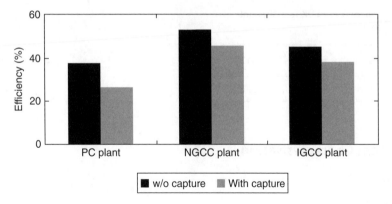

Figure 7.4 Illustrative contemporary power plant efficiencies with and without (w/o) carbon capture on an LHV basis. *Source:* Hammond et al. (2011). Reproduced with permission of Elsevier.

without capture, followed by IGCC technologies and then PC plants. The introduction of capture into the power plants clearly reduces the operating efficiency. The efficiency of the NGCC system is reduced the least, by 14% (Hammond and Ondo Akwe, 2007), while IGCC plant efficiency is reduced by 16%. The greatest fall in efficiency, by 30%, occurs with PC plants. This is a significant reduction in the operating efficiency.

The EPP and EGR were calculated by Hammond and Spargo (2014), based on the sum of energy investments for the PC power plant with a nominal 1 GW_e output over 40 years. The EGR over the lifetime of the reference or non-CCS power plant repays its energy investment in construction, operation and decommission by nearly 11 times, whereas for a contemporary CCS plant this is 9.9. Similarly, the EPP for the non-CCS power plant was estimated to be 3 years and 8 months, which indicates that (following its entry into service) the reference plant will have produced more energy than required for its construction in less than 4 years. The corresponding figure for a nominal 1 GW PC plant with coupled CCS facilities is 4 years, and the 4-month difference between the results is the energy that is required to produce the additional CCS hardware, along with associated transportation (e.g., pipeline) and storage. It should be noted that assumptions that have been made regarding material quantities and additional energy used over the life cycle result in some uncertainty over the EGR and EPP values. Based on current CCS values, this is believed to be in the range ±10% (Hammond and Ondo Akwe, 2007). Even a ±20% range of input values yields only a small change in EGR and EPP, the variation being significantly less than ±20%.

The highest energy penalty associated with PC plants is illustrated in Figure 7.5 (Hammond et al., 2011); it is roughly double the value for NGCC or IGCC systems. The lowest energy penalty is associated with NGCC plants. It should be noted that the energy penalties in this assessment are lower than those in some previous studies. This suggests that the plants studied are particularly suited to the capture systems implemented.

7.6.1.3 Carbon Emissions
NGCC systems produce the lowest operational or 'stack' CO_2 emissions with and without capture (see Figure 7.6), followed by PC plants, and then IGCC systems, even

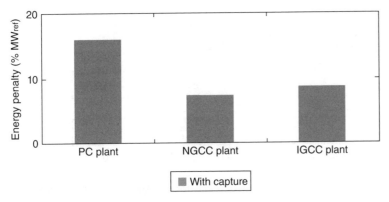

Figure 7.5 Illustrative energy penalties of contemporary power plant associated with carbon capture. *Source:* Hammond et al. (2011). Reproduced with permission of Elsevier.

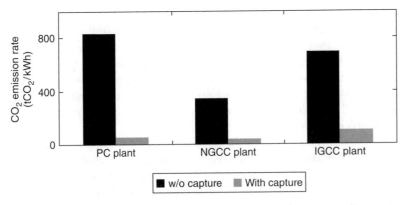

Figure 7.6 Illustrative contemporary power plant CO_2 emissions with and without (w/o) carbon capture. *Source:* Hammond et al. (2011). Reproduced with permission of Elsevier.

though emissions from PC plants are higher than those of IGCC technologies without capture (Hammond et al., 2011). This could be because the physical absorption process used in IGCC plants, being less effective than chemical absorption, compromises the reduction of emissions. In any case, all the emissions are reduced significantly: in PC plants by 93%, NGCC systems by 88% and IGCC technologies by 85%.

Upstream environmental burdens arise from the need to expend energy resources in order to extract and deliver fuel to a power station or other users. They include the energy requirements for extraction, processing/refining, transport and fabrication, as well as methane leakages from coal-mining activities – a major contribution – and natural gas pipelines. These were not taken into account in the above calculations, which allowed only for operational and embodied emissions. The upstream CO_{2e} emissions associated with various power generators and UK electricity transition pathways towards a low-carbon future have recently been evaluated on a whole-system basis (Hammond et al., 2013; Hammond and O'Grady, 2014). The associated stages are

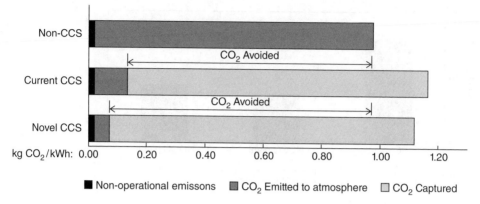

Figure 7.7 CO_2 emissions avoided by adopting CCS technologies compared to conventional (non-CCS) power plants. *Source:* Hammond and Spargo (2014); adapted from the IPCC Special Report on CCS (2005).

shown by the bars in the chart presented as Figure 7.7, with non-operational emissions depicted by the black bars. Here, emissions from a plant with capture are less than those of one without (w/o) capture, although not by as much as the amount of CO_2 captured. This is because additional energy is employed to capture and compress the CO_2, which generates extra CO_2 (assuming both plants are sized to deliver the same amount of electricity to the grid). The amount of CO_2 emissions avoided is therefore an important indicator of the system's effectiveness.

The operational (direct or 'stack') emissions associated with the combustion of fuels are compared with GHG emissions from upstream coal-mining and refining activities in Table 7.4. These data indicate the magnitude of the difference between direct combustion and upstream emissions. Such fugitive GHG emissions indicate (see Table 7.5) that coal CCS are about 2/3 lower in terms of GHG emissions in comparison with conventional coal-fired plant (without CCS), i.e. a reduction from 1.09 to 0.31 kg CO_{2e}/kWh. Thus, CO_2 capture is likely to deliver only a 70% reduction in carbon emissions on a whole-system basis (including both upstream and operational emissions), in contrast to the normal presumption of a 90% saving (Hammond et al., 2013). Consequently, there is a broader range of factors to consider when selecting new UK power generation capacity.

Table 7.4 Upstream GHG emissions from UK coal-fired generation.

Fuel	Defra[a] GHG emissions factor from combustion of fuel (kg CO_2/kWh)	GHG emissions from upstream activities (kg CO_2/kWh)	Resulting ratio (increase)
Coal	0.330	0.060	6.5:1 (+18%)

a) The UK Department for Environment, Food and Rural Affairs (Defra), UK National Atmospheric Emissions Inventory (NAEI) maintained by Ricardo-AEA; see http://naei.defra.gov.uk/.
Source: Hammond and O'Grady (2014); adapted from Hammond et al. (2013).

Table 7.5 UK power technologies in ranked order by whole-systems GHG (upstream plus operational or 'stack') emissions.

Technology (mix)	GHG emissions (kg CO_2/kWh)
Coal	1.09
Grid average, 1990	0.90
Grid average, 2008	0.62
Natural gas	0.47
Coal CCS	0.31
Natural gas CCS	0.08
Nuclear	0.02

Source: Hammond and O'Grady (2014); adapted from Hammond et al. (2013).

7.6.1.4 Economic Indicators

The most costly part of the CCS process is capture of CO_2. It typically represents around 75% of the overall costs of building and running a CCS system (Feron and Hendriks, 2005). In order for CCS technology to be adopted and implemented, it must clearly be economical in the liberalised energy market. UK electricity generators and their investors are looking to construct a new generation of power plants in the medium term that will be environmentally friendly. CCS power plants are high on the list of low-carbon options, but the utilities will not invest unless they can be shown to be economical. Indeed, a UK CCS stakeholder workshop reported by Gough et al. (2010) – and held back in May 2007 – identified a potential to reduce CCS costs of 50–75% by 2040. But several of the industry representatives were also concerned that the UK government was failing to provide sufficient enabling technology 'push' across the entire CCS chain. They argued that greater financial incentives for carbon abatement were required in the form of a higher carbon price from the EU Emissions Trading Scheme (ETS): a 'cap-and-trade' system. In the UK Budget of 2014, it was announced that the UK-only element of the Carbon Price Support (CPS) rate per tonne of CO_2 (tCO_2) would be capped at a maximum of £18 from 2016/2017 until 2019/2020. This will effectively freeze the CPS rates for each of the individual taxable commodities across this period at around 2015 to 2016 levels. It will ensure that the carbon price floor is kept at a rate that the UK Treasury feels will maintain British industrial competitiveness.

The COE can be used as an indicator of the impact on plant economics of adding capture equipment. It incorporates the costs of both the three power generation systems and associated CO_2 capture plants. Hammond et al. (2011) used the Integrated Environmental Control Model (version IECM-cs,2005) developed by Carnegie Mellon University for the US Department of Energy's National Energy Technology Laboratory (DOE/NETL), to provide economic data on power plants and CCS equipment. These sources have previously been widely used in US studies, as well as in the earlier appraisal of the thermodynamic and 'exergoeconomic' performance of NGCC plants with and without carbon capture by Hammond and Ondo Akwe (2007). They also formed the basis for the technical and cost data presented in the IPCC Special Report on CCS (2005). The COE (p/kWh) can consequently be determined using equation 7.5 adapted from Abanades et al. (2007):

$$COE = \text{fixed costs} + \text{fuel costs} + \text{other variable costs}$$
$$= \left\{ \left[(TCR)(FCF) + FOC \right] / \left[(CF)(8760)(NPP) \right] \right\} + (HR)(FC) + VOC \tag{7.5}$$

This expression includes the following factors:

TCR, total capital requirement (UK sterling pence equivalent (p)); FCF, fixed charge factor (fraction); FOC, fixed operating costs (p); CF, capacity factor (fraction); 8760, total hours in a typical year; NPP, net plant power (kW); HR, net heat plant rate (kJ/kWh); FC, unit fuel cost (p/kJ); and VOC, variable operating costs (p/kWh). Thus, the levelised COE (LCOE) for all the potential UK power plants with and without capture can be obtained *via* Eq. (7.5) on a pence-per-kilowatt-hour basis (see Figure 7.8). These data (Hammond et al., 2011) indicate that NGCC systems provide the cheapest electricity, both with and without capture. IGCC technologies exhibit an LCOE that is more expensive compared to PC systems without capture but, when capture is implemented, it becomes relatively less expensive. All of the plants show that when capture is added to the system the LCOE rises. PC plant LCOE increases by 84%, NGCC by 98% and IGCC by 36%. Percentage changes for PC and IGCC systems are in line with those found in earlier studies (for example, by Rubin et al. (2007)), and this should be expected because the PC plant studied uses chemical absorption. In contrast, the IGCC system uses less energy and economically intensive physical absorption *via* an MEA solvent. However, the NGCC plant results in a very high percentage change, most likely because in the study by Hammond et al. (2011) it was also assumed to utilise a chemical (MEA) absorption process, whereas it is possible to use cheaper physical processes (Boyle et al., 2003).

It has been recognised that significant differences and inconsistencies have been exhibited in various methods and metrics employed within international studies of CCS costs (Rubin, 2010; EPRI, 2013). These have included key technical, economic and financial assumptions, such as differences in the plant size, fuel type, capacity factor and cost of capital. It is also true that the underlying methods and cost components that are

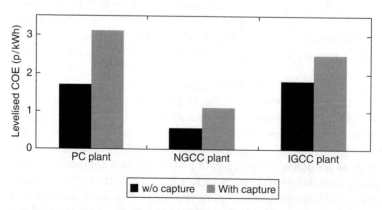

Figure 7.8 Illustrative power plant LCOE with and without (w/o) carbon capture {£(2005)}. *Source:* Hammond et al. (2011). Reproduced with permission of Elsevier.

included (or excluded) in a given study can have a major impact on the results that are reported publicly or in the technical literature (Rubin, 2010). For example, measures that have very different meanings, such as the costs of CO_2 avoided, CO_2 captured and CO_2 abated (Hammond et al., 2011), are often reported using the same units, such as \$/tCO$_2$ (€/tCO$_2$ or £/tCO$_2$). Similarly, there can be major differences between cost estimates associated with FOAK plants and more mature technologies. As a consequence, there is likely to be some degree of confusion, misunderstanding and possible misrepresentation of CCS costs. Rubin (2010) identified a hierarchy of ways to estimate CCS costs. These included expert elicitations, the use of published data, modified published values, new model results and the findings of detailed engineering analysis, such as front-end engineering design (FEED) studies.

7.6.2 CO_2 Transport and Clustering

The International Energy Agency (IEA) in their CCS Technology Roadmap (IEA, 2009) suggest that pipelines will be the main method for CO_2 transportation, with ships and trains being used in the short term in some demonstration projects worldwide (see also Hammond et al., 2011; Hammond and Spargo, 2014). Shipping becomes more economical than piping for the transport of CO_2 over long distances (>1000 km). Liquefied CO_2, which has similar properties to LPG (IPCC, 2005), can be shipped overseas at a pressure of around 0.7 MPa on a commercially attractive basis. The transportation of natural gas and other liquids and gases is well established in the United Kingdom, where natural gas and oil have been piped from North Sea reservoirs since the early 1970s. Consequently, it may be possible to use the existing pipeline infrastructure in the United Kingdom operated by National Grid (Gough et al., 2010) to reduce the investment required to set up a new CO_2 network. However, existing oil and natural gas pipelines out into the North Sea are reaching the end of their engineering life and were designed for rather different operating conditions. CO_2 pipelines would therefore need to be designed to withstand high pressure and the resultant leak risks. The CO_2 behaves differently in various phases, and these can influence the development of corrosion in pipes. Thus, the pipe material would have to be carefully chosen and engineered to minimise the risk of pipeline failure, especially if the pipe is located on the ocean floor. In addition, it will be necessary to devise new metering devices to monitor the quality of the dense-phase CO_2 (Gough et al., 2010). It may also be necessary to insert recompression stages into the pipeline.

In the medium term, it is believed that the CO_2 pipeline network would work most effectively using a number of onshore hubs that would compress and clean the CO_2 transported in smaller pipes from several power stations and industrial capture plants. At these hubs, the more highly compressed and cleaned CO_2 would be transported through one or two larger, stronger pipes to its offshore storage reservoir (APGTF, 2009). This would not only reduce costs in installation and the length of pipeline required but would also allow for an interconnected system that could be shared. Such a network would have the potential to evolve into a network of pipes with redundancy and security should a leak or failure occur (APGTF, 2009; IEA, 2009). A hub–network system also allows better managing of the CO_2 transport with third parties leasing pipe use to the power generators to spread the costs and reduce the maintenance and operation strain on single electricity generator and industry users. Locations for potential CCS cluster regions in the United Kingdom, as well as their proximity to existing power

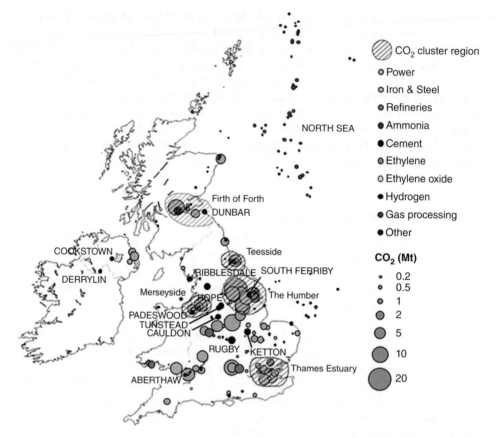

Figure 7.9 Illustrative distribution of CO_2 point sources and CCS cluster regions in the United Kingdom. *Source:* Griffin et al. (2016).

stations and industrial sites, are indicated in Figure 7.9 (Griffin et al., 2016). If the cluster regions identified for storage under the North Sea and the Irish Sea (Element Energy, 2010; Griffin et al., 2016) are considered, existing candidate sites for CCS include Padeswood, South Ferriby and Dunbar. However, only South Ferriby has a large enough raw material reserve to warrant such a long-term investment (Element Energy, 2010). Clearly, the location of future CCS-enabled plants will have to be considered carefully, balancing the cost of transporting CO_2 with the cost of transporting raw materials. Either way, this presents substantial economic barriers (DECC, 2012a) along the UK roadmap for CCS deployment.

The CO_2 transportation needs of the United Kingdom will benefit from the fact that its storage reservoirs in the North Sea are typically located only 200 or 300 km away from the power stations. Stakeholders feel that there are no long-term technical barriers to the development of a CO_2 pipeline network in the United Kingdom (Gough et al., 2010). But a CO_2 pipeline operator runs a significant financial risk (Gough et al., 2010), because of the high cost and low returns associated with the assets. Indeed, Gough et al. (2010) suggested that the increase in the cost between a network and alternative transmission means could be as high as £3 per tonne ($4.5 or €4.0/tCO$_2$).

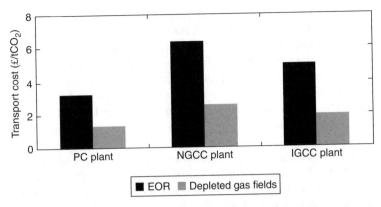

Figure 7.10 Illustrative estimates of CO_2 transport costs {£(2005)}. *Source:* Hammond et al. (2011). Reproduced with permission of Elsevier.

The cost of transporting the CO_2 captured to the storage location is a combination of the distances involved and, more importantly, the quantities transmitted. The distance to the EOR storage site is considerably longer than that for the depleted gas fields; this would be a similar situation for the majority of power stations in the United Kingdom. The transport costs can be separated for two different storage sites, but the costs for onshore and offshore transport for various quantities depending on the power plant have been averaged. Figure 7.10 shows that the transport cost associated with EOR is higher than with the depleted oil fields (Hammond et al., 2011); this is expected due to the notional location of the power stations. They would obviously vary depending on the actual location of individual power stations, if CCS were implemented. It can be deduced from Figure 7.10 (Hammond et al., 2011) that the two storage options for PC plants have the lowest transport costs, followed by IGCC systems and then NGCC technologies. This is because the quantities involved in PC systems are the highest, due to the large amount of CO_2 gas captured. The same argument can be applied in the context of IGCC compared to PC plants. Piping CO_2 on a larger scale reduces the costs.

7.6.3 CO$_2$ Storage

7.6.3.1 Storage Options and Capacities
Naturally occurring geological formations provide potential locations for the storage of captured CO_2: oil or gas recovery, unmineable coal beds, saline aquifers and depleted oil or gas fields (IPCC, 2005; APGTF, 2009; Hammond et al., 2011; Hammond and Spargo, 2014). These are favoured because of the maturity of the technology involved. A global assessment of carbon emissions from contemporary power and industrial sites against the availability of carbon stores was recently performed by Scott et al. (2015). They make some useful observations about the potential temporary (<1 000 years) and permanent (>100 000 years) CO_2 storage sites. CO_2 has been sequestered in geological formations, for example, for over 35 years in both Norway and the United States (APGTF, 2009). Such permeable layers are typically found at least 800 m below the ocean floor. Other potential storage options include ocean storage and CO_2 mineralisation. EOR and enhanced gas recovery (EGR), whereby CO_2 is employed to extract oil

and gas from geological formations, have been widely used in Canada and the United States since the early 1970s (Anderson and Newell, 2004; IPCC, 2005; APGTF, 2009; Hammond et al., 2011; Hammond and Spargo, 2014). The injection of CO_2 into an oil reservoir mixes the gas with the crude oil and thins the resulting mixture. It is then easier to extract from the reservoir. These techniques are presently only employed in inshore applications, and, therefore, they appear to have very limited applications on the UK continental shelf. The Advanced Power Generation Technology Forum (APGTF) in the United Kingdom has argued that EOR and EGR are not currently economical for offshore applications (APGTF, 2009).

A saline aquifer is an underground geological formation in which a large quantity of salt water has become trapped during the formation of the rock layers that surround it. The CO_2 can be pumped down into the deep saline aquifers, where the CO_2 will be stored in the natural gas pockets, where it will dissolve in the water to some extent. Saline aquifers are the most promising long-term CO_2 storage globally according to the SRCC (IPCC, 2005). There has been one major storage project undertaken in a saline formation in the Norwegian sector of the North Sea: the Sleipner field (Boyle et al., 2003; IPCC, 2005; Hammond et al., 2011; Hammond and Spargo, 2014). Monitoring suggests that no CO_2 has currently escaped. However, the monitoring of saline formations is much less developed than that of oil and gas wells (Hammond et al., 2011; Hammond and Spargo, 2014). The confidence in the permanence of storage is consequently lower, especially because the majority of the potential storage is in open saline formations that provide an eventual escape path for CO_2. More development is required in these cases to simulate options and determine whether the CO_2 will be held over hundreds to thousands of years in order to mitigate climate change (Hammond et al., 2011; Hammond and Spargo, 2014). The final option for CO_2 storage, and the one that is most attractive for the United Kingdom, is to store it in geological formations that naturally occur under the seabed of the North Sea. The CO_2 storage capacity in North Sea depleted oil and gas reservoirs is estimated to be around $10\,190\,MtCO_2$ (Hammond and Spargo, 2014). This is equivalent to roughly 59 years of storage, based on 2008 CO_2 emission data. This is supplemented by a further 14 466 Mt. CO_2 (Hammond and Spargo, 2014) of storage capacity available in UK saline aquifers. The Scottish Centre for Carbon Storage (SCCS, 2009) estimated that a total of up to $46\,000\,MtCO_2$ of storage capacity could be available in ten saline aquifers in and around Scotland. This would represent 266 years of UK storage requirements, based on UK CO_2 emissions from power generation in around 2010 (Hammond and Spargo, 2014).

7.6.3.2 Storage Site Risks, Environmental Impacts and Monitoring

Several risk-based studies have identified the critical role of storage within the CCS life cycle or chain (Hammond, 2012). Kimmance and Rogers (2012) employed a risk-management framework based around the ISO 31000:2009 standard. They argued that it has the merit of being able to manage multiple risks associated with many stakeholders and can provide the basis for a Monte Carlo-type simulation of costs and revenues. The authors suggest that the most critical element of CCS chain is the storage component. Failure at initial injection or during longer-term containment would make the project commercially non-viable. A hypothetical CCS demonstration project schedule was employed by Carpenter and Braute (2012) in order to test the risk-management implications of front-loading project costs as a means of meeting an imposed 2015 deadline for the start of operations. Increased commercial or financial risk exposure is also caused by the inability to find adequate CO_2 reservoirs. These were evaluated in the CO2QUALSTORE joint industry guidelines. Carpenter and Braute (2012) acknowledged that finding a balance between deadline risk and site qualification risk for a real project would 'require careful modelling of project activities and costs at a greater level of detail' than in their simplified example.

The IEA-sponsored Greenhouse Gas R&D Programme commissioned an in-depth review by the Alberta Research Council (ARC) in Canada of CCS storage site selection and characterisation methods (IEA GHG, 2009). It drew, in part, on previous work undertaken as part of the SRCCS (IPCC, 2005). The ARC report (IEA GHG, 2009) suggests that the selection should be based on three fundamental requirements. The first of these related to the CO_2 storage capacity of the site over its operational lifetime. Second, its ability to accept emitted CO_2 from the capture plant or transport hub, known as 'injectivity'. Finally, the containment of the CO_2 stored so that it does not migrate and/or leak. It further proposes (IEA GHG, 2009) that site screening should take place in two stages: (i) the elimination of sites that are unsuitable for geological storage and (ii) the determination as to whether the remaining sites pass a set of eligibility criteria. The recommended storage site suitability criteria were grouped into several categories for use in the screening process (IEA GHG, 2009):

- Capacity and injectivity
- Safety and security
- Economics
- Legal and regulatory issues
- Public acceptance.

The characterisation of CO_2 storage sites typically involves (IEA GHG, 2009) geology and rock properties, the history of the hydrocarbon reservoirs and wells (as well as their conditions), the composition and phase behaviour of the indigenous fluids and the injected CO_2 stream and any pre-existing faults or fractures in the formation. A number of qualifiers and threshold values were developed during the ARC study (IEA GHG, 2009) for the safety and security of storage sites in saline aquifers and miscible CO_2-EOR operations. These reflected the expert opinion of the reviewers, which were, in turn, the subject of scrutiny by six external reviewers. Many of the latter cautioned that the guidance presented in the ARC report (IEA GHG, 2009) should always be employed by expert practitioners, and they were wary of using seismic surveying for monitoring purposes.

The SRCCS report (IPCC, 2005) noted that CO_2 can harm marine organisms, including reduced rates of reproduction and growth, and longer-term mortality. There can be immediate mortality near injection points, although some organisms will respond to small additions of CO_2 with consequent impacts on biodiversity and ecosystem services. There is clearly a need to monitor over the long term the integrity of geological reservoirs, notwithstanding the views of the reviewers of the ARC report (IEA GHG, 2009). Early attempts have been made to test leakage risks and monitoring techniques (Hammond, 2012, 2013). Synthetically modelled data, for example, were adopted by Verdon et al. (2012) to examine one specific leakage risk – injection-induced pressure increases that may lead to fractures in the caprock and therefore the leakage of buoyant

CO_2. Passive seismic monitoring (PSM), using 'geophones' placed in boreholes around a reservoir or in larger arrays at ground level, has been shown to yield a relatively inexpensive means of permanently surveying this type of phenomenon. The technical basis of PSM is described, along with its previous usage in the hydrocarbon sector, before its potential use for CCS site monitoring is outlined. Verdon et al. (2012) discussed several circumstances where PSM has been adopted to evaluate subsurface CO_2 injection. These include the CCS site at Weyburn in Saskatchewan, Canada, where CO_2 has been injected since 2000 for the purpose of EOR and storage. They also noted that the approach has again been utilised at the In Salah site. Hannis (2012) provided specific examples of fit-for-purpose monitoring techniques that 'can be used to validate pre-injection predictive methods, image plume development and detect surface anomalies'. A range of monitoring techniques previously designed to address leakage risks associated with the In Salah CO_2 storage site in the Algerian Sahara desert are examined. Hannis drew on experience from various geological storage sites at the Sleipner field in the Norwegian North Sea; Laacher, Germany; Latera, Italy; Frio, Texas, USA; Nagaoka, Japan and Otway, Victoria, Australia. This suggests that monitoring data can provide an early warning system of surface CO_2 leakage, improving the understanding of leakage pathways and the nature of leaks in order that appropriate mitigation measures may be put in place.

7.6.3.3   Storage Economics

The cost of storage within the UK continental shelf will be very similar to that of EOR and gas field projects covered in earlier studies (Rubin et al., 2007; Hammond et al., 2011). Therefore, the storage costs can be taken directly from such studies as they provide experience of actual costs incurred. Storage costs for both gas field and EOR, the opportunities deemed currently viable in the UK North Sea, are depicted in Figure 7.11 (Hammond et al., 2011). EOR can provide financial return from storage, due to increased extraction of valuable oil. This revenue is dependent on the price of oil and therefore can deviate greatly, as has recently been observed post 2014.

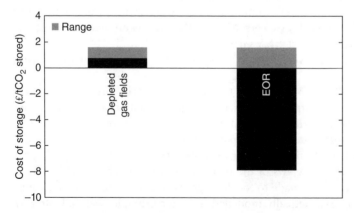

Figure 7.11 Illustrative costs of CO_2 gas storage {£(2005)}. *Source:* Hammond et al. (2011). Reproduced with permission of Elsevier.

7.6.4   Whole CCS Chain Assessment

Any power generation system that has higher operating and construction costs compared to non-CCS fossil-fuelled power stations will lead to an increase in electricity costs for the end-use consumer (unless the financing gap is not met by the government concerned). This is believed to represent a 30–50% increase in the COE produced per kilowatt hour (POST, 2005; HMG, 2009). This would lead to higher electricity bills for UK households, although fossil-fuelled CCS plants have lower operating costs per unit of output than renewable energy sources, such as wind turbines (POST, 2005; Hammond and Waldron, 2008), and a much lower construction cost per unit of output compared to nuclear power stations (POST, 2005). It has been suggested that offshore wind might give rise to an electricity price per kilowatt hour that could be three times that of a non-CCS coal-fired power station, or more than twice that of a power plant with CCS (POST, 2005; Hammond and Spargo, 2014).

In order to assess fully the potential of CCS in the United Kingdom over the whole chain or life cycle, the individual costs from capture, transport and storage were collated by Hammond et al. (2011). The energy requirements of CCS increase the amount of fuel input (and consequently CO_2 emissions) of the entire chain (IPCC, 2005). A commonly used performance parameter for the effectiveness of capture systems is therefore the cost associated with the CO_2 emissions avoided. It reflects the net reduction of emissions and provides a cost for this environmental benefit. This is a widely used measure and indicates the average cost of reducing atmospheric CO_2 emissions using one CCS plant, while providing the same amount of useful product as a reference plant without CCS. The cost of CO_2 avoided {£/tCO_2} can be determined (see Figure 7.7) using the following equation (Rao and Rubin, 2002, 2006; Abanades et al., 2007):

$$\text{Cost of } CO_2 \text{ avoided} = \left[(COE)_{cap} - (COE)_{ref} \right] / \left[\text{Emissions}_{ref} - \text{Emissions}_{cap} \right] \quad (7.6)$$

COE {£/kWh} is taken from the results estimated using Eq. (7.5) for capture (cap) and the reference plant (ref), as well as the mass emission rate (tCO_2/kWh) The values of COE from Figure 7.8 can therefore be used to determine the cost of CO_2 avoided *via* Eq. (7.6) when capture is introduced. The results are presented in Figure 7.11.

Another indicator of capture performance is the cost associated with the CO_2 captured (see Figure 7.13). It is also widely used, but is based on the mass of CO_2 captured as opposed to the emissions avoided. The cost of CO_2 captured (£/tCO_2) is determined *via* the following expression:

$$\text{Cost of } CO_2 \text{ captured} = \left[(COE)_{cap} - (COE)_{ref} \right] / \left[CO_2 \text{ emissions captured} \right] \quad (7.7)$$

It includes the levelised COE (£/kWh) from Eq. (7.5) for capture (cap) and the reference plant (ref), as well as the mass of CO_2 captured (tCO_2/kWh). Thus, the values of LCOE shown in Figure 7.8 can be used in conjunction with Eq. (7.7) to determine the cost of CO_2 captured from the CO_2 separation processes. Costs of capture can be used to evaluate the potential of CCS against possible CO_2 emissions penalties implemented by the government. If the possible emissions penalties reach the levels of carbon-capture cost, then the COE would be same as for the reference plant. This study shows an average carbon-capture cost of approximately £15/tCO$_2$ (see Figure 7.13). The levels of potential CO$_2$ penalties may vary depending on the perceived social cost of carbon, an evaluation of the economic damage that climate change could cause to the earth. However, the CO$_2$ penalties are likely to be determined by government bodies. Penalties of £27/tCO$_2$ (50$/tCO$_2$) have been suggested by the IEA. This would cover the costs of CO$_2$ capture in all UK power plants and capture technologies examined here. Figure 7.12 and Figure 7.13 show that the cost of avoiding CO$_2$ and the cost of capture follow similar trends. The cost of CO$_2$ captured is lower than the cost of CO$_2$ avoided, as a rule. Costs of CO$_2$ avoidance/capture are lowest for IGCC technologies, followed by PC plants, and are most expensive for NGCC

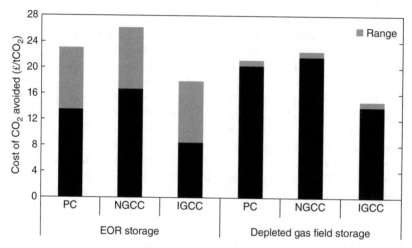

Figure 7.12 Illustrative costs of CO$_2$ avoided by CCS plants on a whole chain basis {£(2005)}
Source: Hammond et al. (2011). Reproduced with permission of Elsevier.

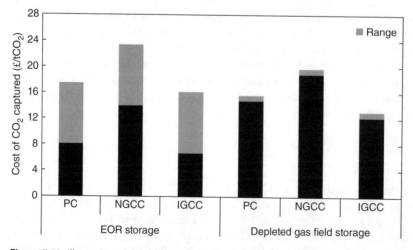

Figure 7.13 Illustrative costs of CO$_2$ captured by CCS plants on a whole chain basis {£(2005)}.
Source: Hammond et al. (2011). Reproduced with permission of Elsevier.

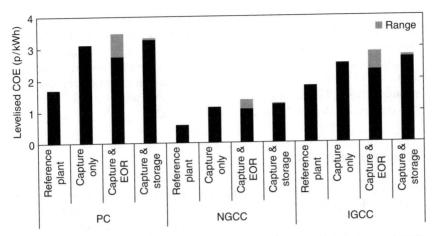

Figure 7.14 Illustrative levelised cost of electricity (LCOE) on a whole chain basis {£(2005)}
Source: Hammond et al. (2011). Reproduced with permission of Elsevier.

systems. The significant range for EOR depends upon the financial revenues incurred from additional oil extracted.

A very useful indicative assessment metric for CCS schemes is how they affect the end product in terms of the COE on a whole CCS chain basis. Figure 7.14 depicts the levelised COE for various power plant/capture technology combinations in the UK context. In the case of PC plants, CCS with EOR results in an increase of COE compared to the reference system of 62–106%, whereas CCS in gas wells leads to a 93–97% increase. For NGCC technologies, CCS with EOR results in an increase of 91–142% and CCS with gas well storage gives rise to a 118–122% increase. IGCC plants with EOR result in an increase of 27–60% and CCS with gas well storage incurs a 45–48% increase. The average price increase for all scenarios is about 84%. Thus, the COE of NGCC remains the lowest with and without CCS, even though it has the highest percentage rise in cost. A key point to note is that even though IGCC has a higher reference system COE than PC plants, the COE of IGCC with CCS is lower than PC with CCS. The rises in COE are significant, but could be reduced as the technologies develop in the future reflected in learning or experience curves (IEA, 2000; Riahi et al., 2004; Mukora et al., 2009).

The UK government established a CCS Cost Reduction Task Force (CCRTF) as an industry-led joint venture to assist with the challenge of making CCS a commercially viable operation by the early 2020s. They estimated nth of a kind (NOAK) CCS levelised costs, in real 2012 money, as being £161/MWh in 2013, falling to £114/MWh in 2020 and £94/MWh in 2028 (CCRTF, 2013). Here, the main cost reduction opportunities were seen as being (i) transport and storage scale and utilisation; (ii) improved finance-ability for the CCS chain and (iii) improved engineering designs and performance. These figures can be contrasted with those for alternative power generators (DECC, 2012b) as follows: NGCC plant –£80/MWh; PC plant (without CCS) –£90/MWh; nuclear power –£85/MWh; onshore wind (>5 MW) – £107/MWh and offshore wind (>5 MW) – £164/MWh. Thus, CCS currently exhibits a significant cost premium over its competitors and will rely on cost reduction to become commercially viable. Greater financial incentives for carbon abatement could, in principle, be secured through a higher carbon price from the EU ETS (see Section 7.6.1.4), although that has been a significant disappointment in terms of the carbon price. It fell from about €20 per tonne in 2005 to around €5 per tonne in 2015. European Ministers, including the UK Secretary of State for Energy and Climate Change, have repeatedly called for decisive action to overcome the travails of the EU ETS. In order to bolster this mechanism, the UK government introduced its CPS, although this was effectively capped in the 2014 UK Budget at a maximum of £18 from 2016/2017 until 2019/2020.

7.7 Concluding Remarks

Sustainability-related performance criteria for the system characterisation of CCS facilities have been identified to lay the groundwork for the evaluation of forthcoming BECCS technologies aimed at providing a net negative emissions option within a low-carbon transition pathway to 2050 and beyond. These are based on well-established, often quantifiable, economic, energy-related and environmental (including climate-change) indicators. They are an essential component in the preparation for a new phase of BECCS deployment in the power sector and industry. Carbon-capture facilities will hinder the performance of power plants and industrial processes as they give rise to an energy penalty, which, in turn, lowers the system (thermodynamic) efficiency. The LCOE can then be used as an indicator of the impact on plant economics of adding capture equipment. However, given the present state of play with CCS research, development and demonstration, many of the system performance characteristics need to be determined on a FOAK basis (Hammond et al., 2011; Hammond and Spargo, 2014). All such cost figures should therefore be viewed as indicative or suggestive (Hammond and Spargo, 2014). They are nonetheless helpful to various CCS stakeholder groups, such as those in industry, policymakers (civil servants and the staff of various government agencies), and civil society and environmental NGOs to enable them to assess the role of this technology in national energy strategies and its impact on local communities. Finally, the environmental performance of CCS developments can be assessed in terms of climate-change impacts (including parameters such as the carbon intensity and the cost of carbon avoided or captured), as well as the effects on biodiversity, land use and water resources. Values for the various CCS performance criteria are illustrated by reference to data for contemporary fossil-fuelled systems (Hammond et al., 2011). They represent the challenges against which BECCS will need to be evaluated. Carbon dioxide geological storage will, in many cases, have potential consequences for the marine environment, including reduced rates of reproduction among organisms, their growth and longer-term mortality. A number of organisations have taken the lead in developing geological storage site selection criteria (IPCC, 2005; IEA GHG, 2009).

Public opposition could prove a showstopper for CCS deployment unless various stakeholders are engaged in an appropriate consultation. Public acceptance of CCS was identified as one of the key site suitability criteria for CO_2 geological storage in the Alberta Research Council report (IEA GHG, 2009). A recent book on the social dynamics of CCS by Markusson et al. (2012) drew attention to what is known as the 'issue-attention cycle' (see also Hammond, 2014), a concept originally applied in the ecological domain. It suggests that disruptive technologies (such as CCS) cycle between discovery and initial enthusiasm, to the realisation of the true costs, and then a gradual decline in public interest. Chalmers et al. (2009) adopted an innovative way to draw lessons for the development of CCS by examining previous major UK energy transitions (Hammond et al., 2011): the post-World War II development of nuclear electricity; the increase in size of PC power stations in the decade around 1960; the opening up of North Sea oil and natural gas fields in the 1960s and 1970s and flue gas desulphurisation in the late 1980s and 1990s. These historical transition studies (Chalmers et al., 2009) provided a number of insights into critically important underpinning actions, which included the importance of active public engagement. In this context, Pidgeon et al. (2014) recently examined some of the critical issues concerning the design and conduct of public deliberation processes on energy policy matters of national importance. In order to develop their argument, they employed as an illustrative case study some of their earlier work on public values and attitudes towards future UK energy system change. They note that national-level policy issues are often inherently complex, involving multiple interconnected elements and frames, analysis over extended scales and different (often high) levels of uncertainty. It is their view that facilitators should engage the public in whole-systems thinking at the problem scale, provide balanced information and policy framings and use different approaches that encourage participants to reflect and deliberate on the issues. Further elaboration of such surveys, employing the deliberative framework proposed by Pidgeon et al. (2014), might go some way towards securing better awareness and understanding of CCS, including BECCS, by the public in general.

Acknowledgments

The work reported here forms part of a programme of research at the University of Bath on the appraisal of energy (including bioenergy and biofuel) systems, industrial energy use and carbon reductions and low carbon transition pathways that has been supported by a series of UK research grants and contracts awarded by various bodies. In addition to his main post as Professor of Mechanical Engineering and Founder Director of the *Institute for Sustainable Energy and the Environment* (*I·SEE*) at the University of Bath, the author holds an *Honorary Professorship in Sustainable Bioenergy* at the University of Nottingham. In the present context, he is the Principal Investigator and Co-Leader of a large consortium of university partners funded by the UK *Engineering and Physical Sciences Research Council* (EPSRC) entitled 'Realising Transition Pathways: Whole Systems Analysis for a UK More Electric Low Carbon Energy Future' (under Grant EP/K005316/1). He is also the Co-Director of the *Centre for Industrial Energy, Materials and Products* (CIE-MAP) (under Grant EP/N022645/1), which forms part of EPSRC 'End Use Energy Demand' (EUED) Centres Programme. The former grant has involved indicative technology assessments of power plant CCS within the context of UK transition pathways to a low-carbon future, whilst the latter involves the study of industrial CCS. The author is grateful to many graduate students and researchers who have contributed to various aspects of these studies, particularly Samuel J.G. Cooper, Paul W. Griffin, Craig I. Jones, Jonathan B. Norman, Áine O'Grady, Serge S. Ondo Akwe, Jack Spargo and Stephen Williams. Finally, the author has benefited from an interchange with external colleagues who are part of the above consortia of university research partners. However, the views expressed in this paper are those of the author alone and do not necessarily reflect the policies of the funders or partners.

References

Abanades, J.C., Grasa, G., Alonso, M. et al. (2007). Cost structure of a postcombustion CO_2 capture system using CaO. *Environmental Science & Technology* **41** (15): 5523–5527.

Advanced Power Generation Technology Forum [APGTF] (2009). *A Technology Strategy for Carbon Capture & Storage*. London, UK: APGTF.

Allen, S.R., Hammond, G.P., Harajli, H.A. et al. (2008). Integrated appraisal of micro-generators: methods and applications. *Proceedings of Institution of Civil Engineers: Energy* **161** (2): 73–86.

Allen, S.R., Hammond, G.P., and McManus, M.C. (2008). Prospects for and barriers to domestic micro-generation: a United Kingdom perspective. *Applied Energy* **85** (6): 528–544.

Anderson, S.T. and Newell, R.G. (2004). Prospects for carbon capture and storage technologies. *Annual Review of Environment and Resources* **29**: 109–142.

Boyle, G., Everett, B., and Ramage, R. ed. (2003). *Energy Systems and Sustainability: Power for a Sustainable Future*, 577–583. Oxford, UK: Oxford University Press.

Brent, R.J. (1996). *Applied Cost-Benefit Analysis*. Cheltenham, UK: Edward Elgar Publishing.

Carbon Capture and Storage Cost Reduction Task Force [CCRTF] (2013). *The Potential for Reducing the Costs of CCS in the UK (Final Report)*. London, UK: The Crown Estate/DECC.

Carpenter, M. and Braute, L. (2012). Can carbon dioxide storage site qualification help meet a 2015 deadline? *Proceedings of Institution of Civil Engineers: Energy* **165** (2): 97–108.

Chalmers, H., Jakeman, N., Pearson, P., and Gibbins, J. (2009). Carbon capture and storage deployment in the UK: what next after the Government's competition? *Proceedings of the Institution of Mechanical Engineers, Part A: Journal of Power and Energy* **223** (3): 305–319.

Cheng, V.K.M. and Hammond, G.P. (2017). Life-cycle energy densities and land-take requirements of various power generators: a UK perspective. *Journal of the Energy Institute* **90** (2): 201–213.

Clift, R. (1995). The challenge for manufacturing. In: *Engineering for Sustainable Development* (ed. J. McQuaid). Royal Academy of Engineering: London, UK.

Clift, R. (2007). Climate change and energy policy: the importance of sustainability arguments. *Energy* **32** (4): 262–268.

Department for the Environment, Food and Rural Affairs [Defra] (2005). *One Future – Different Paths*. London, UK: The Stationery Office Ltd.

Department for Transport [DfT] (2007). *Low Carbon Transport Innovation Strategy*, 19–26. London, UK: DfT.

Department of Energy and Climate Change [DECC] (2009). *A Framework for the Development of Clean Coal: Consultation Response*. London, UK: DECC.

Department of Energy and Climate Change [DECC] (2012a). *CCS Roadmap: Supporting Deployment of Carbon Capture and Storage in the UK*. London, UK: DECC.

Department of Energy and Climate Change [DECC] (2012b). *Electricity Generation Costs*. London, UK: DECC.

Electric Power Research Institute [EPRI] (2013). *Towards a Common Method of Cost Estimation for CO_2 Capture and Storage at Fossil Fuel Power Plants: A White Paper*. Palo Alto, CA: EPRI.

Element Energy (2010). *Developing a CCS Network in the Tees Valley Region Final Report.* Cambridge, UK: Element Energy Ltd.

Elghali, L., Clift, R., Sinclair, P. et al. (2007). Developing a sustainability framework for the assessment of bioenergy systems. *Energy Policy* **35** (12): 6075–6083.

Energy Technologies Institute [ETI] and Ecofin Research Foundation [ERF] (2012). *Carbon Capture and Storage: Mobilising Provate Sector Finance for CCS in the UK.* (A Joint Report by the ETI and the ERF.) Loughborough, UK: ETI.

Feron, P.H.M. and Hendriks, C.A. (2005). CO_2 capture process – principles and costs. *Oil & Gas Science and Technology* **60** (3): 451–459.

Foxon, T.J., Gross, R., Chase, A. et al. (2005). UK innovation systems for new and renewable energy technologies: drivers, barriers and systems failures. *Energy Policy* **33** (16): 2123–2137.

Gibbins, J. and Chalmers, H. (2008). Carbon capture and storage. *Energy Policy* **36** (10): 4317–4322.

Global CCS Institute (2015). *Fact Sheet – CCS/CCUS in the Americas.* Docklands, Victoria, Australia: Global CCS Institute.

Gough, C., Mander, S., and Haszeldine, S. (2010). A roadmap for carbon capture and storage in the UK. *International Journal of Greenhouse Gas Control* **4** (1): 1–12.

Griffin, P.W., Hammond, G.P., and Norman, J.B. (2016). Industrial energy use and carbon emissions reduction: a UK perspective. *WIREs Energy and Environment* **5** (6): 684–714.

Hammond, G.P. (2004). Engineering sustainability: thermodynamics, energy systems, and the environment. *International Journal of Energy Research* **28** (7): 613–639.

Hammond, G.P. (2012). Carbon dioxide capture and geological storage (Editorial). *Proceedings of Institution of Civil Engineers: Energy* **165** (2): 47–50.

Hammond, G. (2013). Briefing: carbon dioxide capture and storage faces a challenging future. *Proceedings of Institution of Civil Engineers: Energy* **166** (4): 147.

Hammond, G.P. (2014). Book review: Markusson, N., Shackley, S. And Evar, B. (eds)., the social dynamics of carbon capture and storage: understanding CCS representations, governance and innovation. *Proceedings of Institution of Civil Engineers: Energy* **167** (2): 86–88.

Hammond, G.P. (2016). Carbon and environmental footprint methods for renewables-based products and transition pathways to 2050. In: *Sustainability Assessment of Renewables-Based Products: Methods and Case Studies* (ed. J. Dewulf, S. De Meester and R.A.F. Alvarenga), 155–178. London, UK: Wiley.

Hammond, G.P., Harajli, H.A., Jones, C.I., and Winnett, A.B. (2012). Whole systems appraisal of a UK Building Integrated Photovoltaic (BIPV) system: energy, environmental, and economic evaluations. *Energy Policy* **40**: 219–230.

Hammond, G.P., Howard, H.R., and Jones, C.I. (2013). The energy and environmental implications of UK more electric transition pathways: a whole systems perspective. *Energy Policy* **52**: 103–116.

Hammond, G.P. and Jones, C.I. (2008). Embodied energy and carbon in construction materials. *Proceedings of Institution of Civil Engineers: Energy* **161** (2): 87–98.

Hammond, G.P. and Jones, C.I. (2011). Sustainability criteria for energy resources and technologies. In: *Handbook of Sustainable Energy* (ed. I. Galarraga, M. González-Eguino and A. Markandya), 21–46. Cheltenham, UK: Edward Elgar.

Hammond, G.P., Jones, C.I., and O'Grady, A. (2015). Environmental life cycle assessment (LCA) of energy systems. In: *Handbook of Clean Energy Systems, Vol. 6. – Sustainability of Energy Systems* (ed. J. Yan), 3343–3368. Chichester, UK: Wiley.

Hammond, G.P. and O'Grady, A. (2014). The implications of upstream emissions from the power sector. *Proceedings of Institution of Civil Engineers: Energy* **167** (1): 9–19.

Hammond, G.P. and Ondo Akwe, S.S. (2007). Thermodynamic and related analysis of natural gas combined cycle power plants with and without carbon sequestration. *International Journal of Energy Research* **31** (12): 1180–1201.

Hammond, G.P., Ondo Akwe, S.S., and Williams, S. (2011). Techno-economic appraisal of fossil-fuelled power generation systems with carbon dioxide capture and storage. *Energy* **36** (2): 975–984.

Hammond, G.P. and Pearson, P.J.G. (2013). Challenges of the transition to a low carbon, more electric future: from here to 2050 (editorial). *Energy Policy* **52**: 1–9.

Hammond, G.P. and Spargo, J. (2014). The prospects for coal-fired power plants with carbon capture and storage: a UK perspective. *Energy Conversion and Management* **86**: 476–489.

Hammond, G.P. and Waldron, R. (2008). Risk assessment of UK electricity supply in a rapidly evolving energy sector. *Proceedings of the Institution of Mechanical Engineers, Part A: Journal of Power and Energy* **222** (7): 623–642.

Hammond, G.P. and Winnett, A.B. (2006). Interdisciplinary perspectives on environmental appraisal and valuation techniques. *Proceedings of Institution of Civil Engineers003A Waste and Resource Management* **159** (3): 117–130.

Hannis, S. (2012). Monitoring carbon dioxide storage using fit-for-purpose technologies. *Proceedings of Institution of Civil Engineers: Energy* **165** (2): 73–84.

Helm, D. (2012). *The Carbon Crunch: How We're Getting Climate Change Wrong – And how to Fix it*. New Haven, USA & London, UK: Yale University Press.

HM Government [HMG] (2009). *UK Low Carbon Transition Plan – National Strategy for Climate and Energy*. London, UK: The Stationery Office Ltd.

HM Treasury [HMT] (2003). *The Green Book: Appraisal and Evaluation in Central Government*. London, UK: The Stationery Office Ltd.

House of Commons Energy and Climate Change Committee [ECCC] (2016). Future of Carbon Capture and Storage in the UK. Report *HC 692*. London, UK: The Stationery Office Ltd.

IEA Greenhouse Gas R&D Programme [IEA GHG] (2009). CCS Site Characterisation Criteria. Technical Study Report No. *2009/10*. Cheltenham, UK: IEA GHG.

Integrated Environmental Control Model [IECM] (2005). *The New Integrated Environmental Control Model: Carbon Sequestration Edition (ICEM-Cs)*. Pittsburgh, PA: Center for Energy and Environmental Studies, Carnegie Mellon University.

Intergovernmental Panel on Climate Change [IPCC] (2005). *IPCC Special Report on Carbon Dioxide and Storage*. Cambridge University Press: Cambridge, UK [Prepared by Working Group III of the IPCC].

International Energy Agency [IEA] (2000). *Experience Curves for Energy Technology Policy*. Paris, France: IEA/Organisation of Economic Co-operation and Development (OECD).

International Energy Agency [IEA] (2009). *Technology Roadmap – Carbon Capture and Storage*. Paris, France: IEA/Organisation of Economic Co-operation and Development (OECD).

Kimmance, J.P. and Rogers, D.A. (2012). Estimating and managing risk across the carbon dioxide capture and geological storage lifecycle. *Proceedings of Institution of Civil Engineers: Energy* **165** (2): 109–116.

Kohyama, H. (2006). *Selecting Discount Rates for Budgetary Purposes.* Cambridge, MA: Harvard Law School.

Liu, J., Mooney, H., Hull, V. et al. (2015). Systems integration for global sustainability. *Science* **347** (6225): 1258832.

Markusson, N., Shackley, S., and Evar, B. ed. (2012). *The Social Dynamics of Carbon Capture and Storage: Understanding CCS Representations, Governance and Innovation.* London, UK: Routledge.

Midttun, A. and Gautesen, K. (2007). Feed in or certificates, competition or complementarity? Combining a static efficiency and a dynamic innovation perspective on the greening of the energy industry. *Energy Policy* **35** (3): 1419–1422.

Mukora, A., Winskel, M., Jeffrey, H.F., and Mueller, M. (2009). Learning curves for emerging energy technologies. *Proceedings of Institution of Civil Engineers: Energy* **162** (4): 151–159.

Orr, F.M. Jr. (2009). CO_2 capture and storage: are we ready? *Energy & Environmental Science* **2** (5): 449–458.

Parkin, S. (2000). Sustainable development: the concept and practical challenge. *Proceedings of Institution of Civil Engineers: Energy* **138** (6): 3–8.

Parliamentary Office of Science and Technology [POST] (2005). *Carbon Capture and Storage (CCS).* Postnote No. 238. London, UK: POST.

Parsons, E.L., Shelton, W.W., and Lyons, J.L. (2002). *Advanced Fossil Power Systems Comparison Study – Final Report.* Morgantown, WV: National Energy Technology Laboratory.

Pidgeon, N., Demski, C., and Butler, C. (2014). Creating a national citizen engagement process for energy policy. *Proceedings of the National Academy of Sciences* **111** (Supplement 4): 13606–13613.

Rao, A.B. and Rubin, E.S. (2002). A technical, economic, and environmental assessment of amine-based CO_2 capture technology for power plant greenhouse gas control. *Environmental Science & Technology* **36** (20): 4467–4475.

Rao, A.B. and Rubin, E.S. (2006). Identifying cost-effective CO_2 control levels for amine-based CO_2 capture systems. *Industrial and Engineering Chemistry Research* **45** (8): 2421–2429.

Riahi, K., Rubin, E.S., Taylor, M.R. et al. (2004). Technological learning for carbon capture and sequestration technologies. *Energy Economics* **26** (4): 539–564.

Rubin, E.S. (2010). Understanding the pitfalls of CCS cost estimates. *International Journal of Greenhouse Gas Control* **10**: 181–190.

Rubin, E.S., Chen, C., and Rao, A.B. (2007). Cost and performance of fossil fuel power plants with CO_2 capture and storage. *Energy Policy* **35** (9): 4444–4454.

Scott, V., Haszeldine, R.S., Tett, S.F., and Oschlies, A. (2015). Fossil fuels in a trillion tonne world. *Nature Climate Change* **5** (5): 419–423.

Scottish Centre for Carbon Storage [SCCS] (2009). *Opportunities for CO_2 Storage around Scotland.* Haddington, UK: Scotprint.

Slesser, M. (1978). *Energy in the Economy.* London, UK: Macmillan Press.

UK Carbon Capture and Storage Cost Reduction Task Force [CCRTF] (2013). *The Potential for Reducing the Costs of CCS in the UK (Final Report).* London, UK: The Crown Estate/DECC.

Verdon, J.P., Kendall, J.M., and White, D.J. (2012). Monitoring carbon dioxide storage using passive seismic techniques. *Proceedings of Institution of Civil Engineers: Energy* **165** (2): 85–96.

Watson, J. (2012). Carbon Capture and Storage – Realising the Potential? UKERC Research Report. London, UK: UK Energy Research Centre.

White, S.W. and Kulcinski, G.L. (2000). Birth to death analysis of the energy payback ratio and CO_2 gas emission rates from coal, fission, wind, and DT-fusion electrical power plants. *Fusion Engineering and Design* **48** (3–4): 473–481.

8

The System Value of Deploying Bioenergy with CCS (BECCS) in the United Kingdom

Geraldine Newton-Cross and Dennis Gammer

Energy Technologies Institute, Loughborough, UK

8.1 Background

8.1.1 Why BECCS?

To tackle the causes of climate change, the United Kingdom has committed to an 80% reduction in its greenhouse gas (GHG) emissions by 2050, compared to 1990 levels. Meeting these targets will require a massive transformation in the way energy is generated and used in the United Kingdom.

The United Kingdom's GHG emissions reduction targets are enshrined in the Climate Change Act (2008). Meeting these targets is going to require a step change in the way energy is generated and used in the United Kingdom. The Energy Technologies Institute (ETI) was set up to identify and accelerate the development and demonstration of an integrated set of low-carbon technologies to deliver this step change cost-effectively. We are a public-private partnership, working across several programme areas including offshore renewables, energy storage and distribution, transport, carbon capture and storage, smart systems and heat, nuclear, energy-system modelling and bioenergy.

The ETI has worked closely with its members and with project partners to develop and refine its strategic thinking on the future low-carbon UK energy system. This has been supported by analysis and research activities evidenced against large-scale development and demonstration projects and field trials undertaken in ETI projects. This accumulating evidence base has been incorporated into the ETI's modelling framework, Energy System Modelling Environment (ESME), which enables us to identify the lowest-cost decarbonisation pathways for the UK energy system (see Figure 8.1). This involves running thousands of simulations exploring the variations in cost-optimal designs within a range of assumptions and constraints in order to identify robust strategies against a broad range of uncertainties.

We have tested different energy-system designs by removing and adding certain technologies and adjusting their cost and performance characteristics. Technologies which are shown to be the most valuable and most resilient to different assumptions can be

Biomass Energy with Carbon Capture and Storage (BECCS): Unlocking Negative Emissions, First Edition.
Edited by Clair Gough, Patricia Thornley, Sarah Mander, Naomi Vaughan and Amanda Lea-Langton.
© 2018 John Wiley & Sons Ltd. Published 2018 by John Wiley & Sons Ltd.

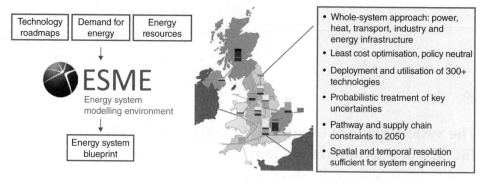

Figure 8.1 Overview of the Energy System Modelling Environment (ESME). (*See colour plate section for the colour representation of this figure.*)

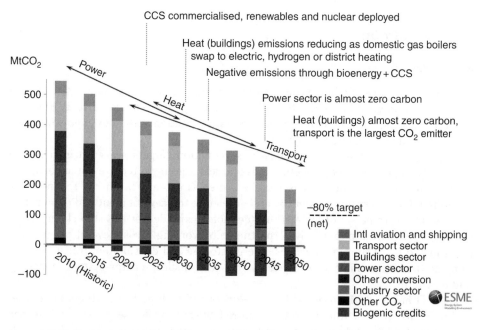

Figure 8.2 Typical decarbonisation pathway modelled in ESME, needed to meet the United Kingdom's 2050 GHG emission reduction targets. Source: Data from v4.2. (*See colour plate section for the colour representation of this figure.*)

identified and assigned relative option values/opportunity costs, i.e. the amount of additional money it would cost to develop an energy system for the United Kingdom which meets its future GHG targets and energy demands, *in the absence of* the technology or sector in question. This enables us to identify priority technologies and sectors and their associated cost and performance characteristics that are required for the United Kingdom to decarbonise its energy system cost-effectively by 2050. An example of the decarbonisation pathway is shown in Figure 8.2, along with a typical 2050 energy-flow Sankey diagram in Figure 8.3.

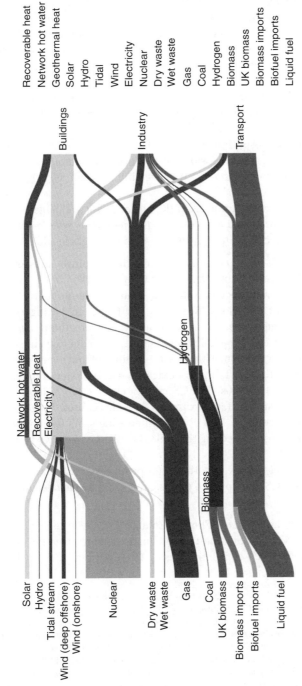

Figure 8.3 Typical (average) energy flows in 2050: Sankey diagram from ESME. Source: Data from v4.2. (*See colour plate section for the colour representation of this figure.*)

Extensive modelling undertaken over the last 5 years, informed by the outputs and insights from ETI projects, has shown the following (ETI, 2015a):

- The United Kingdom can achieve an affordable transition to a low-carbon economy over the next 35 years, with abatement costs ranging from 1–2% of GDP by 2050 – the lower range of which can be achieved through effective planning.
- The United Kingdom must focus on developing and proving a basket of the most promising supply-and-demand technology options, in order to limit implementation risks.
- Key technology priorities for the UK energy system include bioenergy, carbon capture and storage, new nuclear, offshore wind, gaseous systems, efficiency of vehicles and efficiency/heat provision for buildings.
- CCS and bioenergy are especially valuable. The most cost-effective system designs require zero or even negative emissions[1] in sectors where decarbonisation is easiest (power, industry), alleviating pressure in more difficult sectors (such as international aviation and shipping).
- High levels of intermittent renewables in the power sector and large swings in energy demand can be accommodated at a cost, but this requires a system-level approach to storage technologies, including heat, hydrogen and natural gas in addition to electricity.

Bioenergy technologies when combined with carbon capture and storage (BECCS) can deliver negative emissions (net removal of CO_2 from the atmosphere) while producing energy in the form of electricity, heat, gaseous and liquid fuels. Negative emissions provide important emissions 'headroom' as the United Kingdom transitions towards a low-carbon energy system, since the additional 'breathing space' afforded by negative emissions reduces the need for rapid emissions reductions in sectors such as heavy-duty transport and aviation, which are more difficult and expensive to decarbonise. Evidence from ESME, the ETI's peer-reviewed energy-system modelling environment, suggests that by the 2050s, BECCS could deliver c. 55 million tonnes of net negative emissions per annum (approximately half the United Kingdom's emissions target in 2050), while meeting c. 10% of the United Kingdom's future energy demand. This would reduce the cost of meeting the United Kingdom's 2050 GHG emissions target by up to 1% of GDP.

In addition to ETI analysis, several other high-profile organisations have highlighted the importance of BECCS in helping to tackle global climate change, including the following organisations:

- The Intergovernmental Panel on Climate Change (IPCC). In their Fifth Assessment Report (IPCC, 2014) on mitigating future climate change, over 100 of the 116 scenarios associated with CO_2 concentrations of 430–480 ppm in 2100 (the level that is

---

1 The term 'negative emissions' describes the process by which carbon dioxide may be permanently removed from the atmosphere, resulting in lower levels of CO_2 relative to when you started. There are two key steps to this: firstly, CO_2 is 'captured' as the biomass grows, and the carbon is either then stored in its living structures (above and below ground), or in the soil itself. Secondly, when the biomass conversion process is combined with CCS, the CO_2 that is released during the conversion stage can be captured and permanently stored underground. This results in a net reduction in the amount of CO_2 in the atmosphere over the life cycle of the biomass plant. Having an ability to facilitate negative emissions offers an accelerated route to reducing atmospheric CO_2, and this can be used to provide what some call 'emissions headroom' in certain sectors that are challenging to decarbonise cost-effectively, such as transport.

likely to limit average temperature rises to 2 °C) depend on BECCS to deliver global net negative emissions. The IPCC found that many climate models could not limit global warming to below 2 °C if the use of bioenergy, CCS and their combination (BECCS) had limited deployment.

- The Low Carbon Innovation Coordination Group (LCICG). The 2012 Bio-TINA (Technology Innovation Needs Assessment) (LCICG, 2012) stated that, deployed properly, bioenergy and BECCS have the potential to help secure energy supplies, mitigate climate change and create significant green growth opportunities.
- The Committee on Climate Change (CCC). In their report on setting the fifth Carbon Budget[2] (CCC, 2015), the CCC recommended that bioenergy should be used with CCS, and where alternative low-carbon options were not feasible or cost-effective.
- Lord Oxburgh's report on the critical role of CCS (Oxburgh, 2016). This report re-emphasises that BECCS plays a very significant role in both 2 °C and 1.5 °C modelling scenarios for global warming, which are consistent with the Paris Agreement (agreed at COP21). It also echoes the ETI's views that (i) CCS technology is ready for deployment without any more fundamental research and that it is already competitive against other forms of clean technology; and (ii) the capacity to deliver negative emissions has the potential to reduce the overall cost of decarbonisation by compensating for emissions from some hard-to-mitigate sectors, and adds flexibility into any decarbonisation plan. The report also notes that while there are a number of other potential negative emissions technologies (NETs), including NET fuel cells and direct air capture, none can be deployed cost-effectively at scale today in the same way BECCS could be. Some may develop in the future, but they would require an established CCS infrastructure to already be in place, and therefore BECCS is a natural technology to progress first.
- The CCC's 'UK Climate Action following the Paris Agreement' report (2016a). This report (and its two sister reports (CCC, 2016b, 2016c) make it very clear that sustainable bioenergy and BECCS both play a critical role in enabling the United Kingdom to meet its 2050 GHG emissions reduction commitments, and an even more central role in realising the net zero emissions ambitions arising from the Paris Agreement. Specifically, it echoes ETI's views that (i) BECCS could be cost competitive by the 2030s, but requires urgent UK government support; and (ii) the BECCS supply chain would need to draw on both domestically produced and imported feedstocks. CCC suggest a similar domestic planting rate of 30 000 ha/yr to the ETI (2016a), and they highlight the future importance of hydrogen and CCS – initially for heat via injection into the gas grid (ETI has also explored the potential of utilising the United Kingdom's significant salt caverns for hydrogen storage (ETI, 2015b), ultimately providing base- and peak-load electricity generation via hydrogen turbines).

The International Energy Agency Clean Coal Centre (IEACCC), whilst not focusing on BECCS specifically, list in their fifth 'Co-firing Biomass with Coal' conference papers[3] several demonstrations and trials by major producers, e.g. E.ON, Dong, Drax and GDF

---

2 The government accepted the CCC's recommendations and set the fifth carbon budget at 1725 Mt CO_2e for the period 2028–2032 in the Carbon Budget Order 2016. Available from: http://www.legislation.gov.uk/uksi/2016/785/made

3 IEA Clean Coal Centre, 5th Co-firing Biomass with Coal Conference (16–17 September 2015), conference papers available from: http://cofiring5. coalconferences.org/ibis/cofiring5/home

Suez, and scientific institutions working on the practical issues some biomass and waste fuels pose, including fouling, corrosion and downstream catalyst deactivation. This demonstrates the current commercial interest in developing biomass conversion technologies, which is a vital part of BECCS.

8.1.2 Critical Knowledge Gaps

The ETI was established in 2007, and at that time, there were considerable uncertainties and knowledge gaps around BECCS, as well as many individual elements of bioenergy and CCS development themselves. The ETI established CCS and bioenergy programmes (with spends of £32m and £20m, respectively) to address these knowledge gaps, in order to better assess and understand the potential for, and suitability of, BECCS deployment in the United Kingdom.

Specifically, ETI identified a series of questions that encapsulated uncertainties surrounding the use and commercial deployment of bioenergy and CCS in the United Kingdom. By seeking to answer them, ETI has identified and progressed the priority activities needed to quantify and reduce these uncertainties.

Collective project insights gained through ETI's bioenergy and CCS programmes, and informed by the work of others, have enabled us to address four key questions in relation to BECCS deployment in the United Kingdom:

1) Can a sufficient level of BECCS be deployed in the United Kingdom to support cost–effective decarbonisation pathways for the United Kingdom out to 2050?
2) What are the right combinations of feedstock, preprocessing, conversion and carbon-capture technologies to deploy for bioenergy production in the United Kingdom?
3) How can we deliver the greatest emissions savings from bioenergy and BECCS in the United Kingdom?
4) How much CO_2 could be stored from UK sources and how do we monitor these stores efficiently and safely?

The intention of this chapter is to:

1) Highlight the progress that has been made in understanding the key uncertainties associated with BECCS through ETI's projects, which have been delivered in partnership with industrial, academic and research partners over the last 10 years, and outline the insights gained.
2) Demonstrate the potential for UK deployment and the system value in supporting BECCS now in order to meet GHG targets cost-effectively.

A more detailed description of project activities and insights can be found in the individual project insight publications referenced throughout this document.

8.2 Context

8.2.1 Bioenergy

Bioenergy is the largest source of renewable energy in the United Kingdom. In 2015, bioenergy contributed 73% of all renewable energy inputs (151 TWh/yr) and 59% of

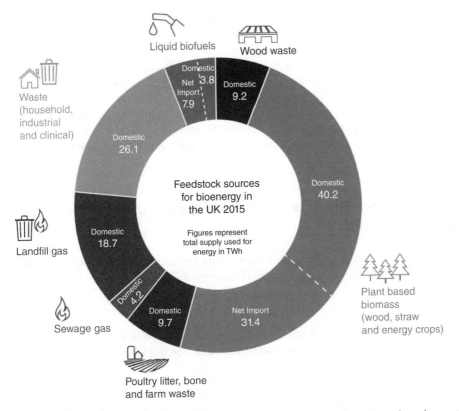

Figure 8.4 2015 biomass and waste feedstock sources. *Source:* BEIS, (2016a). (*See colour plate section for the colour representation of this figure.*)

final renewable energy consumption. This met 5% of all UK final energy demand, of which 38% was electricity, 14% transport fuels and 48% heat (BEIS, 2016a).

The majority of biomass and waste feedstocks are sourced in the United Kingdom, but imports of biomass and biofuels have increased in recent years to meet demand (see Figure 8.4). Most UK bioenergy plants to date either provide power or heat. By the end of July 2016, there was 5.2 GW of bio-power capacity in the United Kingdom, and 3.1 GW of heat, plus 226 MW of biogas capacity supported under the Renewable Heat Incentive (RHI) (BEIS, 2016b).

8.2.2 Bioenergy with CCS

Both biomass combustion and biomass gasification lend themselves to CCS through proven CO_2 capture technologies (on coal and oil residues). The scale of biomass in the United Kingdom today, and in particular its use in a large unit like Drax, produces CO_2 in sufficient quantities to deliver economies of scale in the capture of CO_2. On a smaller, but still important scale, the United Kingdom has the opportunity to fit CCS to existing bioethanol plants, as has been demonstrated in the United States. Although the United Kingdom has no large CCS projects, large-scale underground storage in

North Sea aquifers has been practised in Norway since 1996, and by the end of 2017, 22 plants globally will be running CCS technology applications, spanning post-combustion and pre-combustion coal, natural gas steam reforming, bioenergy (corn to ethanol) and applications from power, gas production, refining, chemicals and steel. Indeed, Toshiba Corporation has recently been selected by Japan's Ministry of the Environment to construct a carbon-capture facility to capture over 500 tons of CO_2 a day from the Mikawa Power Plant (49 MW) (Toshiba, 2016).The plant aims to be operational by 2020, and it will become the world's first power plant capable of capturing carbon from a biomass power plant and, therefore, the first to deliver negative emissions.

8.3 Progressing our Understanding of the Key Uncertainties Associated with BECCS

Section 8.1.2 highlights the critical knowledge gaps around BECCS deployment in the United Kingdom as four key questions. This section highlights the progress made under each.

8.3.1 Can a Sufficient Level of BECCS Be Deployed in the United Kingdom to Support Cost–Effective Decarbonisation Pathways for the United Kingdom out to 2050?

The ETI's ESME modelling consistently selects decarbonisation pathways for the United Kingdom that generate approximately 130 TWh/yr of bioenergy (~10% final energy demand) in 2050 and deliver more than 50 million tonnes of negative emissions a year through the combined deployment of bioenergy and CCS. Using ETI's more detailed Bioenergy Value Chain Model (BVCM) to understand future bioenergy sector development scenarios, we know that this 130 TWh/yr final energy output requires approximately 190 TWh/yr of biomass feedstock (a combination of imported and domestically grown feedstocks) and ~45 TWh/yr of waste feedstocks[4] (BEIS, 2016a). Table 8.1 below sets out estimates for how much additional feedstock is required to meet these 2050 pathways.

ETI analysis indicates that the additional domestic biomass feedstock production needs could be met by converting 1.4 million hectares of UK agricultural land into bioenergy crops and forestry by the 2050s. ETI has taken a conservative approach to assessing the amount of land that could potentially be available to produce domestic biomass feedstocks – limiting the amount, type and location of land to be converted based on a series of assumptions set out in earlier insight reports (ETI, 2015c; ETI, 2016a).

At the start of ETI's bioenergy programme in 2008, much uncertainty and concern existed around the availability and sustainability of bioenergy at the scales required.

---

4 These figures exclude imported biofuels, since they are 'ready to use' and not actually converted to a final vector within the United Kingdom. The United Kingdom currently imports ~678 ktoe (7.88 Twh/yr) of biofuels.

Table 8.1 Estimates of feedstock requirements to meet 2050 pathways.

	A. Amount of feedstock *currently* being used for bioenergy	B. 2050 requirement	C. Additional needs from 2015 level of use to be met over next 35 years (C = −B−A)	D. ETI estimates of *additional* feedstock potentially available in 2050	ETI assessment of whether the 2050 requirement can be met (is B < A + D?)
UK residual waste arising	22 TWh/yr[a]	45 TWh/yr	23 TWh/yr	29 TWh/yr (~8mT p.a.)	YES
Domestically grown biomass	40 TWh/yr (plus ~4 TWh/yr being used to produce liquid biofuels) (BEIS, 2016a)	75–115 TWh/yr	35–75 TWh/yr	70–105 TWh/yr (based on conversion of 1.4mHa)	YES
Imported biomass feedstocks	31 TWh/yr	75–115 TWh/yr	44–84 TWh/yr	69–319 TWh/yr (DECC, 2012)	YES

a) This excludes landfill gas and sewage gas, which has a current input value of 23 TWh/yr., producing energy largely via engines or anaerobic digestion.

Considerable work has been completed by ETI and others in this space over the last few years. The ETI's 4 year Ecosystem Land Use Modelling (ELUM) field trials project has significantly advanced the evidence and understanding of the sustainability of biomass feedstock production in the United Kingdom, especially around soil carbon sequestration (see 'summary findings' box in Figures 8.5 and 8.6) and the ability to deliver genuine carbon savings across bioenergy value chains (see Figure 8.8). Most importantly, it showed that given the right choice of land-use change, crop type and location, substantial emissions savings can be delivered through bioenergy, and many opportunities exist to optimise the wider ecosystem service benefits from biomass feedstock production in the United Kingdom.

Financial sustainability can also be achieved through more strategic approaches to agricultural land use in the United Kingdom, and specifically optimising local productivity by taking account of economic, environmental and wider accessibility factors. As part of ETI's evidence collection and assessment of land available for biomass, we have also assessed the drivers for land-use change to bioenergy (Enabling UK Biomass project (ETI, 2016b)) and the economic counterfactuals, and collated example case studies where farmers/land owners have successfully diversified part of their land to include biomass production (ETI, 2016c).

Summary findings from ELUM (ETI, 2016a):

- The ELUM project was commissioned to provide more data and understanding of soil carbon and GHG fluxes arising as a result of land-use change to bioenergy feedstocks, with a primary focus on the second-generation bioenergy crops Miscanthus, short rotation willow (SR-W).

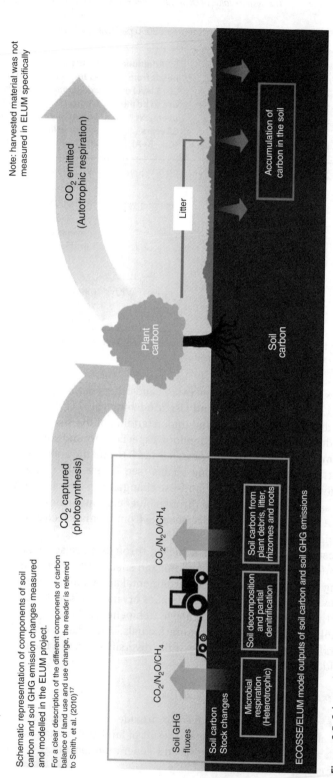

Schematic representation of components of soil carbon and soil GHG emission changes measured and modelled in the ELUM project.

For a clear description of the different components of carbon balance of land use and use change, the reader is referred to Smith, et al. (2010)[17]

Note: harvested material was not measured in ELUM specifically

CO_2 emitted
(Autotrophic respiration)

CO_2 captured
(photosynthesis)

Plant carbon

Litter

Soil carbon

Accumulation of carbon in the soil

Soil GHG fluxes

Soil carbon Stock changes

$CO_2/N_2O/CH_4$

$CO_2/N_2O/CH_4$

Microbial respiration (Heterotrophic)

Soil decomposition and partial denitrification

Soil carbon from plant debris, litter, rhizomes and roots

ECOSSE/ELUM model outputs of soil carbon and soil GHG emissions

Figure 8.5 Schematic representation of components of soil carbon and soil GHG emissions changes measured and modelled in the ELUM project.

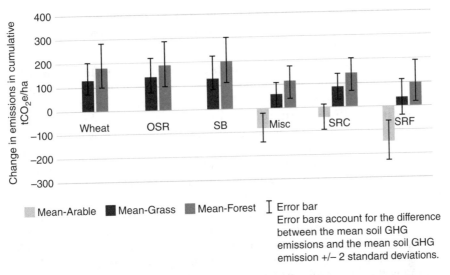

Figure 8.6 Estimates derived from the ELUM model on mean soil GHG emissions over 40 years (relative to counterfactual land use), expressed as net GHG emissions per hectare across the United Kingdom. The model was validated using empirical data collected during the ELUM project (Smith et al., 2010). (*See colour plate section for the colour representation of this figure.*)

- Second-generation (2G) biomass, such as Miscanthus, short rotation coppice (SRC) willow and short rotation forestry (SRF) grown on arable land, or grassland sites where appropriate, offers the greatest potential yield and GHG emissions savings, but also could deliver wider biodiversity and ecosystem service benefits, including hazard regulation (e.g. flood prevention), disease and pest control, improving water and soil quality and acting as wildlife/game cover.
- The GHG benefits of increased planting of 2G bioenergy crops (including forestry) is apparent since crops that are well matched to sites can start acting as net carbon sinks as soon as they start growing (see Figure 8.6 for national average).
- Short rotation forestry is likely to offer the greatest GHG savings in bioenergy value chains, particularly when grown on arable or grassland, due to its ability to deliver greatest soil carbon sequestration.
- It is important to note that the ELUM project findings are only part of the value chain (i.e. soil carbon changes from biomass production) and, therefore, need to be read in conjunction with Figure 8.5, which contextualises the importance of soil carbon changes and wider direct land-use change emissions relative to the GHG emissions across the whole system-level value chain (e.g. taking into account transport and conversion emissions too). Only this value chain assessment enables us to test whether genuine carbon savings could be delivered relative to fossil fuel baselines.
- ETI analysis demonstrated that where land-use change resulted in an increase in soil emissions (i.e. soil carbon losses), these were more than offset by the CO_2 captured and stored when that biomass feedstock was used in a BECCS value chain (ETI, 2016a).

8.3.2 What are the Right Combinations of Feedstock, Preprocessing, Conversion and Carbon-Capture Technologies to Deploy for Bioenergy Production in the United Kingdom?

8.3.2.1 Optimising Feedstock Properties for Future Bioenergy Conversion Technologies

In parallel with understanding whether bioenergy value chains deliver genuine carbon savings and understanding the availability of different feedstocks, it is critically important to understand the characteristics of biomass and waste feedstocks and their variability, in order to understand how energy conversion technologies may perform when utilising these feedstocks. Biomass and waste feedstocks can raise new issues with conversion that have not been encountered with fossil fuels, such as moisture, ash (content and fusion temperature), minor constituents such as silica, calcium, potassium and chlorine, and – especially for waste – issues with tars.

Building the evidence base around physical and chemical composition of different feedstocks enables process designers to assess the relative opportunities for optimising bioenergy value chains – whether it be through selection of feedstocks with particular traits; pretreating the feedstocks in some way; blending the feedstocks to dilute any issues; adapting the conversion technology itself to deal with any fouling, slagging or performance issues better and/or bolstering clean-up technologies to negate any changes in emissions resulting from the use of particular feedstocks. ETI has commissioned a series of projects to build this evidence base, including the Characterisation of Feedstocks; Energy from Waste and Techno-economic Assessment of Biomass Preprocessing (TEABP) projects.[5,6,7] Insights from their combined outputs will be published as an ETI Insights Report in 2017 and have informed the scope of the Biomass Feedstock Improvement Process (BioFIP) demonstrator project currently being commissioned by ETI.

Two ETI projects have looked at the future relevance of different bioenergy conversion technologies, and both have indicated that advanced gasification appears to have the greatest potential for delivering flexible low-cost energy. The Energy from Waste project compared combustion, anaerobic digestion, advanced gasification and pyrolysis and concluded that advanced gasification (with syngas clean-up) offered the greatest potential benefits for converting waste to energy at the town scale (an optimal scale in terms of balancing economies of scale with logistical costs and associated emissions). The BVCM project assessed the current maturity, cost and performance levels of all bioenergy technologies and consistently highlighted advanced gasification as one of the key low-cost means of delivering the required carbon savings at the UK energy system level out to the 2050s.

---

5 ETI's Techno-economic Assessment of Biomass Pre-processing Technologies (TEABP) project is assessing the cost–effectiveness of preprocessing, taking into account different feedstocks, storage, logistics, preprocessing and conversion technologies. Insights informed the commissioning of the Biomass Feedstock Improvement Process (BioFIP) demonstrator project. More information available from: http://www.eti. co.uk/project/techno-economic-assessment-of-biomass-pre-processing/

6 ETI's Energy from Waste project assessed the energy-bearing content and composition of different waste feedstocks in the United Kingdom, and modelled availability out to 2050. More information available from: http://www.eti.co.uk/project/energy-from-waste/

7 ETI's Characterisation of Feedstocks (CoF) project is assessing the physical and chemical properties of a selection of different UK-derived biomass feedstocks. More information available from: http://www.eti. co.uk/project/characterisation-of-feedstocks/

Advanced gasification technology, although not the only option, is a key enabler of flexible energy-system solutions, since the syngas produced can be converted into electricity (either directly or via hydrogen production), CHP, bio-methane and transport fuels, making it one of the most flexible, scalable and cost-effective bioenergy technologies available. ETI has progressed this important technology by supporting the development and demonstration of advanced waste gasification, i.e. where the syngas quality is sufficiently increased and cleaner, such that it could be used consistently in an engine or turbine.

8.3.2.2 BECCS Value Chains: What Carbon-Capture Technologies Do we Need to Develop?

BECCS technologies represent one of the very few practical, scalable and economic means of removing large quantities of CO_2 from the atmosphere relevant for the United Kingdom, and the only approach which generates a useful by-product – power, heat or hydrogen.

ETI work on CCS has examined key aspects of its deployment for both fossil (coal and gas) and biomass (co-firing and dedicated) feedstocks, including costs, risk and technology maturity. The bioenergy programme has focused on the costs and challenges of BECCS technologies,[8] and the CCS programme has examined the risks and cost reduction in CO_2 transportation and storage and the cost of CO_2 capture.

- The cost of CO_2 capture is the largest single cost element in CCS (see Figure 8.7) and can be comparable to that of the original power station. Additionally, the process of capturing CO_2 itself can use up to 20% of the power station output.
- Plant scale remains the principal driver of capex, rather than choice of technology, with larger plants having lower specific capital costs. The weighted feedstock energy

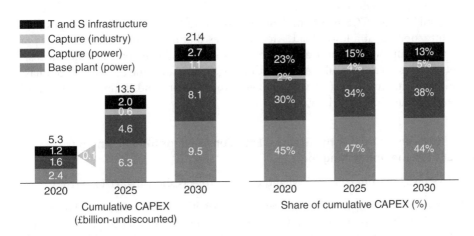

Figure 8.7 Capital costs of building a 50 Mt/year CO_2 (~10 GWe) CCS network (£bn 2014 undiscounted). *Source:* FromETI, (2016d). Reproduced with permission of ETI. (*See colour plate section for the colour representation of this figure.*)

---

8 Techno-Economic Study of Biomass to Power with CCS (TESBiC) project commissioned and funded by the ETI. Further information available at: http://www.eti.co.uk/programmes/bioenergy/biomass-to-power-with-carbon-capture-and-storage

content and cost is one of the key drivers of LCOE, with biomass pellets currently being more expensive.

- A number of capture technologies warrant further development. Whilst some technologies are more likely to be demonstrated first (post-combustion amine), it is not yet possible to identify next-generation technologies which could end up being the dominant CCS/BECCS technology by 2050. It is likely that as technologies develop, such as gasification to generate more flexible fuels such as SNG and H_2, pre-combustion CCS becomes more attractive.
- Dedicated (100% biomass) BECCS technologies offer significant opportunity to deliver substantially more negative emissions than co-firing technologies and would be more attractive in the longer term if and when financial incentives are applied to negative emissions and avoided carbon. In the absence of such incentives, co-firing could be a more cost-effective way of minimising penalties for positive emissions and is likely to play an important transition role.
- The TESBiC project findings indicate that the most significant barriers to the deployment of BECCS technologies will not be technical ones, but rather the scale of investment required, the limited price of carbon and hence the limited value of negative emissions.
- Most cost savings in the next two decades will be delivered through reducing costs by deployment, rather than fundamental technology breakthroughs. Combined with the low growth rate of CCS and the absence of a commercially ready game changer today, this means amines and pre-combustion technologies will continue to be the technology of choice in power production for several years.
- It is important that technologies offering breakthrough performance are funded through to demonstration level so that they can enter the market when other aspects of risk have been reduced.
- By the mid-2030s, CCS plants may have to respond to daily demand changes and, therefore, operate at a lower load – technologies that offer reduced capital costs or system flexibility will be more attractive. ETI system modelling indicates new investments for this market should favour both natural gas plants due to cost and biomass gasification due to the system value of negative emissions and flexibility in terms of the end product (i.e. electricity, SNG or H_2).

8.3.3 How can we Deliver the Greatest Emissions Savings from Bioenergy and BECCS in the United Kingdom?

Collective insights from ETI's ELUM and BVCM projects show that (ETI, 2016a):

- CCS is a game changer. Bioenergy value chains with CCS, as shown in Figure 8.8, render direct land-use change (dLUC) emissions of second-order importance. Most 2G biomass feedstocks grown in the United Kingdom and used in BECCS would deliver substantial negative emissions to the United Kingdom, as well as flexibility across all key vectors of power, heat, liquid and gaseous fuels.
- If bioenergy is deployed without CCS, dLUC emissions can be material, either contributing GHG emissions savings via soil carbon sequestration or by producing additional emissions at the value-chain level, depending on the choice of crop type,

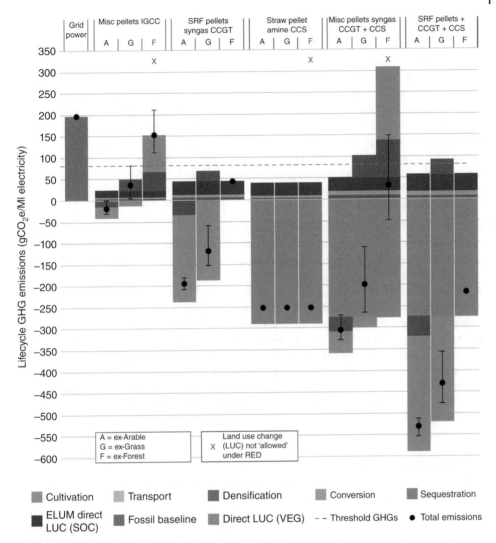

Figure 8.8 Quantifying the impact of dLUC emissions and CCS on UK bio-electricity value chains (life-cycle GHG emissions: gCO₂e/MJ). (*See colour plate section for the colour representation of this figure.*)

location and ultimate use in the energy system. ETI's work has reinforced the need to assess emissions across *the whole value chain* in order to judge the scale of carbon savings achieved, and not just to view feedstock carbon debt, land-use change emissions or conversion technologies in isolation.

- The greatest emissions savings are when bioenergy is used with CCS. This combination means that biomass and waste feedstocks are best deployed in conversion technologies which result in power or hydrogen – since neither have any carbon content in the resultant energy vector, and CCS can be used to capture the maximum amount of CO_2 from the conversion process.

8.3.4 How Much CO_2 Could Be Stored from UK Sources and How Do we Monitor These Stores Efficiently and Safely?

8.3.4.1 Storage Potential

ETI has commissioned a number of projects to assess different aspects of storage of CO_2 around the United Kingdom. It started with the development of an atlas of UK offshore stores – a high-level appraisal of about 600 potential stores (UKSAP Project), and this database is now managed by The Crown Estate and the British Geological Survey under the title 'CO$_2$Stored'. In 2015, DECC funded the ETI to specify, commission and manage appraisal work (Strategic UK Storage Appraisal Project) on five geological storage sites in more detail (both depleted oil and gas fields and saline aquifers), in order to identify the amount of storage that was confidently exploitable in the short and medium term.

- The United Kingdom is endowed with a rich and diverse national offshore CO_2 storage resource, key components of which can be brought into service readiness without extensive appraisal programmes thanks to decades of petroleum exploration and development activity.
- The portfolio of five sites selected in the Strategic UK Storage Appraisal Project (see Figure 8.9) is geographically and technically diverse and presents options for clean energy and industrial development around the United Kingdom.
- The ETI work, together with the three sites (Hewett, Goldeneye and Endurance) which completed FEED studies through the DECC CCS Programme, has enabled a mature and well-qualified UK storage proposition to be developed, such that more than 1.5 Gte worth of stores could be fully operational by 2030. This is enough to service around 10 GW of power generation and other industrial sources fitted with CCS, as highlighted in ETI's CCS scenarios work (ETI, 2015d).
- It is important to site new power stations with CCS close to storage sites and emissions sources (e.g. on populous estuaries close to potential offshore stores, such as Thames, Mersey, Tees and Humber).
- The discounted life-cycle costs (10% discount, 2015) for this offshore pipeline and storage would add only about £5–9 £/MWh to the UK electricity price.

The insights from this work were incorporated into BVCM and enabled us to identify the optimal locations for, and nature and scale of, clusters of BECCS technologies in the United Kingdom, taking into account the likely sources of UK-grown biomass or waste feedstocks, and the port locations for imported biomass. Modelling optimised on minimal cost and GHG emissions revealed strong preferences for clusters of large, highly efficient BECCS plants utilising gasification or combustion technology to produce hydrogen or electricity at two main shoreline hub locations (onshore points at which captured CO_2 is compressed and piped to the injection point of the store): Thames and Teesside. These locations are also closest to key port facilities capable of handling and distributing imported feedstock. Peterhead, Barrow and Easington were also sites of considerable CO_2 sequestration.

8.3.4.2 Managing the Risks of Storage

Risks in storage are assessed through the process of storage appraisal. For those stores which are depleted gas fields, the activities of oil and gas operators have often

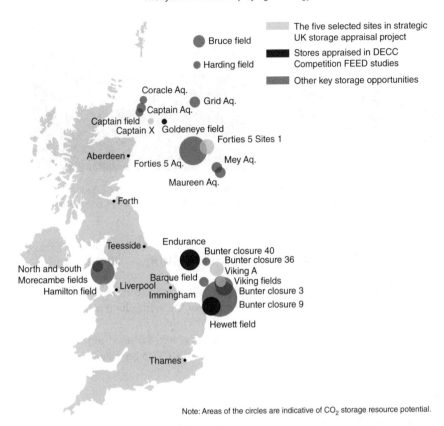

Note: Areas of the circles are indicative of CO_2 storage resource potential.

Figure 8.9 Selected potential CO_2 storage sites in the United Kingdom. *Source:* From ETI Strategic UK Storage Appraisal Project. Reproduced with permission of ETI.

collected enough data to fully evaluate the store. Much of the bulk of the United Kingdom's storage resource, however, located in saline aquifers, has not been explored fully and will require new appraisal wells to check the properties of the storage and sealing cap-rocks. In 2013, the ETI co-funded an appraisal well – the Aquifer Appraisal Project, the United Kingdom's first – for the huge Endurance store off Yorkshire. The results were excellent and National Grid progressed design of the store; much of the findings are published in Key Knowledge Deliverables (KKDs) on the government website (BEIS, 2016c).

Current research and evidence shows that leakage from stores is highly unlikely. Behaviour of the CO_2 deep below the seabed (e.g. 2 km) is expected to be observed by a suite of monitoring tools. However, if CO_2 did escape, it would be difficult to predict with certainty exactly where and how it would reach the seabed. To enable low-cost, reliable marine monitoring, the ETI initiated a project to develop a long-range mobile autonomous vehicle for measurement, monitoring and verification of storage integrity (the MMV Project), which will patrol the sea floor over large areas. The prototype (see Figure 8.10) will be trialled in the North Sea in 2017.

Figure 8.10 Autonomous marine vehicle prototype being developed through the ETI MMV project (The Autosub Long Range AUV, Stephen McPhail, National Oceanography Centre, the United Kingdom).

8.4 Conclusion: Completing the BECCS Picture

It is often helpful to view the whole-system demonstration and commercial deployment of BECCS as a jigsaw puzzle, with each core element of the value chain being a separate jigsaw piece (Figure 8.11).

Critical evidence and understanding has now been created for most of the individual jigsaw pieces, which has enabled key insights around BECCS to be drawn.

8.4.1 Next Steps

The aim of this chapter was to describe the importance of BECCS and negative emissions and to highlight the progress that has been made over the last 8 years. The collective insights and progress delivered have shown that all the core component parts of a BECCS system have been significantly de-risked and advanced, with very few technical

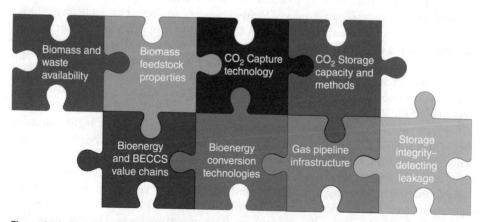

Figure 8.11 Core elements of the BECCS value chain.

or sustainability barriers being identified. The next steps needed are to put the components together in a full chain and in parallel develop the domestic bioenergy feedstock supply.

Overall, the United Kingdom is well placed to utilise BECCS as a means of helping to meet 2050 GHG emissions reduction targets. However, to realise the benefits of negative emissions, support is needed over the next five to 10 years to deploy BECCS technology and CO_2 storage at a commercial scale.

UK government support of BECCS technology is key to the United Kingdom fulfilling its GHG carbon reduction commitments by 2050, since the final decision is a political and financial one, not fundamentally technical (Oxburgh, 2016).

This progress in the technical, environmental and financial evidence and understanding, and the commercial demonstration steps being taken by others globally, should give the UK government confidence to commit to and support the deployment of this vital technology in the United Kingdom. The Oxburgh report suggests that full-chain CCS costs at c. £85/MWh are feasible under the right circumstances, a figure which ETI's analysis can corroborate. The report concludes that, under the right conditions, even the first CCS projects can compete on price with other forms of clean electricity. Given the evidence and progress highlighted in this report, we would urge the government to give consideration to ensuring that the United Kingdom's CCS strategy encompasses demonstration of BECCS technology and delivering negative emissions within the next decade.

All BECCS jigsaw pieces are now clear and on the table. Others have started to put them in place internationally, and the United Kingdom should do the same.

References

BEIS (2016a). Digest of UK Energy Statistics (DUKES) 2016. Department for Business, Energy & Industrial Strategy. Available at: https://www.gov.uk/government/collections/digest-of-uk-energy-statistics-dukes

BEIS (2016b). July 2016 RHI Deployment Data. Department for Business, Energy & Industrial Strategy. Available at: https://www.gov.uk/government/statistics/rhi-deployment-data-july-2016

BEIS (2016c). Carbon Capture and Storage Knowledge Sharing. Department for Business, Energy & Industrial Strategy. Available at: https://www.gov.uk/government/collections/carbon-capture-and-storage-knowledge-sharing

CCC (2015). The Fifth Carbon Budget: The Next Step Towards a Low Carbon Economy. Committee on Climate Change, November 2015. Available at: https://www.theccc.org.uk/wp-content/uploads/2015/11/Committee-on-Climate-Change-Fifth-Carbon-Budget-Report.pdf

CCC (2016a). UK Climate Action Following the Paris Agreement. Committee on Climate Change, October, 2016. Available at: https://www.theccc.org.uk/publications/uk-action-following-paris/

CCC (2016b). Next Steps for UK Heat Policy. Committee on Climate Change, October 2016, https://www.theccc.org.uk/publication/next-steps-for-uk-heat-policy/

CCC (2016c). Meeting Carbon Budgets – Implications of Brexit for UK climate policy. Committee on Climate Change, October 2016. Available at: https://www.theccc.org.uk/wp-content/uploads/2016/10/Meeting-Carbon-Budgets-Implications-of-Brexit-for-UK-climate-policy-Committee-on-Climate-Change-October-2016.pdf

DECC (2012). UK Bioenergy Strategy. Department of Energy and Climate Change. Available at: https://www.gov.uk/government/publications/uk-bioenergy-strategy

ETI (2015a). *Options, Choices, Actions: UK Scenarios for a Low Carbon Energy System Transition*. Loughborough: Energy Technologies Institute Available at: http://www.eti.co.uk/insights/options-choices-actions-uk-scenarios-for-a-low-carbon-energy-system/.

ETI (2015b). *The Role of Hydrogen Storage in a Clean Responsive Power System*. Loughborough: Energy Technologies Institute Available at: http://www.eti.co.uk/insights/carbon-capture-and-storage-the-role-of-hydrogen-storage-in-a-clean-responsive-power-system/.

ETI (2015c). *Insights into the Future UK Bioenergy Sector, Gained Using the ETI's Bioenergy Value Chain Model (BVCM)*. Loughborough: Energy Technologies Institute Available at: http://www.eti.co.uk/bioenergy-insights-into-the-future-uk-bioenergy-sector-gained-using-the-etis-bioenergy-value-chain-model-bvcm/.

ETI (2015d). *Building the UK Carbon Capture and Storage Sector by 2030 – Scenarios and Actions*. Loughborough: Energy Technologies Institute Available at: http://www.eti.co.uk/insights/carbon-capture-and-storage-building-the-uk-carbon-capture-and-storage-sector-by-2030/.

ETI (2016a). *Delivering GHG Emission Savings through UK Bioenergy Value Chains*. Loughborough: Energy Technologies Institute Available at: http://www.eti.co.uk/insights/delivering-greenhouse-gas-emission-savings-through-uk-bioenergy-value-chains/.

ETI (2016b). *Enabling UK Biomass*. Loughborough: Energy Technologies Institute Available at: http://www.eti.co.uk/bioenergy-enabling-uk-biomass/.

ETI (2016c). *Bioenergy Crops in the UK: Case Studies of Successful Whole Farm Integration*. Loughborough: Energy Technologies Institute Available at: http://www.eti.co.uk/library/bioenergy-crops-in-the-uk-case-studies-on-successful-whole-farm-integration-evidence-pack.

ETI (2016d). *Reducing the Cost of CCS, Developments in Capture Plant Technology*. Energy Technologies Institute, Loughborough. Available at: http://www.eti.co.uk/insights/reducing-the-cost-of-ccs-developments-in-capture-plant-technology/

IPCC (2014). Climate change 2014: mitigation of climate change. In: *Contribution of Working Group III to the Fifth Assessment Report of the Intergovernmental Panel on Climate Change* (ed. O. Edenhofer, R. Pichs-Madruga, Y. Sokona, et al.). Cambridge and New York: Cambridge University Press.

LCICG (2012). Technology Innovation Needs Assessment (TINA). Bioenergy Summary Report. Low Carbon Innovation Coordination Group, September. Available at: http://www.lowcarboninnovation.co.uk/working_together/technology_focus_areas/bioenergy/

Oxburgh (2016). Lowest cost decarbonisation for the UK: the critical role of CCS. Report to the Secretary of State for Business, Energy and Industrial Strategy from the Parliamentary Advisory Group on Carbon Capture and Storage (CCS). Available at: http://www.ccsassociation.org/news-and-events/reports-and-publications/parliamentary-advisory-group-on-ccs-report/

Smith, P., Lanigan, G., Kutsch, W.L. et al. (2010). Measurements necessary for assessing the net ecosystem carbon budget of croplands. *Agriculture, Ecosystems and Environment* **139**: 302–315. http://dx.doi.org/10.1016/j.agee.2010.04.004.

Toshiba (2016). *Toshiba and Mizuho Information & Research Institute to Lead Japan's Largest CCS Project*. Press release Available at: https://www.toshiba.co.jp/about/press/2016_07/pr2601.htm.

Part III

BECCS in the Energy System

9

The Climate-Change Mitigation Challenge

Sarah Mander[1], Kevin Anderson[1], Alice Larkin[1], Clair Gough[1] and Naomi Vaughan[2]

[1] *Tyndall Centre for Climate Change Research, School of Mechanical, Aerospace and Civil Engineering, University of Manchester, UK*
[2] *School of Environmental Sciences, University of East Anglia, Norwich, UK*

9.1 Introduction

The concept of linking bioenergy with carbon capture and storage began to emerge in peer-reviewed literature in the early 2000s. This work included combining CCS and biomass for power generation to produce negative emissions (Keith 2001; Obersteiner et al., 2001) and the potential for CCS to reduce emissions from the paper and pulp industry in Sweden (Mollersten et al., 2003). Initially, biomass energy with carbon capture and storage (BECCS) was conceived as a 'stop-gap' solution that could allow more ambitious targets to be achieved, rather than a mainstay of mitigation policy. From 2005 onwards, BECCS started to appear within the integrated assessment modelling (IAM) work that underpinned the stabilisation pathways of the IPCC (van Vuuren et al., 2007), and, as we will explain in this chapter, it now plays a central role within a significant number of the 2 °C climate futures modelled using IAMs.

This chapter describes how the global carbon budget is diminishing rapidly, given delays to emissions reduction efforts. The concept of negative emissions generally, and BECCS more specifically, is increasingly prominent as it becomes more challenging for IAM runs to generate scenarios that remain within global carbon budgets associated with 2 °C. To complement previous chapters, which provide an overview of the technical and physical aspects of implementing BECCS technologies, we present a summary of the contribution of BECCS within the IAMs used by the climate-change community. We consider the role of negative emissions in compensating for exceeding cumulative emissions budgets that are consistent with global climate-change ambitions, along with the implications of implied levels of deployment for bioenergy supply (including land-use change), infrastructure roll-out and CO_2 storage capacity. The chapter concludes

Biomass Energy with Carbon Capture and Storage (BECCS): Unlocking Negative Emissions, First Edition.
Edited by Clair Gough, Patricia Thornley, Sarah Mander, Naomi Vaughan and Amanda Lea-Langton.
© 2018 John Wiley & Sons Ltd. Published 2018 by John Wiley & Sons Ltd.

with an assessment of whether BECCS can deliver what is expected of it in within climate-change policy.

9.2 Cumulative Emissions and Atmospheric CO_2 Concentration for 2 °C Commitments

The UNFCCC 21st Conference of the Parties held in Paris in 2015 was an historic event; for the first time, nations across the world agreed to address climate change by limiting global mean surface temperature increases to 'well below 2 °C' as well as to 'pursue efforts to limit the temperature increase to 1.5°C'. Those engaged in the climate-change debate will be aware of the significance of avoiding a 2 °C global temperature rise and be familiar with its use as both a focused global commitment and a political anchor point for the debate (Jordan et al., 2013; Sharmina et al., 2013). However, what is perhaps less commonly understood is the significance of the associated framing of the Paris Agreement around 'well below 2 °C', or indeed how much more challenging, or arguably impossible, it will be to 'limit the temperature increase to 1.5 °C'.

As the Paris Agreement is scrutinised, it is becoming increasingly evident that emissions scenarios demonstrating pathways to a 2°C (or a better 1.5°C) future are heavily reliant on negative emissions technologies (NETs) to remove CO_2 from the atmosphere (Fuss et al., 2014; Rogelj et al., 2015). Moreover, these technologies, of which there are as yet only demonstration and pilot-scale operations, are assumed to withdraw, on average, 12 Gt of CO_2 each year, extracting, on average, 550 Gt of CO_2 over the latter half of the century (Smith et al., 2016). To date, most NETs are little more than conceptual designs for which the costs are as yet unknown; no governments have committed to regulatory support mechanisms, and there is poor understanding of the associated risks. Understanding why such unproven technologies are considered necessary to achieve the ambitions of the Paris Agreement is an important part of the climate-change debate.

The concept of a global carbon budget is based on earth-system modelling findings that global temperature change is proportional to cumulative carbon emissions (Matthews et al., 2009; Zickfeld et al., 2012). Whilst the concept of a limited stock of CO_2 (the carbon budget) being linearly related to future temperature changes (within certain bounds) is relatively simple, a clear understanding of the particular carbon budget constraints for a given temperature is muddied by the different timeframes and pathways to reaching this given temperature. For instance, the budget for keeping the global mean temperature rise to below 2 °C by a specific year in the future will be smaller than one associated with passing through that temperature goal, at the same date, en route to a higher temperature (Rogelj et al., 2016; Zickfeld et al. 2012). Assumptions regarding the level and timing of emissions of non-CO_2 greenhouse gases (some of which are inherently harder to mitigate than CO_2 emissions) have an important bearing on the size of the carbon budget. This adds further complexity to determining the available carbon budget associated with any given probability of staying below a 2 °C threshold. Moreover, there remain long-standing uncertainties surrounding how the global climate will respond to future emissions and resulting changes in the atmospheric CO_2 concentration (Freeman et al., 2015).

The IPCC's Fifth Assessment report (AR5) estimates different carbon budgets for given global mean temperatures over the twenty-first century based on an assessment of the likelihood of achieving that temperature. Expressed as probabilities, these are, in effect, the proportion of model runs included in AR5 for which global mean temperature remains below this threshold. Thus, the choice of probability of exceeding a particular temperature threshold further alters any chosen budget. For instance, having a high probability (>66%) of staying below a 2 °C temperature rise relates to a much smaller carbon budget over the century than choosing a lower chance (<33%) of staying below the 2 °C threshold. By the selection of probability, associated carbon budget and hence extent of mitigation and adaptation, society is making a choice as to how the problem of climate change will be approached. The Paris commitments on 2 °C and 1.5 °C and the carbon budgets associated with different probabilities of achieving the specific temperature increases embedded in the most recent IPCC report (IPCC, 2014a) offer a clear route for generating quantitative energy-system pathways commensurate with the Paris Agreement.

The Paris Agreement commitments can be converted into carbon budgets using the IPCC's guidance notes to the authors of their latest report (Mastrandrea et al., 2010). The language of the Agreement is very clear: to stay 'well below 2 °C' – and importantly to 'pursue efforts to limit the temperature increase to 1.5 °C'. We can translate this qualitative language of chance into quantitative probabilities – and therefore into carbon budgets. Based on the IPCC's Synthesis Report guidance, a carbon budget of between 850 and 1000 $GtCO_2$ for the period post-2011 equates to the temperature commitments in the Paris Agreement (IPCC, 2014a). The lower end of this range equates to an 'unlikely' chance of staying below 1.5 °C (i.e. a probability of 0 to 33% of <1.5 °C) with the upper end relating to a 'likely' chance of staying below 2 °C (i.e. a probability of 66–100% of <2 °C).

In order to estimate the global energy-only CO_2 budget for the remainder of the century (2016–2100), we need to calculate emissions for the last 5 years. The 850 to 1000 $GtCO_2$ carbon budget includes all carbon dioxide emissions from all sectors for the post-2011 period. Therefore, in order to understand the size of the budget available from 2016 onwards, it is necessary to subtract those emissions released between 2011 and 2016, since the budget was first published. Based on Carbon Dioxide Information Analysis Center (CDIAC) data (Le Quéré et al., 2016), extrapolated out to include 2015, at least 150 $GtCO_2$ have been emitted since 2011, leaving a range of 700 to 850 $GtCO_2$ for the period 2016 onwards.

In order to focus on the energy sector, it is necessary to remove projected deforestation and industrial process emissions (primarily cement) for the period 2016 onwards from the CO_2 budget. Based on research published in *Nature Geoscience* (Anderson, 2016), an optimistic interpretation of deforestation and cement process emissions for 2016 to 2100 are, respectively, in the region of 60 $GtCO_2$ and 150 $GtCO_2$. Both of these figures are based on the assumption that efforts to reduce emissions from deforestation and industrial processes are broadly in line with those required across the energy sector.

Combining recent emissions (i.e. those between 2011 and 2016) with those from deforestation and cement (process only) leaves an energy-only global CO_2 budget for 2016 onwards of 490 to 640 $GtCO_2$ (i.e. in the region of 500 to 650 $GtCO_2$). What would this mean for mitigation in lower-income (non-Annex 1) and higher-income (Annex 1) nations? This is undoubtedly an area where different interpretations of fairness and equity can give potentially very different results in terms of national carbon budgets. However, the Paris Agreement specifically acknowledges that the peak in emissions from the industrialising and poorer nations will be later than that within the wealthier industrial nations.

Assuming an aggregate peak in the industrialising nations' emissions in 2025, followed by a programme of rapidly ramping up mitigation rates in these countries to deliver mitigation of around 10% per annum by 2035, then the total emissions for post-2016 would be in the region of 550 to 600 $GtCO_2$. Put simply, a mitigation agenda across the industrialising and poorer nations at a level of ambition far beyond anything discussed in Paris would nevertheless leave, at best, only 50 to 100 $GtCO_2$ for the wealthier industrialised nations for 2016 onwards. This equates to a combined mitigation rate for these wealthier nations of between 13% and over 20% per annum, and starting immediately!

Recent data suggest that, since 2014, global emissions of CO_2 from fossil fuels, industrial processes and gas flaring have levelled off (Le Quéré et al., 2016). However, this plateauing has to be seen against the backdrop of rapidly rising emissions since the inception of the UNFCCC in 1992. Despite model and other analysis illustrating that it has been technically and physically possible to mitigate carbon emissions significantly, ambitious and effective mitigation has not materialised. Furthermore, the failure of the climate community to recognise and communicate clearly the scale of the mitigation effort required to deliver on earlier 2 °C targets has seemingly led to still tighter commitments in the Paris Agreement that are unlikely to be achieved. It is therefore perhaps unsurprising that, in order to make the 2 °C goal appear superficially viable, most modelling now assumes the rapid uptake of unproven negative emissions technologies applied at an unprecedented global scale. BECCS is particularly important in this context, and with its reliance on linking, at a large scale, bioenergy production with carbon capture and storage (CCS) with the consequences for food production, energy provision, energy system capacity and the environment (Vaughan and Gough, 2016; Smith et al., 2016), the feasibility of this dependence on BECCS is coming under increased scrutiny, as its significance to the Paris targets becomes better understood.

9.3 The Role of BECCS for Climate-Change Mitigation – A Summary of BECCS within Integrated Assessment Modelling

Given the unprecedented emissions reduction rates required to resolve the cumulative carbon budgets associated with 1.5 °C and 2 °C outlined in Section 9.2, whilst NETs are not explicitly discussed within the Paris Agreement, it implicitly relies on negative emissions to allow an overspend of carbon budgets or to compensate for sectors that are deemed harder to mitigate. Unsurprisingly, future emissions scenarios that achieve 1.5 °C are dependent on BECCS (Rogelj et al., 2015). There is also a growing and significant dependence on BECCS in future emission scenarios that do not exceed 2 °C warming. Only scenarios achieving concentration levels within 430–480 ppm by 2100 (and a small number of the scenarios extending to 530 ppm) were associated with a greater than 66% chance of achieving the policy goal of limiting global atmospheric temperature rise to below 2 °C (IPCC, 2014a); over a hundred of the 116 scenarios associated with concentrations between 430 and 480 ppm CO_2 include BECCS to deliver *global net* negative emissions in the IPCC Fifth Assessment Report (AR5) (Fuss et al., 2014). Global net negative emissions are achieved when the negative emissions associated with BECCS are greater than total emissions from *all* other sources (i.e. anthropogenic and non-anthropogenic) (Fuss et al., 2014). The median value for cumulative carbon removal using BECCS in these IPCC scenarios is around 616 GtCO$_2$ (168 GtC) by 2100 (Wiltshire et al., 2015). Scenarios that did not feature BECCS but still achieved 430–480 ppm CO$_2$ had a peak in emissions in 2010 (Wiltshire et al., 2015) – a benchmark already missed by 7 years, during which emissions have continued to accumulate. The use of BECCS is not confined to emissions scenarios that achieve 1.5 °C or 2 °C, however. Over 1000 emissions scenarios fed into the process through which the Representative Concentration Pathways (RCPs) were developed (Fifth Assessment Report of the IPCC (WG3)), including emissions pathways likely to exceed 1000 ppm CO$_{2eq}$ and, consequently, climate warming up to 5 °C and beyond. In an analysis of these higher-temperature pathways, Fuss et al. (2014) found that roughly half of all the scenarios include a significant contribution from BECCS.

In the context of contracting carbon budgets, the ability to achieve negative emissions by removing carbon dioxide from the atmosphere and storing it underground for extensive periods of time (beyond 1000 years) potentially enables the overspend of a carbon budget. There is a number of methods of carbon dioxide removal (CDR) but only BECCS has the co-benefit of significant energy provision combined with a potential scalability of >5 PgC/yr. CDR methods such as direct air capture and storage have this potential scalability but themselves require significant energy, whilst smaller-scale potential contributory CDR methods, such as biochar, have co-benefits such as improving some soil properties, i.e. water and nutrient retention (Vaughan and Lenton, 2011; Smith 2016). Thus, BECCS is the method of CDR that dominates IAM scenarios. Direct air capture with storage is included in some IAMs (e.g. Chen and Tavoni, 2013). Afforestation is also included in some but, as a CDR method, has limited scalability and is more vulnerable to unintended carbon loss through disease, pests and fire as well as potential impacts of future climate change (Vaughan and Lenton, 2011; Smith et al., 2016).

The option to overspend a carbon budget whilst generating energy makes BECCS an attractive approach for potentially enabling mitigation costs to be reduced, permitting more ambitious targets to become feasible than would otherwise be possible or allowing a delay in the year in which emissions peak and overshooting long-term concentration targets in the near term ('buying time') (Friedlingstein et al.; 2011; Huntingford et al., 2012; van Vuuren et al., 2013; Bernie and Lowe, 2014; Rogelj et al., 2015). That said, BECCS should not be seen as a substitute for direct mitigation measures – analysis by Kriegler et al. (2013) suggests that its key potential lies in providing cost savings in balancing sectors, such as transport, that are particularly challenging to abate. Taking aviation as an example, there exist few technical options for decarbonisation in the short-to-medium term. Longer-term options, such as hydrogen-powered aircraft, are far from commercialisation and require both technical innovation and large-scale deployment of capital-intensive infrastructure. At a global scale, there is continued growth in passenger kilometres, particularly within developing economies, and demand management is likely to be unpopular with travellers and is challenged by the industry. Balanced against these challenges, BECCS offers a theoretically appealing approach to compensating for the continued use of oil within transport sectors such as aviation.

However, the extent to which future negative emissions can compensate for overshoot is unclear and depends on the way the carbon cycle is modelled within IAMs. In many cases, simplified climate-carbon cycle models which are calibrated against more complex earth-system models (ESMs) are used within IAMs. The simple climate carbon cycle models are able to emulate the more complex ESMs (Jones et al., 2016) and are therefore an appropriate tool to model the required contribution of NETs for different emission scenarios. However, uncertainties remain due to the spread of responses from ESMs and processes that are not yet represented within ESMs such as release of carbon from thawing permafrost (Jones et al., 2016). ESMs demonstrate that understanding the behaviour of natural carbon sinks under low or negative emission scenarios is key to developing carbon budgets (Jones et al., 2016) and assessing whether negative emissions can revert global mean temperatures to 2 °C or below (Tokarska and Zickfeld, 2015). When there are global *net* negative emissions, anthropogenic removal of CO_2 from the atmosphere is offset to a degree by the outgassing of CO_2 from marine and terrestrial sinks. The greater the amount of global net negative emissions, the greater the outgassing, which in turn impacts the amount of negative emissions required to achieve a given removal of CO_2 from the atmosphere (Jones et al., 2016; Tokarska and Zickfeld, 2015). Tokarska and Zickfeld (2015) also demonstrate how, unlike global mean temperature, which decreases over centennial timescales after overshoot, sea levels continue to rise for several centuries despite CO_2 removal.

Assessing the options for mitigating climate change is the task of Working Group 3 (WG3) of the IPCC, who in the Fifth Assessment Report (AR5) describe four Representative Concentration Pathways (RCPs). These have been developed to represent the wide range of emissions scenarios from different sources published across the literature; the RCPs are presented as cumulative greenhouse gas (GHG) concentrations over time (1850–2100) and are associated with different levels of radiative forcing. The RCPs provide a consistent set of pathways for subsequent analysis in different areas of climate-change research – for example by climate modellers to analyse potential climate impacts associated with the pathways (including projected global average temperature rise) and in IAMs to explore alternative mitigation scenarios consistent with achieving the concentration pathways (IPCC, 2014a; van Vuuren et al., 2011). The recently published Shared Socio-Economic Pathways (SSPs) allow further exploration of possible futures by combining different global socio-economic futures with each of the RCPs (Riahi et al., 2017).

The RCPs are grouped according to the estimated radiative forcing due to GHG emissions in 2100. There are two stabilisation pathways, RCP6 (~850 ppm CO_{2eq}) and RCP4.5 (~650 ppm CO_{2eq}[1]), a high pathway, RCP8.5 (~1370 ppm CO_{2eq}) and a low pathway RCP2.6 (~450 ppm CO_{2eq} by 2100). This latter pathway, RCP2.6, includes an 'overshoot' whereby concentrations reach a peak of 490 ppm CO_{2eq} before declining by 2100; this peak-and-decline profile is achieved by including a negative emissions component based on deploying BECCS (Figure 9.1) and, to a lesser extent, afforestation measures (van Vuuren et al., 2011; van Vuuren and Riahi, 2011). The baseline scenario assumes

---

1 CO_2 concentrations may be reported as either CO_2 equivalents (CO_{2eq}) (*e.g.* RCPs and some IAMs), or as carbon dioxide (CO_2). CO_{2eq} can represent total greenhouse gas forcing or total anthropogenic forcing, *e.g.* including cooling due to aerosols. Here we quote figures that maintain the format of the original literature under review.

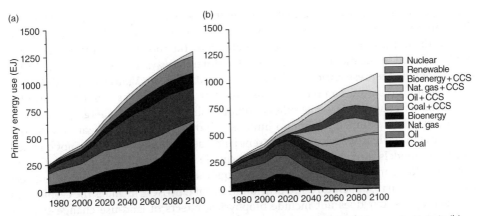

Figure 9.1 Trends in global energy use for (a) baseline scenario and (b) RCP2.6 scenario. Note in (b) BECCS use (bright green) from 2020, non-CCS bioenergy use (dark green) and fossil fuel with CCS use (blue, pale orange and grey). Source: van Vuuren et al., 2011, Figure 2 (p102). *(See insert for colour representation of this figure.)*

historical trends continue in the following decades, with growth in energy demand and the energy supply dominated by fossil fuels. Under this baseline scenario, CCS does not emerge for any fuel (van Vuuren et al., 2011).

The process of relating GHG concentrations to projected temperature rise is complex and carried out by models of the Earth's climate system. The IAMs used to produce emissions pathways are driven by economics-based decision-making approaches – and most deliver cost-minimising scenarios (i.e. emissions profiles that deliver a particular level of mitigation at the least aggregate cost within the set of constraints defined in the model) (IPCC WG3). Those pathways presented in the IPCC assessment that do not include negative emissions are estimated to be significantly more expensive in the model results (Fuss et al., 2014). However, there are very significant uncertainties associated with the use of BECCS as a carbon dioxide removal approach. Emissions pathways derived from IAMs all cover a global scale, and, although much of the data is regionally disaggregated, the assumptions driving the models are nevertheless developed at a top-down level. There is currently one integrated fossil CCS power plant in operation, Boundary Dam in Canada; a coal gasification plant with an expected capture capacity of 3 MTCO$_2$pa is due to open in 2017, and there is a single existing BECCS plant, a storage demonstration project using biogenic CO$_2$ from an ethanol facility, at Decatur, USA. Thus, most modelling assumptions are extrapolated from data often derived from desk-based analyses, or, at best, from small-scale demonstration plants; the models introduce large-scale estimates of globally significant levels of deployment with a vastly expanded biomass energy market. Furthermore, it remains uncertain as to whether the technology will be ready for the ambitious levels of BECCS deployment typically assumed from 2020 onwards (McGlashan et al., 2012; McLaren, 2012). Here, we summarise some of the broader assumptions made in IAMs around the potential role for BECCS.

IAMs create cost-optimal pathways to reach set targets, e.g. 2 °C, constrained by a wide range of assumptions about the global economy and energy systems, including future socio-economic assumptions relating to population, diets and living standards.

The upper limits of BECCS potential are set within each model as constraints, based on literature estimates; the amount of BECCS used in any one scenario is an output of the model run and will not exceed the levels defined by the model assumptions. The way BECCS is used in the models is constrained in different ways, for example some IAMs include detailed representations of land use (e.g. IMAGE, van Vuuren et al., 2011), and with input from more complex land-use models (e.g. LPJml as input to MAgPIE, Humpenöder et al., 2014), while others do not explicitly model bioenergy production, using instead an assumed maximum limit, e.g. 200 EJ/yr (e.g. GET, Azar et al., 2013; Kriegler et al., 2013). Models that represent land use, such as IMAGE, are not coupled to climate models directly, instead they use simple climate models (e.g. MAGICC) to generate global mean temperature and precipitation and pattern scaling to convert these into regional temperature and precipitation data. IMAGE includes the carbon and nitrogen cycles, but does not include biophysical effects of land-use change such as albedo (Bouwman et al., 2006).

The RCP2.6 scenario assumes 430–580 Mha is available for energy crop production; this is restricted to abandoned agricultural land and natural grassland systems (e.g. savannah, scrubland, tundra and grasslands) excluding areas with low potential such as parts of tundra and deserts (Hoogwijk, 2004; van Vuuren et al., 2009). To represent further constraints such as biodiversity protection, a land-cover accessibility factor is applied (De Vries et al., 2007). The biophysical potential is calculated using this land area and potential yields of sugar cane and woody biomass bioenergy crops. The technical potential allows for actual yields being lower than potential yields and limited conversion efficiency in the energy extraction process (van Vuuren et al., 2009). As with other IAMs, up to half of the bioenergy used is assumed to come from agriculture and forestry residues, thus having no direct impact on additional land demand (Daioglou et al., 2015; van Vuuren et al., 2013; Azar et al., 2013; Wise et al., 2009).

9.3.1 Key Assumptions

In this section, we identify the key assumptions that could be critical in modelling the contribution of BECCS to achieving climate-change mitigation targets (Table 9.1). Some assumptions are made explicitly and communicated in published papers, whilst others are implicit – either not published or tacitly made in constructing the models. The IAMs used to create scenarios all construct their models differently, with the inclusion, and more detailed representations, of different components of the global energy system and its contributing factors; as such they have different assumptions and representations of BECCS. Certain assumptions are easy to identify while others can be hard to decipher, especially as the models evolve over time. In Table 9.1 we describe explicit and implicit key assumptions, noting that many of these assumptions are strongly interconnected and interdependent.

9.4 Implications and Consequences of BECCS

When we move beyond the abstracted world of models, and begin to consider the real-world deployment of BECCS, the sheer complexity of the task required to achieve the emissions pathway outlined in Figure 9.1 soon becomes apparent; moreover, this is set

Table 9.1 Summary of the key uncertainties in future BECCS scenarios.

Assumptions	Details
Future climate change	The earth-system response to future climate change, and the impact on the carbon cycle, is implicitly assumed in the cumulative emissions budgets used in the IAM.
Bioenergy potential	
Agricultural efficiency gains	Assumed trends in agricultural efficiency gains impact the amount of available land as well as future bioenergy crop yields.
Land area requirement for BECCS	For models that explicitly represent land use, the land area available for energy crops is crucial. IAMs without this make an implicit assumption given by total bioenergy potential. Most scenarios assume abandoned agricultural land for dedicated bioenergy crops, to avoid issues of food production, biodiversity and land-use change carbon emissions.
Crop yields	Most scenarios focus on dedicated lignocellulosic crops and assume productivity levels in keeping with abandoned agricultural land (i.e. associated with lower yields than agricultural land). Scenarios have differing fertiliser and irrigation assumptions: most assume rain-fed land. Improvements are achieved through the use of irrigation and/or fertiliser, with an associated trade-off with N_2O emissions and embedded carbon, and technological developments.
Residue availability	Many scenarios include residues as well as dedicated lignocellulosic crops. Residue availability is dependent upon the types and levels of socio-economic activity.
Infrastructure	Transport infrastructure for biomass and purpose-built biomass energy generation plants. Some negative emissions can be achieved through co-firing, but most IAMs assume purpose-built biomass energy generation plants.
CCS capability	
Maximum annual rate of CO_2 stored	This refers to the amount of CO_2 stored annually in geological formations. To equate to CO_2 removal, or project-level negative emissions, this assumes that CO_2 is captured only from dedicated biomass plant.
CCS infrastructure	A strong implicit assumption is that CCS infrastructure is established and available to capture, transport and store CO_2. CCS technology is currently entering the demonstration phase, with very limited experience of dedicated BECCS systems.
CO_2 storage capacity	Total storage capacity in suitable reservoirs. Estimates of technical potential capacity in appropriate geological formations cover a very large range; IAMs typically incorporate an assumption of how much of this will be suitable for secure storage. While storage potential in hydrocarbon fields is relatively well quantified, the size of the potential for large-scale, long-term storage in saline remains uncertain, with large regional variations in uncertainty in capacity estimates.
BECCS	
BECCS as a % of primary energy	IAMs in WG3 cluster around 20–30% of total primary energy from BECCS, although there are extreme outliers.
Cost of BECCS per t CO_2 stored	As IAMs typically optimise on cost, the relative costs of different mitigation options is an important driver. Assumptions lie in the range of 60 to 250 US$/t CO_2 (IPCC, 2014b).

(Continued)

Table 9.1 (Continued)

Assumptions	Details
Policy support	Sufficient and effective policy and governance frameworks and incentives are a prerequisite to developing and establishing BECCS technology.
Net negative emissions	Assumed net negative emissions across the full life cycle of the BECCS system; includes large uncertainties in bioenergy production, e.g. direct and indirect land-use change, fertiliser use and water availability.
Political and socio-economic	
Population, lifestyle, diets	This is a key overarching set of assumptions that feed into the agricultural assumptions that underpin the bioenergy potential as well as establishing global energy demand.
Sustainable land use	IAMs that strive for sustainable bioenergy include assumptions about land areas that are not available for bioenergy, such as primary forest and food production.
Social acceptability	Most IAMs do not consider social acceptability, although some express this in terms of a scaling down of technical potential. This is relevant across the entire BECCS supply chain.
Global participation	Most scenarios assume global participation in emissions reductions.
Carbon price (or equivalent)	IAMs assume that an effective (global) carbon pricing mechanism exists.
Global governance system	A BECCS supply chain will incorporate a diverse mix of nations, regions, technologies and actors, which will require a coordinated regulatory framework in order to deliver, verify and account for negative emissions.

within a programme of significant mitigation across the complete energy system and the associated demand for a skilled workforce, knowledge, funding, expertise, regulation, etc. What is immediately striking is that the deployment of negative emissions at scale, and on which the remaining IAM carbon budget is reliant, is typically very early, with some IAMs assuming BECCS as early as 2020 to 2025. BECCS requires the integration of three distinct elements: (i) a biomass supply chain (production, processing and transport); (ii) energy generation and (iii) a CCS facility and infrastructure (capture, transport and CO_2 storage). As of 2017, however, commercial experience of BECCS is limited to the storage of biogenic CO_2 for the purposes of reservoir demonstration (Global CCS Institute, 2017). Limited commercial operation is not just the case for BECCS, but also for storage of fossil CO_2, where despite significant policy attention paid to the technology there is a legacy of failed and ceased projects (Reiner, 2015), with one operational integrated power plant and a coal gasification facility due to open in the United States in 2017.

Table 9.1 illustrates the challenges from an infrastructure, policy and global governance perspective that need to be overcome if sustainable biomass supply chains are to be established and an extensive CO_2 transport and storage infrastructure developed. It is useful to reflect as well on the inherent uncertainty within these assumptions (Vaughan and Gough, 2016). While the emissions pathways derived from IAMs rely, in part, on regionally disaggregated data, they nevertheless reflect a more global and top-down framing of mitigation. That said, the assumptions made within more bottom-up studies (IEAGHG, 2011) are equally open to examination. In the IEA report 'Potential for Biomass with Carbon Capture and Storage' (IEAGHG, 2011), the technically feasible potential for BECCS is estimated at 10 $GtCO_2$ per annum (and 47 EJ of primary energy from biomass) by 2050 for a dedicated biomass and CCS route with biomass-integrated gasification combined cycle (BIGCC) and circulating fluidised bed (CFB) technology. The economic potential, however, is assumed to be much less at 3.5 $GtCO_2$ in 2050 (20 EJ of primary energy from biomass) for gasification routes. These figures are based on IEA estimates of deployment rates of CO_2 transport infrastructure and known storage reservoirs as well as projections of the sustainable supply of biomass. Contrasting the IEA projections of BECCS with those not untypical in the IAMs demonstrates how sensitive emissions pathways are to assumptions of the uptake of what is currently very much a conceptual technology. The assumed primary energy for BECCS under the RCP2.6 pathway in Figure 9.1b suggests in the region of 70 EJ are required by 2050, higher than even the technical potential from the IEA study.

There is technological diversity across each element of the BECCS chain with a variety of options for combining biomass energy with CCS, although, as already highlighted, dedicated biomass electricity generation offers the greatest potential for delivering negative emissions (IEAGHG, 2011). Routes to dedicated 100% biomass electricity generation are offered via combustion in fluidised beds, gasification and biomass chemical looping. Of these and at relatively small scale (up to 100 MW), biomass combustion (CFB) is already commercial. Gasification of either coal with biomass or dedicated biomass is at a smaller scale and less commercially proven. Biomass-based chemical looping remains a longer-term technology. In the context of BECCS applied to power generation, systems may be either co-fired (with conventional fossil fuels) or dedicated biomass; however, to achieve net negative emissions, co-firing applications above 20–30% are required (IEAGHG, 2011), which require technical challenges to be overcome. Scaling up and integration not only of technology but also across the wider socio-technical system including the skills, industries and standards adds further complexity and uncertainty.

Taking the requirement of 70 EJ per annum of dedicated BECCS by 2050, the assumption that deployment of dedicated BECCS power plants of a capacity of 1 to 2 GW, and with a high annual load factor could begin in 2025, would suggest construction and commissioning up to two large BECCS plants each week for a quarter of a century. This ambitious scale of roll-out has to be set against the backdrop of the Marshall-style construction programme that RCP2.6 assumes across the complete energy system. For example, the fossil CCS uptake outlined in Figure 9.1 is over twice that assumed for BECCS – equivalent to perhaps a coal or gas CCS plant coming online every day or two for 25 years – as well as two BECCS plants per week. Clearly this pace of deployment has to be considered in light of the recent withdrawal of financial support by the governments of countries initially seen as pioneering CCS technology, such as the United Kingdom and the United States, with perhaps a maximum of three integrated CCS projects likely by 2020 (Reiner, 2015).

Moving on to consider the storage of CO_2, global estimates of CO_2 storage capacity vary hugely from 100 to 10 000s $GtCO_2$, (Bradshaw et al., 2007), a range it is hard to base policy upon without further assessment, with our knowledge of potential carbon sinks lagging far behind the understanding of carbon stores (Scott et al., 2015). There are numerous bottom-up assessments of national capacity (for example British Geological

Survey & The Crown Estate, 2013 for the UK; Stewart et al., 2014 for the EU), though as with global assessments, there are large uncertainties associated with these estimates. These uncertainties arise from the availability of data (for example geological data on the reservoir characteristics), the assumptions required to make these estimates (for example about trapping mechanisms) and the extent to which theoretical resources can be practically or economically exploited (Bradshaw et al., 2007). The two most promising types of storage formation for CO_2 storage are depleted oil and gas reservoirs and saline aquifers. Depleted oil and gas reservoirs are, for the most part, portions of saline formations which have stored oil and gas for millions of years, therefore have proven trapping mechanisms and reliable seals. As a result of exploration for, and production of, oil and gas, most of these reservoirs are better characterised compared to saline aquifers. There are deep saline aquifers across every part of the world, and although they are less well characterised than depleted oil and gas reservoirs, because of their large volume and wide distribution, deep saline formations have the greatest potential storage capacity.

Whilst the availability of safe and secure CO_2 storage capacity is independent of the source of the CO_2, to date much discussion related to total potential storage has taken place within a fossil CCS discourse, where CO_2 storage is a bridge towards a long-term goal of decarbonisation and the move away from fossil fuels. BECCS, however, is not typically framed in such time-limited terms (Vaughan and Gough, 2016), and, therefore, total usable storage capacity becomes important in terms of how long a significant reliance on BECCS can be sustained, and the possibility that deployment rates for BECCS are limited by the rate at which CO_2 storage can be exploited.

The reliability of reservoir capacity assessments notwithstanding, providing enough storage infrastructure fast enough to meet the storage requirements at the levels suggested within the IAM modelling will also be challenging. Bellona (2014) models the deployment of storage for 12.8 $GtCO_2$ to be captured and stored, in Europe, by 2050, with capture starting in 2025. In order to meet these storage requirements, in the region of 19 storage sites are required in 2030, 85 in 2040 and 127 in 2050. To achieve these levels, characterisation of storage sites, which is estimated to take 9.5 years, needed to have commenced in 2014 (Bellona, 2014). This work highlighted the importance of the injectivity of reservoirs (rate at which CO_2 can be injected) and the potential limits on rates of exploitation – with a corresponding increase in the cost of storage – should more wells need to be drilled to maintain injectivity. Injectivity will be affected by a number of real-world challenges including the geometry of traps, pressure increases and interactions between the operations of different storage operators injecting into the same reservoir, which may lead to pressure wave interferences. Even the amount of CO_2 storage infrastructure required to store a total of 12.8 $GtCO_2$, which is dwarfed by the requirements to deliver the 2–10 $GtCO_2$/yr within AR5 scenarios, is estimated to need annual investments in the region of £500 million by 2020 (Bellona, 2014) and a functioning investment environment to encourage CO_2 storage investors. Furthermore, in the United Kingdom, storage will be offshore and will require new pipeline infrastructure between capture and storage sites in addition to any transboundary transport required for any CO_2 storage network (Stewart et al., 2014). Transport will also be a problem: although CO_2 is currently transported by pipeline (notably in the United States), large-scale CO_2 storage would require a new pipeline infrastructure to be developed with a potential capacity to transport up to

10 $GtCO_2$/yr – the current gas pipeline infrastructure transports the equivalent of around 1.5 GtC (van Vuuren et al., 2010).

Whilst acknowledging that BECCS is not a blue-skies mitigation option, it has to be recognised that it is in the early or conceptual stages of development. It is also clear that in order to address the climate mitigation challenge a wholesale and rapid transition to a new, low-carbon energy system is required. This raises the question of whether BECCS can deliver what is anticipated of it within existing climate-change policy frameworks, and, indeed, of the desirability of the potential environmental, social and ethical challenges associated with BECCS, as outlined in Chapter 12.

9.5 Conclusions: Can BECCS Deliver what's Expected of it?

Given the gap between Paris Agreement ambitions and the reality of the remaining carbon budget, as outlined in Section 9.2, it is perhaps unsurprising that the IAMs, which are influential in informing policy, are strongly reliant on negative emissions technologies as a means of expanding the available carbon budget space. It is in part the outputs of these, typically cost-optimised, IAMs that have persuaded policy-makers (amongst many others) that green growth, win-win opportunities and incremental, but nonetheless important, changes, rather than more radical societal changes and associated emissions reduction, can deliver on the Paris commitments.

Within IAM outputs, one particular negative emissions technology, namely BECCS, makes a significant contribution (10 $GtCO_2$/yr from 2050, van Vuuren et al., 2011); this is adopted at a global scale, capturing and transporting hundreds of billions of tonnes of CO_2 from the atmosphere into long-term and secure storage. Although as of 2017 there are only a handful of demonstration plants and one commercial plant in operation, it is in many respects a sociotechnical imaginary (Jasanoff and Kim, 2009), an approach to mitigation which is at a very early stage of maturity, yet one which policy-makers are relying upon and one which legitimises the view that climate change can be solved within our current energy-system paradigm and without more fundamental changes to our current economic and societal structures. In this sense, BECCS is a technical fix for the complex social, political and economic challenge of climate-change mitigation. Whilst such a solution may appear to offer hope that the ambitions of the Paris Agreement can be achieved, the challenge of its implementation may mean that this hope is offered on a false promise, with the high risk that mitigation and adaptation plans are made on an unrealistic basis as a consequence.

BECCS entails making more efficient use of existing biomass resources, at the same time utilising agricultural and forestry residues and increasing production of bioenergy crops, which absorb carbon dioxide through photosynthesis. Given the global nature of the bioenergy economy, once harvested and processed, it is likely that biomass energy resources for BECCS would be shipped all around the world for use in power stations. Following CO_2 capture, CO_2 is compressed (such that it has the properties of a liquid), pumped through pipeline networks or shipped over potentially very long distances and finally stored deep underground in various geological formations (depleted oil and gas reservoirs and saline aquifers) for a millennium or so. As articulated in Chapters 5 and 7 of this book, there is an energy penalty associated with CO_2 capture and for other stages of the process; these, and other assumptions concerning variation in crop yield or capture efficiency, have to be accounted for through the LCA approaches described in Chapter 6 to determine realisable negative emissions. Given that a BECCS supply chain will incorporate a diverse mix of nations, regions, technologies and actors, a globally coordinated regulatory framework will be required in order to deliver, verify and account for negative emissions, with a value placed on negative emissions in order to support deployment (see Chapter 11).

The sheer scale of BECCS assumed in many of the IAM scenarios, and consequently the Paris Agreement temperature commitment, is enormous. The demand for bioenergy for BECCS must be met alongside anticipated future demand for bioenergy from other sectors as a renewable energy source, noting that both the aviation and shipping industries anticipate using biofuels, and biomass is a potential feedstock for the chemical sector. Whilst, as highlighted in Table 9.1, IAMs assume a sustainable use of biomass to limit impacts on food production and natural habitats, this will require appropriate governance mechanisms with sustainability safeguards to avoid additional pressure on natural habitats.

Despite its implicit reliance on negative emissions to deliver its stated goals, there is no reference to either NETs or BECCS throughout the 32 pages of the Paris Agreement. Only very recently has this assumption, which underpins the political and economic optimism on climate change, been subject to discussion and debate (see for example Fuss et al., 2014; Lomax et al., 2015). Consequently, there is a need for a candid, transparent and scientifically informed dialogue that includes a diversity of perspectives across policy, scientific, business and civil-society actors; this needs to be open and honest about the profound political and economic changes required if BECCS is to be deployed at the scale implied by the technical futures that are currently shaping climate-change policy.

References

Anderson, K. (2016). Duality in climate science. *Nature Geoscience* **8**: 898–900.

Azar, C., Johansson, D.J.A., and Mattsson, N. (2013). Meeting global temperature targets: the role of bioenergy with carbon capture and storage. *Environmental Research Letters* **8**: 034004.

Bellona (2014). *Scaling the CO_2 storage industry: a study and a tool.* http://bellona.org/assets/sites/4/Scaling-the-CO2-storage-industry_Bellona-Europa.pdf (accessed 31 May 2017).

Bernie, D. and Lowe, J.A. (2014). *Future Temperature Responses Based on IPCC and Other Existing Emissions Scenarios. AVOID2 Report WPA1.*

Bouwman, L., Kram, T., and Klein-Goldewijk, K. (2006). *Integrated Modelling of Global Environmental Change. An Overview of IMAGE 2.4.* Bilthoven: Netherlands Environmental Assessment Agency.

Bradshaw, J., Bachu, S., Bonijoly, D. et al. (2007). CO_2 storage capacity estimation: issues and development of standards. *International Journal of Greenhouse Gas Control* **1**: 62–68.

British Geological Survey & The Crown Estate (2013). CO_2 *stored.* http://www.co2stored. co.uk (accessed 31 May 2017).

Chen, C. and Tavoni, M. (2013). Direct air capture of CO_2 and climate stabilization: a model based assessment. *Climatic Change* **118**: 59.

Daioglou, V., Stehfest, E., Wicke, B. et al. (2015). Projections of the availability and cost of residues from agriculture and forestry. *GCB Bioenergy* **8**: 456–470. doi: 10.1111/gcbb.12285.

De Vries, B.J.M., van Vuuren, D.P., and Hoogwijk, M.M. (2007). Renewable energy sources: their global potential for the first half of the 21st century at a global level: an integrated approach. *Energy Policy* **35**: 2590–2610.

Freeman, M.C., Wagner, G., and Zeckhauser, R.J. (2015). Climate sensitivity uncertainty: when is good news bad? *Philosophical Transactions of the Royal Society A: Mathematical, Physical and Engineering Sciences* **373**: 2055.

Friedlingstein, P., Solomon, S., Plattner, G.K. et al. (2011). Long-term climate implications of twenty-first century options for carbon dioxide emission mitigation. *Nature Climate Change* **1**: 457–461.

Fuss, S., Canadell, J.G., Peters, G.P. et al. (2014). Betting on negative emissions. *Nature Climate Change* **4**: 850–853.

Global CCS Institute (2017). *Illinois Basin Decatur Project (CO_2 injection completed, monitoring ongoing)*. https://www.globalccsinstitute.com/projects/illinois-basin-decatur-project (accessed 31 May 2017).

Hoogwijk, M. (2004). *On the Global and Regional Potential of Renewable Energy Sources*. Utrecht University. ISBN: 90-393-3640-7.

Humpenöder, F., Popp, A., Dietrich, J.P. et al. (2014). Investigating afforestation and bioenergy CCS as climate change mitigation strategies. *Environmental Research Letters* **9**: 6.

Huntingford, C., Lowe, J.A., Gohar, L.K. et al. (2012). The link between a global 2°C warming threshold and emissions in years 2020, 2050 and beyond. *Environmental Research Letters* **7**: 014039.

IEAGHG (2011). *Potential for Biomass with Carbon Capture and Storage in 2011/06*, July 2011 IEA Greenhouse Gas R & D Programme.

IPCC (2014a). *Climate Change 2014: Mitigation of Climate Change. Contribution of Working Group III to the Fifth Assessment Report of the Intergovernmental Panel on Climate Change* (ed. O. Edenhofer, R. Pichs-Madruga, Y. Sokona, et al.). Cambridge: Cambridge University Press.

IPCC (2014b). *Climate Change 2014: Synthesis Report, Contribution of Working Groups I, II and III to the Fifth Assessment Report of the Intergovernmental Panel on Climate Change* (ed. Core Writing Team, R.K. Pachauri and L.A. Meyer), 151. Geneva: IPCC.

Jasanoff, S. and Kim, S.-H. (2009). Containing the atom: sociotechnical imaginaries and nuclear power in the united states and South Korea. *Minverva* **47**: 119–146.

Jones, C.D., Ciais, P., Davis, S.J. et al. (2016). Simulating the earth system response to negative emissions. *Environmental Research Letters* **11**: 095012.

Jordan, A., Raynor, T., Schroeder, H. et al. (2013). Beyond 2 degrees: the risks and opportunities of different options. *Climate Policy* **13** (6): 751–769.

Keith, D.W. (2001). Sinks, energy crops and land use: coherent climate policy demands an integrated analysis of biomass. *Climatic Change* **49**: 1–10.

Kriegler, E., Edenhofer, O., Reuster, L. et al. (2013). Is atmospheric carbon dioxide removal a game changer for climate change mitigation? *Climatic Change* **118**: 45–57.

Le Quéré, C., Andrew, R.M., Canadell, J.G. et al. (2016). Global carbon budget 2016. *Earth System Science Data* **8**: 605–649. doi: 10.5194/essd-8-605-2016.

Lomax, G., Lenton, T.M., Adeosun, A., and Workman, M. (2015). Investing in negative emissions. *Nature Climate Change* **5**: 498–500.

Mastrandrea, M.D., Field, C.B., Stocker, T.F., et al. (2010). Guidance note for lead authors of the IPCC fifth assessment report on consistent treatment of uncertainties. *Intergovernmental Panel on Climate Change (IPCC)* https://www.ipcc.ch/pdf/supporting-material/uncertainty-guidance-note.pdf (accessed 14 December 2017).

Matthews, H.D., Gillett, N.P., Stott, P.A., and Zickfeld, K. (2009). The proportionality of global warming to cumulative carbon emissions. *Nature* **459**: 829–832.

McGlashan, N., Shah, N., Caldecott, B., and Workman, M. (2012). High-level techno-economic assessment of negative emissions technologies. *Process Safety and Environmental Protection* **90**: 501–510.

McLaren, D. (2012). A comparative global assessment of potential negative emissions technologies. *Process Safety and Environmental Protection* **90**: 489–500.

Mollersten, K., Westermark, J., and Yan, M. (2003). Potential and cost-effectiveness of CO_2 reductions through energy measures in Swedish pulp and paper mills. *Energy* **28**: 691–710.

Obersteiner, M., Azar, C., Kauppi, P. et al. (2001). Managing climate risk. *Science* **294** (5543): 786–787.

Reiner, D. (2015). Where can I go to see one? Risk communications for an "imaginary" technology. *Journal of Risk Research* **18**: 710–713.

Riahi, K., van Vuuren, D.P., Kriegler, E. et al. (2017). The shared socio-economic pathways and their energy, land use, and greenhouse gas emissions implications: an overview. *Global Environmental Change* **42**: 153–168.

Rogelj, J., Luderer, G., Pietzcker, R.C. et al. (2015). Energy system transformations for limiting end of century warming to below 1.5°C. *Nature Climate Change* **5**: 519–527.

Scott, V., Haszeldine, R.S., Tett, S.F.B., and Oschlies, A. (2015). Fossil fuels in a trillion tonne world. *Nature Climate Change* **5**: 419–423.

Sharmina, M., Anderson, K., and Bow-Larkin, A. (2013). Climate change regional review: Russia. *Wiley Interdisciplinary Reviews: Climate Change* **4** (5): 373–396.

Smith, P. (2016). Soil carbon sequestration and biochar as negative emission technologies. *Global Change Biology* **22**: 1315–1324.

Smith, P., Davis, S.J., Creutzig, F. et al. (2016). Biophysical and economic limits to negative CO_2 emissions. *Nature Climate Change* **6**: 42–50.

Stewart, R.J., Scott, V., Haszeldine, R.S. et al. (2014). The feasibility of a European-wide integrated CO_2 transport network. *Greenhouse Gases: Science and Technology* **4** (4): 481–494.

Tokarska, K.B. and Zickfeld, K. (2015). The effectiveness of net negative carbon dioxide emissions in reversing anthropogenic climate change. *Environmental Research Letters* **10**: 9.

Vaughan, N.E. and Gough, C. (2016). Expert assessment concludes negative emissions scenarios may not deliver. *Environmental Research Letters* **11** (9): 095003.

Vaughan, N. and Lenton, T. (2011). A review of climate geoengineering proposals. *Climatic Change* **109**: 745–790.

van Vuuren, D.P., Deetman, S., van Vliet, J. et al. (2013). The role of negative CO_2 emissions for reaching 2°C – insights from integrated assessment modelling. *Climatic Change* **118** (1): 15–27.

van Vuuren, D.P., den Elzen, M.G., Lucas, P.L. et al. (2007). Stabilizing greenhouse gas concentrations at low levels: an assessment of reduction strategies and costs. *Climatic Change* **81**: 119. doi: 10.1007/s10584-006-9172-9.

van Vuuren, D., Edmonds, J., Kainuma, M. et al. (2011). The representative concentration pathways: an overview. *Climatic Change* **109**: 5–31.

van Vuuren, D. and Riahi, K. (2011). The relationship between short-term emissions and long-term concentration targets. *Climatic Change* **104**: 793–801.

van Vuuren, D., Stehfest, E., den Elzen, M.J. et al. (2011). RCP2.6: exploring the possibility to keep global mean temperature increase below 2°C. *Climatic Change* **109**: 95–116.

van Vuuren, D.P., Stehfest, E., den Elzen, M.G.J., van Vliet, J., and Isaac M. (2010). Exploring IMAGE model scenarios that keep greenhouse gas radiative forcing below 3 W/m2 in 2100. *Energy Economics* **32**(5): 1105–1120.

van Vuuren, D.P., van Vliet, J., and Stehfest, E. (2009). Future bio-energy potential under various natural constraints. *Energy Policy* **37**: 4220–4230.

Wiltshire, A., Davies-Bernard, T., and Jones, C.D. (2015) *Planetary limits to Bio-Energy Carbon Capture and Storage (BECCS) negative emissions. AVOID2 Report WPD2a.*

Wise, M., Calvin, K., Thomson, A. et al. (2009). Implications of limiting CO_2 concentrations for land use and energy. *Science* **324**: 1183–1186.

Zickfeld, K., Arora, V.K., and Gillett, N.P. (2012). Is the climate response to CO_2 emissions path dependent? *Geophysical Research Letters* **39**: L05703.

10

The Future for Bioenergy Systems: The Role of BECCS?

Gabrial Anandarajah[1], Olivier Dessens[1] and Will McDowall[2]

[1] UCL Energy Institute, University College London, UK
[2] UCL Institute for Sustainable Resources, University College London, UK

10.1   Introduction

The countries that met at 21st Conference of Parties (COP21) in Paris agreed to hold the increase in the global average temperature to well below 2°C above pre-industrial level and also make efforts to limit the temperature increase to 1.5°C. These are ambitious targets, and while many low-carbon resources and technologies are available, there are major limitations in the plausible rate at which human societies can deploy them, due to a range of technical, financial, political and social constraints.

The challenges of meeting 2 or 1.5°C targets are sufficiently great that many believe that low- or zero-carbon technologies are not enough: negative emissions technologies (NETs) are required. NETs are a disparate group of methods that have been proposed for removing CO_2 from the atmosphere with the objective of limiting climate change. Some have argued that NETs are crucial to meet climate targets (Rogelj et al., 2015; Dessens et al., 2016) as the global carbon budget left during the next 85 years is very limited, and a large portion will be used up during the next few decades, due to locked-in investments. Several NETs are discussed in the literature, such as afforestation, biochar, biomass with CCS (BECCS), direct air capture (DAC), oceanic and terrestrial enhanced weathering, land-use management and ocean fertilisation, among others (McLaren, 2012).

Among NETs, BECCS has received most attention. However, uncertainties remain in two main areas that affect the cost-effectiveness and role of this technology to meet global climate targets: the availability of biomass, which is affected by many inter-linked factors, including availability and suitability of land for biomass production; and sustainability of bioenergy production. Sustainability of the biomass is not limited to greenhouse gas (GHG) emissions: there are several issues around its sustainability such as environmental emissions, natural capital, social values, ecosystem services, biodiversity. But these are beyond the scope of this study. The large diversity of options and feedstocks available for bioenergy accentuates the complexity of the issue of

Biomass Energy with Carbon Capture and Storage (BECCS): Unlocking Negative Emissions, First Edition.
Edited by Clair Gough, Patricia Thornley, Sarah Mander, Naomi Vaughan and Amanda Lea-Langton.
© 2018 John Wiley & Sons Ltd. Published 2018 by John Wiley & Sons Ltd.

sustainability. Biomass can only be a useful element of the energy system in the future if it is economically, socially and environmentally sustainable. Sustainability indicators include soil quality, water quality and quantity, GHG emissions, biodiversity, air quality, food competition, productivity, economic competitiveness and global equity. However, we develop several scenarios by varying the availability of biomass under 2 °C and 1.5 °C climate-change mitigation targets at a global level to analyse the possible limitation on biomass production under sustainable goals.

This chapter uses the TIAM-UCL global energy-system model to analyse the role of BECCS to meet the global climate-change mitigation targets in line with the Paris Agreement. The chapter is divided into four sections. Following the introduction, Section 10.2 presents the tool used for modelling, and the representation of biomass and CCS in the model and defines the scenarios; Section 10.3 presents and discusses the model results, especially the role of BECCS to meet 2 °C and 1.5 °C scenarios and Section 10.4 concludes the findings.

Optimisation modelling of this kind provides a clear and relatively simple conceptual framework with which to explore the techno-economics of alternative possible technology options. It should be clear that neither the dynamics of the model (which assumes perfect foresight and implicitly represents the view of a single global social planner) nor the input assumptions (relating to long-term forecasts of the characteristics of technologies, which are clearly deeply uncertain) support crisp quantitative predictions. Rather, the aim of such tools is to enable insights into the potential orders of magnitude and relative importance of techno-economic factors.

10.2 Methodology

10.2.1 TIAM-UCL

TIAM-UCL is a whole energy-system model covering energy resources, conversion, infrastructure and end-use sectors (Loulou and Labriet, 2007). The model has been developed under the UK Energy Research Centre (UKERC) Phase II project (Anandarajah et al., 2011) and enhanced under different projects (McGlade et al., 2015; McGlade and Ekins, 2015). It is a linear-programming-based partial-equilibrium model that maximises societal welfare (defined as the sum of consumer and producer surplus). The model thus identifies the optimal energy-system pathway subject to constraints such as carbon targets (and 'optimal' is defined in techno-economic terms). Within the model, the world is divided into 16 geopolitical regions of different size (from single nation state to groups of different countries). Base-year energy-service demand is exogenous and future projections are based on drivers such as GDP, population, household size and sectoral outputs. The base-year (2005) primary energy consumption, energy conversion and final consumptions are calibrated to the IEA Energy Balance at sector and sub-sector levels. A simplified representation of the TIAM-UCL model structure is presented in Figure 10.1. The world regions are linked through the trade in crude oil, hard coal, pipeline gas, liquefied natural gas (LNG), petroleum products (diesel, gasoline, naphtha, heavy fuel oil), energy crops, solid biomass and emissions credits.

Biomass is modelled in TIAM-UCL from resources to conversion to end-use devices. Regions in the model can trade energy crops, solid biomass, biodiesel and other

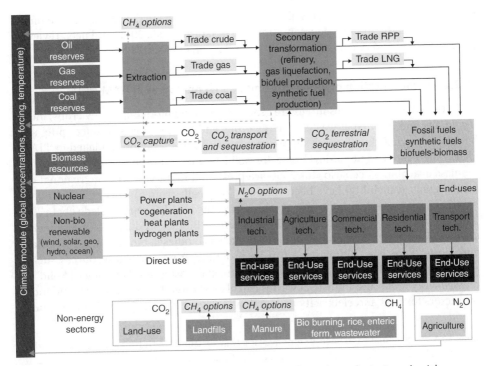

Figure 10.1 TIAM-UCL energy-system structure (for more information refer to Anandarajah et al., 2011).

bioproducts in addition to fossil fuels. Biomass is available for electricity and heat production with and without carbon capture and storage (CCS) and also in biofuel production with CCS. As presently represented in the model, biomass with CCS will always yield a negative net emission from the process when it is dedicated biomass with CCS for electricity or heat production.

End-use technologies are modelled at a detailed level in each end-use sector. For example, the transport sector is divided into passenger versus goods transport, as well as by modes (car, bus, train, air, ship, HGV, LGV, etc.). Vehicle technologies such as internal combustion engines (gasoline, diesel, CNG, LPG), hybrids, plug-in hybrids, electric vehicles and fuel-cell vehicles are modelled in each transport mode where appropriate. TIAM-UCL also has a climate module, which calculates impacts on atmosphere – CO_2 and other GHG emissions concentrations, radiative forcing and temperature changes – and can be constrained to a particular maximum temperature rise, such as 2 °C.

In addition to the global social discount rate of 3.5%, TIAM-UCL includes various hurdle rates. TIAM-UCL uses hurdle rates to calculate the equivalent annual capital cost (annualised capital cost) of a technology during its lifetime. The higher the hurdle rate, the higher the equivalent annual capital cost. The model will use the global discount rate to calculate the annualised capital cost for a technology if a hurdle rate was not modelled for the technology. Hurdle rates are higher than the social discount rate to represent market failures, barriers, consumer preferences, etc., for sector-specific technologies. These hurdle rates also vary across regions. Further details of the model are available in the model documentation (Anandarajah et al., 2011) and peer-reviewed papers (Kesicki and Anandarajah, 2011; Anandarajah et al., 2013).

10.2.2 Representation of Bioenergy and CCS Technologies in TIAM-UCL

In TIAM-UCL, biomass resources are grouped into solid biomass, energy crops, industry wastes and municipal wastes. Energy crops and solid biomass resource potentials have been modelled with three supply-cost curves for each region modelled in TIAM-UCL. Aggregated regional-level data for solid biomass and energy crops data (high biomass availability scenario) for 2050 are presented in Table 10.1. The regional data (costs versus availability) have been extracted originally from Smeets et al. (2004), but have been reassessed with the global values from the UKERC report (Slade et al., 2011) *'Energy from Biomass: the Size of the Global Resource'.* Biofuel productions (bio-refineries) are modelled to produce a range of biofuels such as biodiesel, bio-kerosene, bioethanol and bio-jet kerosene. Biomethane can be produced from various resources such as energy crops, solid biomass, industry and municipal wastes and landfill gas. Solid biomass, energy crops and industrial wastes can also be directly used in power and industry sectors for heat and electricity production involving various technologies including

Table 10.1 Solid biomass and energy crops availability in 2050 in TIAM-UCL (high biomass scenario).

TIAM_UCL region	Biomass feedstock type (PJ)	
	Solid biomass	Energy crops
AFR	19 300	9000
AUS	1974	13 000
CAN	2020	6000
CHI	10 323	5000
CSA	15 130	17 000
EEU	1196	5700
FSU	2961	43 000
IND	7030	5000
JPN	75	100
MEA	276	1000
MEX	1462	2000
ODA	1953	6000
SKO	105	100
UK	270	400
USA	5044	16 400
WEU	4956	6400

Results are presented in PJ (10^{15} J) for 16 regions: Africa (AFR), Australia (AUS), Canada (CAN), China (CHI), etc. (for a full list of regions, refer to Anandarajah, 2011).

combined heat and power (CHP) technologies. Biomass technologies compete directly at energy-service demand level with fossil-fuel technologies to meet energy services (such as residential, industrial and commercial heating demand, transportation and residential cooking).

The CCS technology has been modelled with various biomass technologies upstream for biofuel production (Fischer–Tropsch process) and hydrogen production, in the power sector for electricity and heat production, and in the industry sector with large-scale heat production. Biomass is also modelled as a co-firing fuel (with coal) with CCS for electricity, heat, hydrogen and biofuel production with various technologies. CCS technologies become available in the scenarios from 2025 onwards. The assumptions about costs of the technologies as well as the storage potential are extracted from the IEA *Technology Roadmap* report (IEA, 2013) and the IPCC special report on carbon dioxide capture and storage (IPCC, 2005). Finally, a growth constraint has been introduced in the model at a maximum 10% on the roll-out of the CCS technologies.

10.2.3   Scenario Definitions

Scenarios have been developed in order to examine the implications of BECCS within global decarbonisation pathways. These scenarios address the following question: what are the implications of BECCS for cost-optimal global decarbonisation pathways?

The first set of scenarios explore optimal pathways to a 2 °C target, with and without BECCS. These scenarios provide a way of understanding how the use of BECCS can influence the options available to the rest of the energy system and the cost and investment implications of BECCS availability. It is worth explaining the rationale for these scenarios and their relationship with plausible or possible envisaged futures. In particular, one may question the realism of a scenario in which CCS is available for fossil fuels, but not for bioenergy. The point here is that these scenarios are experiments that explore the dependence of model outcomes on the availability of BECCS.

The two basic scenarios (2 °C, with and without BECCS) are further explored using a range of sensitivity scenarios, which explore key uncertainties:

1) The availability of sustainable biomass for the energy system. The TIAM-UCL model adopts relatively conservative[1] baseline assumptions on the availability of bioenergy that can be sustainably used, and the literature shows a considerable range of estimates for global bioenergy availability. The relative availability of bioenergy can be expected to be an important determinant of the modelled potential for BECCS. A scenario has thus been developed to explore the implications of a more generous availability of bioenergy than is assumed in the base case, by doubling the availability of bioenergy.
2) The year in which globally coordinated decarbonisation efforts begin (2015, 2020 or 2025). These scenarios are a highly abstracted representation of complex global political processes. The aim of these delayed-action scenarios is to test how the relative role of BECCS changes in the model results as effective global action to reduce emissions is delayed. The scenarios are thus designed to test the consequences of delay. They are not intended to be realistic 'storylines', as the geopolitical and

---

1 This is in comparison with the range of estimates presented in Slade *et al.* (2011).

practical realities of delivering global emissions reductions on such scales render such neat turning points unlikely.

The scenarios in this chapter include key assumptions on climate-change mitigation policies, including perfect international cooperation. This international cooperation starts in the scenarios presented in 2015, 2020 or 2025, and as a consequence produces a homogeneous international carbon price from this year onwards. The years chosen for the starting of such global cooperation are dictated by the structure of the TIAM-UCL time representation with periodic outputs every 5 years from 2005. The year 2015 is a point already in the past. However, large uncertainties in emissions and climate representation in the model should be taken into account and the starting dates for the mitigation policies presented should be regarded as a present-day start versus delay by 5 or 10 years. It is also worth emphasising that the scenarios presented in the chapter are not forecasts, as each result is built on a large number of uncertain assumptions including, for example, the socio-economic development of the world. The structure and philosophy of TIAM-UCL, a bottom-up cost optimisation energy-system model with perfect foresight, enables the extraction of informed insights into the long-term energy system under specific constraints. As a consequence, decisions and policies can be developed as long as the limitations of such models are acknowledged and understood.

A further set of scenarios, as presented in Table 10.2, explore the importance of BECCS in achieving a global 1.5 °C target, both with and without BECCS. These scenarios have been developed in light of the agreement reached at the UNFCCC COP21 in Paris, in December 2015.

Table 10.2 Scenario descriptions.

Scenario	Scenario shorthand	Scenario description
2 degree scenario	2D	The climate module in the TIAM-UCL model has been constrained to limit the temperature changes to 2 °C.
2D without BECCS	2D-NoBECCS	BECCS technologies have been made unavailable in this scenario. The climate policy is same as the one applied in the 2D scenario.
2D biomass sensitivity scenarios	2D-HBio; 2D-NoBECCS-HBio	The 2D and 2D-NoBECCS scenarios have been run with increased biomass availability.
2D delayed-action sensitivity scenarios		The 2D and 2D-NoBECCS scenarios have been run with global emissions abatement effort beginning in 2015, 2020 and 2025.
1.5 degree scenario	1.5D	The climate module in the TIAM-UCL model has been constrained to limit global temperature changes to 1.5 °C.
1.5 degree without BECCS	1.5D-NoBECCS	As the 1.5D scenario, but with BECCS made unavailable
1.5 degree biomass availability sensitivity scenarios	1.5D-HBio; 1.5D-NoBECCS-HBio	The 1.5D and 1.5D–NoBECCS scenarios are both run with increased biomass availability.

10.3 Results and Discussions

10.3.1 2 °C Scenarios With and Without BECCS

Figure 10.2 shows CO_2 emissions and the marginal mitigation costs for the two 2 °C scenarios (2D and 2D-NoBECCS). The results clearly show that the global energy system requires a deeper CO_2 reduction in the near and medium term when BECCS is unavailable, if the 2 °C target is to be achieved. The importance of BECCS is particularly clear in the near term: the 2D-NoBECCS scenario requires an annual average CO_2 reduction rate of 2% between 2015 and 2035, whereas the 2D scenario sees emissions remaining at 2015 levels until 2030. This shows that the availability of BECCS reduces the total discounted energy-system cost for the same climate target and relaxes the early mitigation requirements. However, while BECCS availability delays the need for near-term reductions, total emissions in 2055 are more or less the same under both scenarios. This is possible as BECCS can help to capture CO_2 from the atmosphere in the long term, especially during the second half of the century, leading to near-zero net global emissions by 2100. In other words, the late start is compensated by negative emissions in later years in the scenario in which BECCS is available. Unavailability of BECCS not only requires deeper early reductions but also results in much higher CO_2 mitigation costs (carbon price). BECCS can lower the marginal CO_2 mitigation cost in the 2D scenario to less than half of that in the 2D-NoBECCS scenario.

CO_2 mitigation in the near term takes place largely in the power sector, which reduces its emissions substantially under both scenarios, with a sharper reduction in the 2D-NoBECCS scenario between 2015 and 2030 (Figure 10.3). In contrast, end-use sectors such as transport, residential and industry undergo decarbonisation during the latter period of this scenario. This pattern of decarbonising the power sector first, and subsequently switching end uses to low-carbon electricity, is common to a wide range of decarbonisation scenarios at national and global scales, and the scenarios here are no exception. Power-sector emissions decrease from 9.6 $GtCO_2$ in 2015 to 5.6 $GtCO_2$ in 2035 in the 2D scenario. The early mitigation requirements in the 2D-NoBECCS

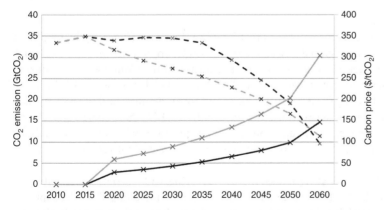

Figure 10.2 Global CO_2 emissions (dashed lines, left axis, $GtCO_2$) and marginal CO_2 abatement costs (solid lines, right axis, \$/tCO_2) under 2D (with BECCS; black lines) and 2D-NoBECCS (grey lines) scenarios.

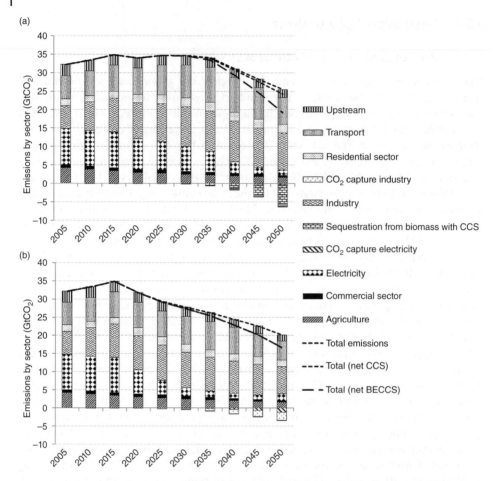

Figure 10.3 Sectoral emissions to 2050 in the 2D (a) and 2D-NoBECCS (b) scenarios generated using TIAM-UCL.

scenario reduce power sector emissions to 1.7 $GtCO_2$ in 2035. This is equivalent to an average of 8% reduction in global power sector emissions annually from 2015 to 2035, a rate considerably in excess of the rate of decarbonisation in the United Kingdom during the well-known 'dash for gas' in the early 1990s, during which time annual reductions were less than 5% annually (CCC, 2010). The length (over 20 years) and the global character of the power sector decarbonisation in the 2D-NoBECCS scenario stretch the plausibility of this scenario and raise the following question: what kinds of political, economic or technical global developments can be envisaged that render such a scenario possible?

CCS plays a role in mitigating emissions by capturing CO_2 in both electricity and industry sectors under both scenarios. In 2050, CCS (not including BECCS) captures 1.1 $GtCO_2$ in the electricity sector and 2.2 $GtCO_2$ from the industry sector in the 2D-NoBECCS scenario. When BECCS is available, as in the 2D scenario, the availability of lower-cost abatement via BECCS reduces pressure on industrial emissions, and as a result, CCS captures only 0.9 GtCO$_2$ in 2050 in the industry sector. BECCS alone captures and stores 5.1 GtCO$_2$ in 2050. This demonstrates the significance of BECCS for technology development priorities: if BECCS, under specific policy decisions (related to sustainability) or due to biomass production limitation (food prioritising), was certain to be unavailable, it would greatly strengthen the case for significant investments in the development of CCS technologies adapted to industrial processes and emissions.

Early mitigation in the power sector translates into very rapid reductions in power sector CO$_2$ intensity during the next 20 years. This is true in both scenarios, but the optimal rate of change in the 2D-NoBECCS scenario is extremely fast. In the 2D scenario, CO$_2$ intensity of electricity halves from 520 g/kWh in 2015 to 250 g/kWh in 2035 and goes into negative values from 2045 due to BECCS (Figure 10.4). When BECCS is not available, as in the 2D-NoBECCS scenario, CO$_2$ intensity of electricity must decrease sharply to 73 g/kWh in 2035, which is a seventh of the 2015 value, and 17 g/kWh in 2050. It perhaps goes without saying that this represents a global transformation that is unprecedented in the combined rate and scale of technological substitution.

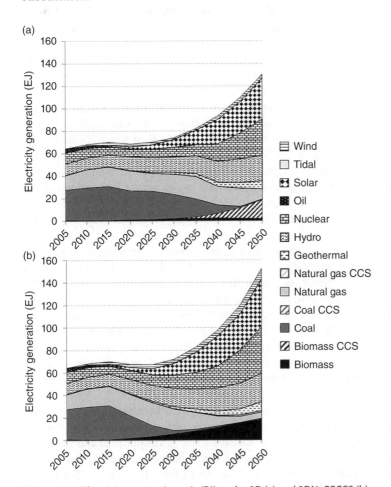

Figure 10.4 Electricity generation mix (PJ) under 2D (a) and 2DNoBECCS (b) scenarios.

In order to accomplish this transition and corresponding reductions in CO_2 intensity, the power sector needs investment in a range of low-carbon technologies in the near and medium term. Figure 10.4 presents generation mix under 2D and 2D-NoBECCS scenarios. At present, the global electricity system heavily depended on fossil-fuel generation capacity, with coal and gas providing two-thirds of total generation (2015 data). Coal is the dominant fuel, accounting for 43% of the total generation in 2015. Fossil-fuel-based generation decreases from 69% in 2015 to 44% in 2035 to only 7% in 2050 in 2D scenario. The fossil-fuel generation is replaced mainly by hydro and solar in 2035 and by nuclear and solar in 2050. There is a rapid increase in the share of solar PV generation between 2015 and 2035 from just 1% to 16% in the 2D scenario. Under the 2D-NoBECCS scenario, the share of fossil fuels in generation further shrinks to 20% in 2035 and 5% in 2050. The fossil fuels are replaced by nuclear and renewable generation, especially solar PV. Nuclear and solar PV are the dominant technologies in 2050 under both scenarios, contributing together more than half of the generation, of which solar PV has a slightly higher share than nuclear.

In the two scenarios presented, the proportion of electricity generation from biomass reaches 18 500 PJ in 2050 (from 500 PJ in 2015), representing 15% and 12% of the total generation for 2D and 2D-NoBECCS, respectively. However the growth in capacity between 2015 and 2050 is different. In the absence of BECCS, biomass-based technologies generate 10% of the electricity generated in 2035; at that time under BECCS availability biomass represents only 3% of generation: 1% from BECCS and 2% from biomass without CCS. From 1% in 2035, the BECCS share increases rapidly to 13% in 2050 in the 2D scenario. The two scenarios present a huge addition of biomass-based generation over a relatively short time. In the case of BECCS, the addition occurs after 2030 when the model invests in 3 GW of BECCS capacity in 2030 and increases it to 39 GW in 2035, which corresponds to an annual capacity addition of 7 GW. The installed capacity increases to 623 GW in 2050. Under the 2D-NoBECCS scenario, the development of biomass use in power generation occurs from 2015 as the model needs to mitigate earlier to compensate the lack of negative emissions with the additional annual capacity increasing from 4 to 7 GW between 2015 and 2035.

There is not much difference in the total electricity demand between the scenarios until around 2025. However, in the absence of BECCS, the global energy system needs 17% more electricity in 2050 under the 2D-NoBECCS scenario compared to that in the 2D scenario. This clearly shows that, in order to offset the emissions captured in power and industry sectors by BECCS in 2D, the 2D-NoBECCS scenario decarbonises end-use sectors by means of electrification – that is, by shifting to low-carbon electricity and away from fossil fuels. This happens in all major end-use sectors: transport, residential, industrial and commercial. The transport sector consumes 32% more electricity in the 2D-NoBECCS scenario compared to that in 2D in 2050. The respective figure for all other end-use sectors is about 10%.

Differences in the near-term and long-term generation mix between the scenarios are reflected in the primary energy consumption of the respective scenarios (Figure 10.5). There are notable differences in coal and natural gas consumption in the near and medium term between the scenarios. In the long term, total primary energy production in the 2D-NoBECCS scenario is slightly lower than that in the 2D scenario. Lower primary energy consumption in the 2D-NoBECCS scenario in 2050 is driven by a larger

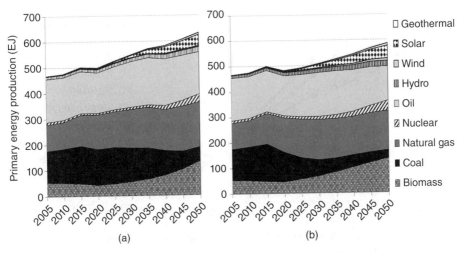

Figure 10.5 Primary energy production under 2D (a) and 2D-NoBECCS (b) scenarios.

share of nuclear and solar in the generation mix replacing gas and also end-use sector electrification.

10.3.2 Sensitivity Around Availability of Sustainable Bioenergy

As discussed previously, the relative availability of bioenergy can be expected to be an important determinant of the perceived importance of BECCS. However, it is not intuitively obvious how changes in bioenergy availability might influence the importance of BECCS. Greater availability of bioenergy might be expected to increase the apparent benefits of BECCS in the model results, since it is possible that the deployment of BECCS might be limited by the available bioenergy. On the other hand, greater availability of bioenergy could directly offset the consumption of fossil fuels, reducing the need for more expensive NETs such as BECCS.

The sensitivity scenarios presented here explore the implications of a more generous bioenergy resource than assumed in the earlier runs, with a doubling of the available potential biomass. The scenario is named 'high biomass'. The change in resource has been applied to the scenarios with BECCS and without BECCS.

A second sensitivity experiment is conducted looking at the impact of delaying the start of mitigation policies (or the peaking year for CO_2 emissions) between 2015 (as in the previous section) and 2020 or 2025. This has created a set of 12 different scenarios summarised in Table 10.2.

Figure 10.6 presents the CO_2 emissions pathways (net of all carbon capture and sequestration) of the 12 different sensitivity scenarios for the 2 °C target. Within the pathways, three different behaviours in CO_2 emissions can be described. The six no-BECCS scenarios are grouped in one lower emissions trajectory after 2030; the 2035 emissions in these scenarios are limited to 25 Gt(CO_2). A second set corresponds to the scenarios combining BECCS with low biomass availability (with 2035 emissions 20% higher than those without BECCS). Finally, the case of BECCS with high biomass availability allows emissions to be around 10% higher during the first part of the pathway

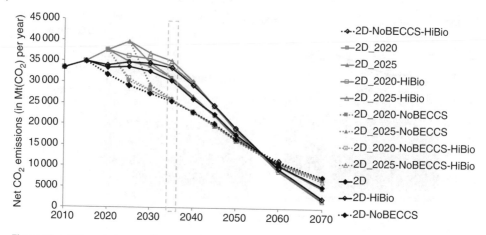

Figure 10.6 CO_2 emissions until 2070 under different scenarios for the 2 °C target (for the list of scenarios see Table 10.2).

(2015 to 2035) compared to the previous set, and almost 30% higher than in the scenarios in which BECCS is unavailable. The higher early emissions that occur in the BECCS scenarios are compensated later in the century (after 2070) by larger negative emissions, keeping the total carbon budget within the overall constraint. The inflexion point (from higher emissions during the first half of the century to lower during the second half) occurs in 2055. A key message from Figure 10.6 is that, when BECCS is unavailable and global action is delayed until 2020 or 2025, reductions must happen with great speed if the carbon budget is not to breach the 2 °C target. This occurs regardless of the amount of bioenergy available (though of course a much less generous amount of bioenergy would bring the BECCS cases closer to the no-BECCS result, as opportunities for negative emissions via BECCS would be constrained by resource limits).

Figure 10.7 presents the annual mean reduction in emissions needed between the peaking year and 2035 in the four scenarios created by combining the high and low biomass with the BECCS availability. The figure shows the rate at which emissions need to be reduced during the period 2015 to 2035. The no-BECCS scenarios all reach a common level of 25 Gt(CO_2) in 2035, but the CO_2 reduction rate needed to reach this level is, of course, much higher for the delayed case, reaching 3.5% per year. In these no-BECCS scenarios, the amount of biomass available does not affect the rate of reduction. When BECCS is available, the amount of bioenergy does influence the optimal decarbonisation rate: with high biomass, the mean annual rate of change to 2035 decreases to a more manageable −0.2% if global action begins in 2015, rising to only −1% per year if global action is delayed. With low biomass, the optimal rate of decarbonisation needed to achieve the 2 °C target rises from −1.5% to −3.5% for the early and late policy, respectively, as shown in Figure 10.7. It is useful to compare these rates of decarbonisation to estimates of what might be a plausible upper bound on the rate at which global decarbonisation could be achieved. Den Elzen et al. (2011), for example, have suggested that 3.5% is the maximum possible annual global reduction rate of CO_2 that could be achieved, taking into account assumptions about technological

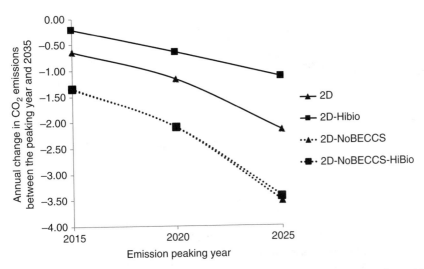

Figure 10.7 Year 2035 annual mean reduction in global CO_2 emissions between the peaking year (2015, 2020 and 2025) for the four families of scenarios incorporating the amount of biomass and the availability of BECCS.

development, economic costs and socio-political factors. This makes clear the importance of early global mitigation action in the case of BECCS failure in the future as in this case any delay would bring the reduction rate close to the maximum value that Den Elzen et al. view as plausible.

A common pathway in the three major cases discussed above (no-BECCS, BECCS low biomass and BECCS high biomass) is reached after 2035. The pathways diverge again after 2070 (not shown), with respectively lower emissions for the delayed mitigation scenarios (a difference reaching 10% for the year 2080 emissions between pathways peaking in 2025 and 2015). This difference compensates the added emissions of CO_2 from the start of the simulation in the delay in mitigation policy case.

The carbon prices calculated for the 12 different scenarios are presented in Figure 10.8. In almost all cases the peaking year for emissions is important in determining the

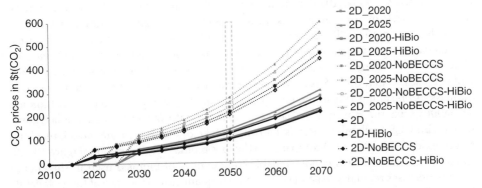

Figure 10.8 Carbon prices until 2070 (in $/t($CO_2$)) under different scenarios as in Figure 10.6.

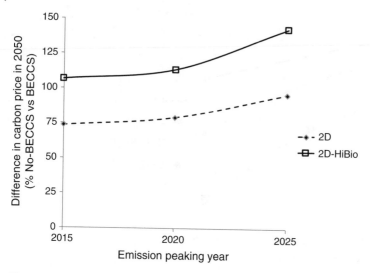

Figure 10.9 Difference in carbon prices between no-BECCS and BECCS scenarios for the year 2050, as a function of emissions peaking year (2015 to 2025) under low and high biomass availability.

carbon price level. When BECCS is available in the scenarios, the carbon price stays below \$150 per tonne of CO_2 in 2050. However, when BECCS is available and bioenergy is abundant, the peaking year makes little difference to mitigation costs in 2050: such scenarios provide a large potential for removing emissions via BECCS during the second part of the century. In the case of low biomass availability, late action matters: the increase in carbon price compared to the 2015 scenario is 5% if starting the mitigation in 2020 and 15% if only starting in 2025 as negative emissions are becoming more restricted by biomass production.

As expected, the no-BECCS scenarios show a significant impact on carbon prices. When BECCS is unavailable, carbon prices range from \$200 to \$280 per tonne of CO_2 in 2050. The exclusion of BECCS technology in the scenarios increases the carbon prices by 75%, 80% and 95% (mitigation policy starting in 2015, 2020 and 2025, respectively) under normal biomass production and 105%, 115% and 145% under high biomass (this is because high biomass with BECCS scenarios present the lowest carbon prices of the 12 pathways calculated), as seen in Figure 10.9. Finally, the changes in carbon price from the impacts of not having access to BECCS increase as global action is delayed.

The discussion now will focus on the use of available bioenergy across the scenarios. In a global energy-system model, the optimisation procedure identifies the least-cost allocation of bioenergy feedstocks across a wide range of possible technologies, sectors and end uses. When BECCS is available starting from year 2025 there is an increase in biomass use reaching 10% in the energy system in 2050 compared to the low biomass availability scenario. However, at a later period in the pathway under high biomass combined with CCS availability there is a strong uptake of biomass in the model, reaching an increase of 60% after 2070 compared to the low-biomass configuration (Figure 10.10). In contrast, when BECCS is not available the quantity of biomass used in the energy system increases regularly to achieve 45% higher quantities in the case of high availability after 2070.

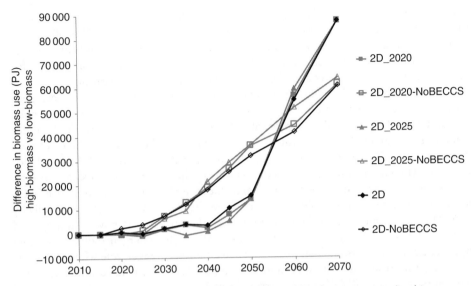

Figure 10.10 Change in bioenergy use in 2 °C scenarios until 2070: high biomass vs low biomass availability.

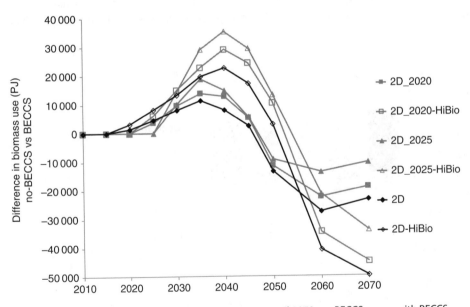

Figure 10.11 Change in bioenergy use in 2 °C scenarios until 2070: no-BECCS versus with BECCS scenarios.

In terms of peaking year, the change from no-BECCS to BECCS availability is plotted in Figure 10.11. As seen before, the implication of a failure in BECCS increases the biomass use pre-2050 (positive difference), peaking in 2035 to 2040, depending on the scenario, and reduces it after 2050 when biomass can contribute to negative emissions using BECCS. In all cases, delaying the peaking year from 2015 to 2025 increases the amount of biomass under the no-BECCS scenarios around 2035 from 20% to 30% in the low-biomass case and 30% to 45% in the high-biomass case. These increases in conjunction with other low-carbon sources (nuclear and variable renewable) compensate for the reduction in fossil fuels (coal and gas) involved in electricity generation.

Figure 10.12 presents the change in primary energy resources used for electricity generation between the high and low biomass availability scenarios for 2 years of interest: 2035 and 2050. Negative values represent a decrease in the use of the resource under high biomass availability while positive values display a higher share in the electricity generation mix. The cases of BECCS available or no-BECCS are treated separately for each year. The larger availability of biomass during the scenario is translated into an increase in biomass use in the generation mix. Under the 2 °C constraint and no-BECCS availability scenario, coal is rapidly removed from the electricity mix, almost totally by 2035. This corresponds to an immediate global effort to convert power stations and establish global bioenergy supply chains. How realistic the emergence of such a rapid supply chain would be is unclear. Coal can have a more prolonged role when BECCS is available (still present in 2035 mix) and biomass is almost unchanged under the high availability scenarios. BECCS creates stronger carbon removal in the later part of the pathway, allowing a prolonged use of coal and fewer stranded assets. In this scenario, the persistence of coal is accompanied by a slower implementation of natural-gas-fired power stations that is noticeable only after 2050. The increase in biomass use around the mid-century mostly reduces the more expensive wind, solar, hydro and nuclear

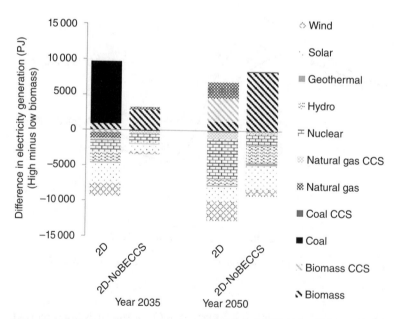

Figure 10.12 Difference in electricity generation by technology under 2 °C scenarios between high- and low-biomass scenarios. Only 2035 and 2050 results are presented from the pathways with no delaying in mitigation policies (peak emissions 2015); negative values convey a reduction and positive values an increase in the use of technology under high biomass availability.

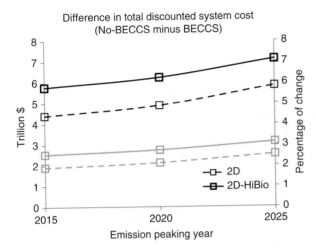

Figure 10.13 Difference in total discounted system cost between no-BECCS and BECCS scenarios to meet 2 °C targets as a function of biomass availability (black line = absolute change; grey line = relative change).

generation (respectively, 1.0 EJ, 3.5 EJ, 2.5 EJ and 1.5 EJ under no-BECCS or 2.7 EJ, 2.1 EJ, 1 EJ and 5.5 EJ under BECCS availability in 2050). It is noticeable that the largest increase in biomass use as feedstock for electricity generation under higher biomass availability (with or without BECCS) occurs only after 2050.

Availability of BECCS can bring down the near-term mitigation costs by up to 2–3% depending on assumptions on biomass availability (Figure 10.13). This translates into a saving, on annual energy-system costs, of up to $5.8 trillion under the 2D-highbio scenario and up to $7 trillion under the 2D-lowbio scenario if actions to reduce emissions are postponed until 2025. Effects of this cost reduction due to BECCS on carbon prices are significant (Figure 10.9), i.e. required carbon prices to meet the 2° target under the no-BECCS scenario will be at least twice as high as those in the corresponding BECCS scenario with low biomass availability. However, there are challenges as BECCS (or CCS) is not yet a commercially viable technology, and a considerable amount of investment for research, development and deployment will be necessary to make CCS technology commercially viable. Further, global cooperation among the leading nations who are developing and deploying CCS technologies is required to make this technology development a success.

10.3.3   1.5 °C Scenarios

Four scenarios have been developed with a climate policy target of achieving 1.5 °C by the end of the century. Figure 10.14 presents CO_2 emissions pathways under different 1.5 °C target scenarios. All scenarios require an even more rapid early reduction of CO_2 between 2015 and 2020 than were observed in the 2 °C scenarios, especially the two scenarios without BECCS. The results show that, in order to meet the 1.5 °C target without BECCS, CO_2 emissions should decrease at a rate of 11% annually between 2015 and 2020, while the scenarios with BECCS require annual reduction rates of 4–7%

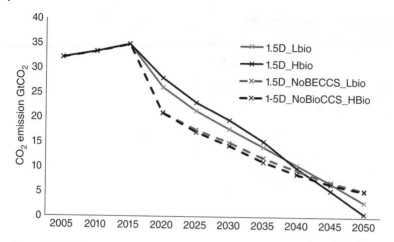

Figure 10.14 Emissions (GtCO$_2$) under different 1.5 °C scenarios.

depending on the scenario. Achieving a reduction rate as high as that in the BECCS scenario appears to be impossible. As Den Elzen et al. (2011) argue, the energy system cannot develop and invest in low-carbon technologies to reduce CO$_2$ emissions at such rates due to the very short lead time required for development and installation of low-carbon technologies for both supply and demand sectors.

Indeed, even ignoring many of the behavioural and social issues that Den Elzen et al. considered, the TIAM-UCL model is only able to meet the 1.5 °C target when 'backstop' technologies are available when BECCS is not available. These are dummy technologies modelled in TIAM-UCL to avoid infeasibility; they produce negative emissions when a certain carbon price level is achieved (for this chapter the level chosen is 5000 $/t(CO$_2$)). Due to this high cost, the processes are only starting off very late in the scenarios after 2070 to keep the global temperature just below the 1.5 °C limit. To a certain degree, these 'backstop' processes could represent unspecified and uncertain NETs, such as DAC, enhanced weathering or ocean fertilisation. This strongly suggests that BECCS appears to be essential to meet the 1.5 °C target in the absence of alternative NETs.

The BECCS scenarios require a relatively low CO$_2$ reduction rate (about half) during 2015–2020 compared to the no-BECCS scenarios, allowing more emissions in the early period, as these emissions can be offset by BECCS at a later period (post-2050), reaching net CO$_2$ emissions negative during the fourth quarter of the century. Net CO$_2$ emissions in 1.5D-LBio and 1.5D-HBio scenarios are –2.5 tCO$_2$ and –6.8 tCO$_2$ in 2100.

The scenarios with BECCS also require a rapid early reduction, which leads to rapid investments in expensive low-carbon technologies in the early period, leading to higher marginal CO$_2$ abatement costs compared to those in the 2D scenarios. Figure 10.15 presents the marginal CO$_2$ abatement costs under different 1.5 °C scenarios. This clearly shows that meeting the 1.5 °C target is infeasible without BECCS as it needs a carbon price of over 1000 USD per tonne of CO$_2$ in 2020.

The model heavily invests in BECCS from 2025 to meet the 1.5 °C target (Figure 10.16). In order to meet the stringent climate policy, depending on the scenario, installed

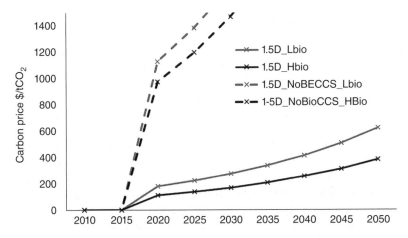

Figure 10.15 Marginal CO_2 prices ($/tCO_2$) under different 1.5 °C scenarios.

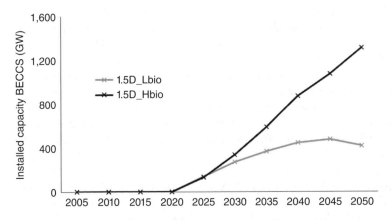

Figure 10.16 Installed capacity of BECCS (GW) under different 1.5 °C scenarios.

capacity of BECCS varies from 134 GW in 1.5D-HBio scenarios to 142 GW in 1.5D-LBio scenario in 2025 (Figure 10.16), generating at least 5% of the total electricity generation from BECCS. Further, a twofold increase in the installed capacity is required during the following 10 years (between 2025 and 2035) in the low-biomass scenario to meet the climate target. The respective figure for the high-biomass scenarios is at least a fourfold increase. This is clearly a very challenging level of investment as BECCS is not a fully commercialised technology yet. On the other hand, without BECCS it is also not possible to meet the stringent target of 1.5 °C.

Electricity generation from BECCS is at least 60% more in 2035 and three times more in 2050 in high-bioenergy scenarios compared to that in the low-bioenergy scenarios. Additional investment in BECCS capacity in high-bioenergy scenarios mainly reduces generation from gas CCS plants, and also nuclear and wind generations to a certain extent, compared to that in low-bioenergy scenarios.

10.4 Discussion and Conclusions

This chapter uses the TIAM-UCL global energy-system model to investigate the extent to which BECCS is critical for meeting global CO_2 reduction targets under different long-term scenarios during 2005–2100. We generated and analysed 16 scenarios by varying the availability of biomass in terms of quantity (resource potential) and time (delayed availability) for GHG emissions mitigation under 2 °C and 1.5 °C climate-change mitigation targets at global level.

Analysis shows that availability of BECCS can reduce the pressure on near-term mitigation requirements under a 2 °C scenario as BECCS can be used to capture CO_2 from the atmosphere in the long term, especially during the second half of the century, leading to near-zero net global emissions by 2100. It especially reduces pressure on the power sector: the CO_2 intensity of electricity halves from 520 g/kWh in 2015 to 250 g/kWh in 2035 under the 2D-BECCS scenarios while it must decrease sharply to 73 g/kWh in 2035 under the 2D-NoBECCS scenario. BECCS alone captures and stores 5.1 $GtCO_2$ in 2050 with an installed capacity of 623 GW. Unavailability of BECCS doubles the carbon price required in 2050 to meet the 2 °C target from 150 to 300 $/t($CO_2$). Sensitivity analysis shows that later action combined with no-BECCS (delaying the peaking year to 2025) can further double the carbon price required to meet the target. Unavailability of BECCS requires the almost complete removal of coal from electricity production by 2035. How realistic such a rapid global supply-chain decline may be is unclear. In return, the availability of BECCS creates stronger carbon removal in the later part of the pathway, allowing a longer use of coal and lower stranded assets.

In order to meet the 1.5 °C target, the model results show that without BECCS, CO_2 emissions should decrease at a rate of 11% annually between 2015 and 2020, resulting in a carbon price of $1000/t$CO_2$ in 2020. Moreover, in this case, NETs, other than BECCS, are still needed at the end of the century to achieve the stringent target. But the BECCS scenario requires a relatively low CO_2 reduction rate (about half, ~5%) during 2015–2020 compared to the no-BECCS scenarios. This requires an installed capacity of at least 134 GW of BECCS in 2025, generating 5% of the total electricity generated globally. This is clearly a very challenging level of investment, as BECCS is not yet a commercialised technology. On the other hand, without BECCS (or other process producing negative emissions), it appears to be impossible to meet the stringent target of 1.5 °C.

Many pathways developed here involve extremely rapid reductions in global emissions, implying a degree of global cooperation that is unprecedented. Such a task requires not only agreement between nations but also national governments capable of affecting such a dramatic energy-system transformation. The plausibility of such pathways – in political and socio-economic terms – is highly uncertain, and there may be valid disagreements about whether or not such pathways represent possible futures. In this context, the model outcomes can be regarded as techno-economically optimal (though their optimality even in narrowly defined techno-economic terms is highly uncertain), but their socio-technical plausibility remains open to question. This generates two possible interpretations of some of the principal results of these scenarios.

First, results appear to suggest that the availability of BECCS provides some 'breathing space' to enable globally coordinated mitigation efforts to be ramped up to the required level. Such an interpretation could be read as reducing the need for near-term action. This would be a mistake. The model formulation, based on optimisation using linear programming, implicitly assumes a world in which coordination barriers are low or non-existent, and technology deployment can proceed without being held up by the behavioural, institutional or political factors that result in slow technology adoption in the real world.

Alternatively, one can understand the scenarios as suggesting that without BECCS, the targets simply become implausible – certainly 1.5° but perhaps also 2°. The social, political and economic conditions under which such a rapid global transformation may occur are difficult to even imagine. Such an interpretation puts BECCS – or other NETs – as being absolutely essential for avoiding dangerous climate change, and it suggests that the urgency of investment and learning around BECCS technology is extreme. Even with BECCS, the temperature targets require unprecedented global action such that this interpretation sees BECCS as neither a 'get-out-of-jail-free card' nor as 'silver bullet' – but as a last hope.

References

Anandarajah, G., McDowall, W., and Ekins, P. (2013). Decarbonising road transport with hydrogen and electricity: long term global technology learning scenarios. *International Journal of Hydrogen Energy* **38**: 3419–3432. doi: 10.1016/j.ijhydene.2012.12.110.

Anandarajah, G., Pye, S., Usher, W. et al. (2011). *TIAM-UCL Global Model Documentation*. UKERC Publication.

CCC (2010). The Fourth Carbon Budget – Reducing Emissions through the 2020s. Committee on Climate Change.

Den Elzen, M.G.J., Hof, A.F., and Roelfsema, M. (2011). The emissions gap between the Copenhagen pledges and the 2°C climate goal: options for closing and risks that could widen the gap. *Global Environmental Change* **21**: 733–743.

Dessens, O., Anandarajah, G., and Gambhir, A. (2016). Limiting global warming to 2°C: what do the latest studies tell us about costs, technologies, and other impacts? *Energy Strategy Reviews* **13–14**: 67–76.

IEA (2013). *Technology Roadmap 2035 2040 2045 2050 Energy Technology Perspectives: Carbon Capture and Storage*. Paris: OECD/IEA.

IPCC (2005). IPCC special report on carbon dioxide capture and storage. In: *Prepared by Working Group III of the Intergovernmental Panel on Climate Change* (ed. B. Metz, O. Davidson, H.C. de Coninck, et al.). Cambridge and New York: Cambridge University Press.

Kesicki, F. and Anandarajah, G. (2011). The role of energy-service demand reduction in global climate change mitigation: combining energy modelling and decomposition analysis. *Energy Policy* **39**: 7224–7233.

R. Loulou and M. Labriet (2007). ETSAP-TIAM: the TIMES integrated assessment model Part I: Model structure. *Computational Management Science*, 10.1007/s10287-007-0046-z. http://www.springerlink.com/content/j8613681347971q5/fulltext.pdf

McGlade, C., Dessens, O., Anandarajah, G., and Ekins, P. (2015, 2015). Global scenarios of greenhouse gas emissions reduction. In: *Global Energy: Issues, Potentials and Policy Implications* (ed. P. Ekins, M. Bradshaw and J. Watson). Oxford University Press.

McGlade, C. and Ekins, P. (2015). The geographical distribution of fossil fuels unused when limiting global warming to 2 °C. *Nature* **517**: 187–190.

McLaren, D. (2012). A comparative global assessment of potential negative emissions technologies. *Process Safety and Environmental Protection* **90** (6): 489–500.

Rogelj, J., Luderer, G., Pietzcke, R.C. et al. (2015). Energy system transformations for limiting end-of-century warming to below 1.5°C. *Nature Climate Change* **5**: 538, June.

Slade, R., Saunders, R., Gross, R., and Bauen, A. (2011). *Energy from Biomass: the Size of the Global Resource*. London: UK Energy Research Centre.

Smeets, E., Faai, A., and Lewandowski, I. (2004). A Quickscan of Global Bio-energy Potentials to 2050: An Analysis of the Regional Availability of Biomass Resources for Export in Relation to the Underlying Factors. *Report NWS-E-2004-109*, 90-393-3909-0.

van Vuuren, D.P., Stehfest, E., den Elzen, M.G.J., van Vliet, J., and Isaac M. (2010). Exploring IMAGE model scenarios that keep greenhouse gas radiative forcing below 3 W/m2 in 2100. *Energy Economics* **32**(5): 1105–1120.

11

Policy Frameworks and Supply-Chain Accounting

Patricia Thornley[1] and Alison Mohr[2]

[1] *Tyndall Centre for Climate Change Research, School of Mechanical Aerospace and Civil Engineering, University of Manchester, UK*
[2] *School of Sociology and Social Policy, University of Nottingham, UK*

11.1 Introduction

The majority of future energy scenarios that facilitate the United Kingdom meeting its greenhouse gas (GHG) reduction targets require the use of negative emissions technologies, and, of these, bioenergy with carbon capture and storage (BECCS) is most often invoked in energy-system models. This is essential because it offers energy with negative emissions at a relatively competitive cost. Other energy technologies generally incur at least some level of carbon emissions, and other GHG removal technologies are less well proven with consequently higher costs and greater performance and economic uncertainties.

However, this characterisation of BECCS as providing net removal of GHGs from the atmosphere is predicated on consideration of the entire supply chain from biomass production through to energy generation and utilisation. It is well established that such a comprehensive supply chain coverage makes sense in most cases in order to ensure that we are assessing the full environmental impact of new energy technologies. However, such a comprehensive scope of system also poses some distinctive challenges. It may provide consistency of comparison to other generation technologies, but does it make sense from an analytical and governance perspective?

The answer to this question involves consideration of four sets of key questions:

1) Is the boundary being drawn around the BECCS system an ideologically plausible and defensible one from an analytic perspective, i.e. is this an actual system or simply a co-alignment of connected stages?
2) Is the scope of the system ethically justifiable, i.e. is there a justification for combining carbon sequestration and emissions incurred in different countries and sectors? The uptake of BECCS in integrated assessment models is driven by models which combine sequestration credits with energy production in the supply chain. This supply chain focus was originally intended to prevent unintended significant upstream GHG releases being excluded from GHG calculations, but is it always appropriate to

Biomass Energy with Carbon Capture and Storage (BECCS): Unlocking Negative Emissions, First Edition.
Edited by Clair Gough, Patricia Thornley, Sarah Mander, Naomi Vaughan and Amanda Lea-Langton.
© 2018 John Wiley & Sons Ltd. Published 2018 by John Wiley & Sons Ltd.

include upstream carbon dioxide sequestration in biomass? If so, can we rely on governance and accounting frameworks to be confident that BECCS systems as proposed will deliver/are actually delivering a net reduction in global GHG emissions and making best use of biomass resources, given that there is no governance rationale for combining these emissions contributions and therefore no incentive to minimise the net GHG emissions of the system?

3) What are the lessons/implications for BECCS governance of the spatial and temporal ordering of bioenergy technologies and the distribution of their impacts across North/South and global/local contexts?

4) What lessons/implications for BECCS governance can be derived from the limited interaction with issues of social sustainability of certification schemes that underpin bioenergy policies? How could an extended analytical focus to encompass the social processes of BECCS deployment and their consequences help shape its governance?

This chapter attempts to answer these questions by exploring the rationale for supply-chain accounting for bioenergy systems, the governance of BECCS and the adequacy of existing institutions and frameworks supporting BECCS development, in order to determine whether negative emissions from BECCS systems can be secured.

11.2 The Origin and Use of Supply-Chain Analysis in Bioenergy Systems

11.2.1 Rationale for Systems-Level Evaluation

Renewable energy systems are often deemed low or zero carbon, with little attention paid to the net GHG balance of such systems. It is simply assumed that that they would be low carbon. However, while most renewable and bioenergy systems are low carbon, there are some where significant GHG emissions are incurred in plant construction, feedstock production, processing or conversion. So, if a feedstock has been cultivated and processed in an energy-intensive manner and/or used inefficiently, then the carbon savings delivered relative to equivalent use of fossil fuels could be low or non-existent. It is therefore very important with bioenergy systems to consider fully the supply chain and account for all parts of the bioenergy system which may be contributing GHG emissions. A key first stage is feedstock production where bioenergy crops will typically sequester carbon dioxide from the atmosphere during growth. It is inclusion of this stage as a negative (sequestered) carbon benefit, which gives rise to the 'negative emissions' connotation of BECCS systems.

Clearly there is justification for including all steps of a process that contribute to its end product in the analysis of its environmental impact, and this is necessary to ensure that full account is taken of all contributions to environmental impact. Figure 11.1 illustrates the scope of typical bioenergy systems with carbon capture and storage for analysis, and it is relatively easy to justify the inclusion of different stages if the objective of the assessment is to gain a true perspective of the environmental impact of the whole system, including its supply chain, by-product disposal and any infrastructure necessary to achieve the system outputs.

So, if we start at the top of the figure with biomass, this will include harvesting of the feedstock which will require machinery using energy, resulting in carbon and other

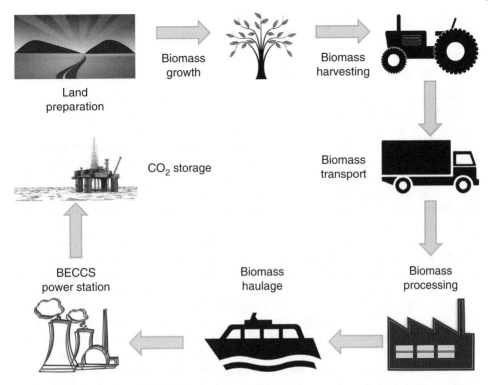

Figure 11.1 Scope of a typical BECCS system.

emissions from the system, which should be taken into account. Transport of the biomass from where it is grown to where it is processed will also have an impact and should be included. Often there is then intermediate processing (e.g. pelletisation which requires energy) and that should be included. Long-distance haulage (including sea freight) usually takes place after such processing and the biomass ends up at a power station where it is converted. Many plant-specific parameters here will influence overall system performance, not least energy efficiency and carbon capture rate.

Finally, the CO_2 is stored, and where the scope of system 'ends' should be considered here. Is it enough to produce pure CO_2 that can be stored or should pumping, storage and leakage all be taken into account? It may be argued that the infrastructure associated with these stages would be primarily used for other (fossil) CO_2 sources and so it does not make sense to include them. Similarly, looking at the top left-hand corner of Figure 11.1, does it make sense to include land preparation? It is difficult to argue that this is part of the process, but without appropriate land the biomass will not grow. During biomass growth, we would certainly want to include any actions taken, e.g. agrochemical application and so this should ordinarily be included in the analysis. The impacts of including or excluding different steps can be significant, and this is discussed in more detail in Chapter 6 on life-cycle assessment (LCA), where the importance of a comprehensive system appropriate to the LCA question being asked is stressed. However, for the purposes of the discussion in this chapter, the question is not so much whether there is a process or physical link between the different systems elements, but how the system behaves as an entity from a governance perspective.

11.2.2 Importance and Significance of Scope of System

As shown in Figure 11.1, a BECCS system is actually a series of linked process steps. The strength of the physical and organisational links between the steps is variable, and this can make analysis and incentivisation of the system challenging.

At one extreme, it could be argued that a BECCS system is nothing more than a number of independent processes combined for the purposes of carbon accounting: a forestry system (sequestering carbon and producing timber, some of which will be sold as saw logs other components for bioenergy); road transport (linked to forest product marketing); a wood-processing plant (producing pellets for international trade, perhaps for a variety of uses); sea transport (likely to a specific BECCS end user, but *via* an international shipping intermediary); conversion to electricity in a BECCS power plant (a dedicated facility that is definitely part of the specific BECCS system); separation and storage of CO_2 (most likely in a larger infrastructure that has been developed for fossil-fuel-based sequestration). The negative emissions associated with the BECCS system are achieved by linking the carbon sequestration credited to the forestry system with the energy delivered in the conversion plant and associated long-term carbon capture system. It could be argued that globally exactly the same net emissions impact would be experienced if we simply maintained an equivalent amount of forest (since this provides a convenient mechanism of extracting CO_2 from the atmosphere) and captured an equivalent amount of CO_2 elsewhere (from fossil fuels or indeed any other sources). In other words: why bother to attempt the engineering challenge associated with bio-CCS when the same net impact on global carbon emissions could be achieved with these much simpler steps?

There are several problems with that argument. First, it does not actually provide any energy and so the avoided fossil fuel emissions provide no benefit; second, there is no framework that would incentivise it to happen, e.g. why would forest carbon sequestration increase without any additional demand for forest products? It is worth bearing in mind here that while maintenance of carbon stock in the biosphere is important, extraction of CO_2 from the atmosphere is also important and this takes place at the greatest rate early in the forest life cycle. So, having a demand that actually uses up the wood produced and turns over the age profile of the global forestry base actually accelerates the rate of CO_2 extraction (up to a limiting point). These dynamics are complex and worthy of careful analysis, but the principle holds that without an additional market/purpose for the wood there will not be additional sequestration.

So while, on paper, BECCS may not be a necessary element to initiate the same net result in global GHG balances, in practice it provides a very attractive tangible system with which to incentivise increases in carbon sequestration *via* biomass growth, careful management and attention to the GHG emissions associated with forest products and trade, and long-term carbon storage – all of which are exactly what is needed!

However, this does not help identify the physical entity associated with a well-defined BECCS system that is recognised from an accounting, regulatory and policy perspective. A key focus of this chapter is therefore to look at the form this system takes, how it sits within existing frameworks and what is needed to ensure appropriate development and governance of BECCS in a way that does actually deliver the objectives.

11.2.3 Importance and Significance of Breadth of Analysis

Supply chains are heterogeneous systems of organisations, people, activities, technologies, infrastructure, markets, information, resources and social and cultural norms and practices involved in the production and distribution of products or services. In the case of BECCS, the products and services are biomass energy and carbon sequestration leading to negative emissions. The dynamic interactions between these heterogeneous components necessitate the consideration of a broader range of impacts beyond the environmental consequences of mitigating GHG emissions and an assessment of the wider socio-economic and ecological impacts of BECCS across the entire supply chain. From this whole-systems perspective, two analytical dimensions are often overlooked in technology and policy assessments of bioenergy and CCS: the spatial dimension and the human dimension.

BECCS supply chains are spatially ordered. They traverse organisational, sectoral and country boundaries at the same time as being embedded in the global bioeconomy. The sustainability of some bioenergy supply chains has been called into question as early visions of bioenergy production embedded at national or local scales gave way to awareness of the spatially uneven impacts of globally integrated bioenergy supply chains on land-use change and GHG emissions (Raman and Mohr, 2014). In a globalised bioeconomy, national supply chains will inevitably interface with adjacent systems, wider networks and markets. The spatial ordering of BECCS supply chains therefore needs to be interrogated to ask more fundamental questions about sustainability beyond those considered by quantitative modelling or sustainability indicators, in order to identify areas where potential conflicts may arise at the intersect of North/South and global/local territorial contexts.

Cutting across most sustainability impact indicators are questions of uneven spatial distribution related to, for example, where biomass has come from, which regions have borne the negative impacts and which have benefited (Raman and Mohr, 2014). While the physical impacts of GHG changes will be felt on a global level, all other impacts will be experienced local to the point of production, conversion, processing and use. The spatial order assumed in visions of BECCS systems is distinctly global. UK targets for bioenergy use assume a system of international trade in biomass commodities. The breaching of territorial visions for bioenergy production has in some circumstances led to poorer Southern countries exporting raw biomass for use as value-added energy products and services in the richer North. The exporting of any negative environmental impacts while reaping the benefits across North/South and local/global (in the case of GHG emissions) territorial boundaries has been referred to as "environmental load displacement" (Hornborg, 2008).

Modelling the potential for BECCS to deliver negative emissions under different scenarios is useful for quantifying both the human drivers of environmental change and the consequences of these changes for environmental systems and their impacts. However, the process of conducting aggregated, target-oriented techno-economic analyses of different systems simplifies the causes, dynamics and scope of societal transformations (Geels et al., 2016). Models typically conceptualise systems as collections of technologies and their interactions and understand transitions as changes in consumption and production patterns, technologies and resources. The role of institutional and socio-economic innovations, human–technology interactions and understanding the effect of social and cultural practices and routines on systems change is overlooked.

The successful implementation of BECCS systems is therefore dependent not only on techno-economic processes but also on social processes and their consequences – its human dimensions. An important aspect of BECCS deployment is thus associated with social attitudes, processes and behaviours related to how we maintain, protect and enhance the natural environment. Environmental change practices and systems often neglect the human and social dimension and fail to address institutional, social and behavioural barriers to change. It is therefore vital to understand the different ways in which humans value, use, interact with and depend on natural resources. Value-based visions of bioenergy have highlighted concerns about difficulties monitoring large-scale supply chains, the potential for distributing impacts unfairly and competition for biomass in the global bioeconomy. On the other hand, it has been argued that public concerns related to CCS are not simply down to knowledge gaps but have to do with differences in values and how the public frames potential risks, compared with experts (Markusson et al., 2012; Mander et al., 2010).

Framing sustainability impacts entirely as factual ones ignores the fact that assessments are wrapped up with value choices over which impacts to measure and how to measure them and how these assessments might differ from a global or local perspective. Understanding the breadth of potential BECCS impacts therefore requires getting to grips with value-based judgements and the conflicts between them to identify potential trade-offs. Assumptions of the feasibility of BECCS may vary with regard to the land area and type used for biomass production, biomass types and yields and proportion of energy supply from biomass to carbon storage capacity, capture rate and technology uptake (Vaughan and Gough, 2016). Clarifying the different value judgements and assumptions regarding the distribution of potential impacts at each stage of the BECCS supply chain will be vital to understanding where value conflicts may emerge, the distribution of beneficial and adverse impacts and associated trade-offs.

At the core of conflicting perspectives on BECCS as either a potential problem or an innovative solution for emissions reductions are different assessments of the breadth of BECCS systems' impacts and whether these extend beyond GHG emissions and environmental impacts to include wider consideration of the spatial and human dimensions of BECCS supply chains. The extent to which BECCS deployment will be able to deliver against emissions targets is therefore not a sufficient sustainability measure from a whole-systems perspective, where considerations by policy analysts of whose values and visions count, which impacts are important and whether they are justly distributed, are also important.

11.3 Policy Options

11.3.1 Objectives of BECCS Policy

BECCS is expected to be an essential component of the strategy to be outlined in the IPCC's special report on the impacts of global warming of 1.5 °C above pre-industrial levels and related global GHG emissions pathways, due in 2018. Addressing the challenge of deep decarbonisation as envisaged by the IPCC and strategies such as the Paris Agreement and the UK's Climate Change Act will be dependent on getting the policy framework right if BECCS is to play an important part in ensuring the decarbonisation of the United Kingdom's energy system by 2050 or earlier. The scale of this policy challenge is significant.

The future potential of BECCS technologies to deliver negative emissions is highly uncertain and will ultimately depend on future land-use policy, sustainable biomass supply chains, improvements in CCS technologies, the availability of CO_2 storage infrastructure and the development of CO_2 transport networks to service dispersed smaller-scale CCS sites (OECD/IEA, 2016). As with bioenergy and CCS policies individually, if BECCS is to be deployed at the scale envisaged, then it is critical that any potential direct and indirect sustainability impacts stemming from deployment are identified to aid responsible political choices regarding which BECCS systems or pathways to support through policy incentives.

We have argued that the relations between humans, the technologies they use and their environments are complex, and this complexity cannot be grasped through techno-economic modelling tools alone. For example, environmental degradation such as loss of biodiversity due to depletion or destruction of natural resources poses a broader cost to society, often felt far into the future. The wider and longer-term impacts of unsustainable production systems are under-represented in bioenergy policy because of a tendency to overlook questions of whose value choices count when deciding which impacts to measure and how to measure them (the human dimension) or the distant degradation of social groups, places and ecologies (the spatial dimension) or the implications of land use and CO_2 storage for future generations (the temporal dimension). The role of policy is to limit negative externalities and forms of degradation but, as we now discuss, existing bioenergy policy instruments have had mixed success in this regard. This implies that policy decisions on the complex and highly uncertain issues raised by BECCS will need to rely on mixed policy measures to account for and manage the range of possible externalities.

The negative externalities of bioenergy production that is otherwise producing useful public goods (renewable energy and carbon dioxide) are the direct and indirect sustainability impacts it generates. The globalisation of bioenergy industries has meant that biomass production systems, and the negative externalities they generate, have shifted towards the global South, aided by weak social and environmental governance (van der Horst and Vermeylen, 2011), while renewable energy consumption systems and the positive externalities derived from using cleaner energy are centred on the global North. While there may be some justification that the global South is more vulnerable to climate impacts and production systems can support resilience, there is a danger of reinforcing exploitative relationships in which affluent countries are able to effectively outsource the negative externalities of realising their domestic biofuel polices.

To manage the externalities arising from the global scale of bioenergy production and consumption systems, sustainability certification was introduced as a mode of global governance that extends the reach of EU policy along supply chains (di Lucia, 2010). However, the interaction of EU certification with the heterogeneity of local contexts has highlighted the limited scope of such schemes when it comes to dealing with social impacts. Tomei's (2015) study of biofuel certification in Guatemala revealed that sustainability criteria can have adverse impacts by facilitating unsustainable social practices. The displacement of subsistence tenant farmers in favour of land-owning elites seeking more lucrative agreements with sugar mills destroyed rural livelihoods and forced farmers into becoming labourers on the plantations that displaced them (Tomei, 2015). Sustainability certification's limited ability to address social concerns and the heterogeneity of local contexts made it complicit in the unjust distribution of negative externalities.

The weakness of bioenergy sustainability criteria is further extended to the issue of indirect land-use change (iLUC, discussed in Chapter 10) as the measurement of carbon emissions is predicated on accounting methodologies that not only struggle to quantify the issue but which also fail to capture the non-carbon impacts of land-use change highlighted above. The EU introduced iLUC factors linked to particular crop groups to estimate the emissions arising from global land-use change due to bioenergy production, but these have been criticised for narrowing the terms of sustainability and related public debate to just carbon, as a means of closing down the human and spatial dimensions of the controversy (Palmer, 2012).

A key lesson for BECCS arising from the overlapping case of bioenergy is that a mixture of policy measures will be needed to comprehensively counter any negative externalities, and these must be sensitive to the heterogeneity of local contexts and to distant (spatial and temporal) impacts. The uptake of BECCS relies heavily on policy incentives. Uncertainties related to the heterogeneity of technology–feedstock combinations and associated sustainability impacts in various bioenergy pathways make it difficult to design policy instruments which incentivise cost-effective contributions of bioenergy to renewable energy and GHG mitigation targets. Different pathways will give rise to different externalities, meaning policy-makers face the challenge of how to weigh external costs and benefits of a given pathway against others to solve associated trade-offs.

A mixture of policy instruments will be needed involving choices between price and quantity instruments, various bioenergy technologies and feedstocks and policy-planning security and adaptive flexibility. The latter is a critical challenge for BECCS. Balancing policy flexibility and the ability to respond to new developments and changes in public attitudes and to correct errors, needs to be weighed against policy stability underpinned by firm expectations and planning security to provide market actors with the safeguards and confidence they need to invest (Parkus et al., 2015). The inclusion of stakeholders in decision-making processes as a core guiding principle for BECCS policy design would help strike that balance as well as balancing other trade-offs. Finally, the focus of policy attention also needs to be appropriate to the technology's nascent stage of commercial maturity and the specific political-economic context in which it is being deployed.

11.3.2 Review of Existing Policy Frameworks

11.3.2.1 International Policy Frameworks

11.3.2.1.1 United Nations Framework Convention on Climate Change

In December 2016, the UK government ratified the United Nations Framework Convention on Climate Change's (UNFCCC) Paris Agreement, committing to pursue efforts to limit the temperature increase to 1.5 °C above pre-industrial levels by cutting emissions to net zero later this century. This shared goal would commit the United Kingdom not only to reducing emissions but to removing from the atmosphere carbon dioxide that has already been emitted. To achieve the net-zero emissions set by the Paris Agreement, the government's Committee on Climate Change acknowledges that GHG removal options such as afforestation and BECCS will be required alongside widespread decarbonisation (Committee on Climate Change, 2015).

Established in 1992, the UNFCCC, and its Conference of the Parties (COP) which takes place annually, enacts global climate policy with the objective of stabilising GHG concentrations in the atmosphere and limiting anthropogenic interference in the climate system. The COP periodically adopts new accords, such as the 1997 Kyoto Protocol and the recent agreements in Paris in 2015. The Kyoto Protocol bound member states to act in the interests of human safety even in the face of considerable scientific uncertainty. In line with its core principle of 'common but differentiated responsibility', as the main GHG emitters industrialised (Annex 1) countries were expected to lead by example by reducing emissions to below 1990 levels by the year 2020. Under the UNFCCC and the Kyoto Protocol, emissions related to biomass use are reported and accounted for under LULUCF (land use, land-use change and forestry).

The Paris Agreement further builds upon the Convention with the aim of establishing a common framework to account for the contribution from land use and forests in reaching the long-term goal of achieving 'a balance between anthropogenic emissions by sources and removals by sinks of GHGs in the second half of this century' (Article 4 [1]). It is hoped a common accountability framework will address criticisms of irregular land-use accounting and reporting under the Kyoto Protocol whereby Annex 1 Parties are able to opt into or out of accounting for certain LULUCF activities. If these irregularities are allowed to continue under the Paris Agreement after the Kyoto Protocol expires at the end of 2020, this raises the possibility of the GHG benefits of BECCS counting towards future GHG commitments while the disadvantages of using unsustainable biomass are ignored. Any damaging effects may therefore outweigh the benefits of negative CO_2 emissions. Assessment of biomass sustainability will not be possible by GHG reporting alone if a BECCS system is fuelled by biomass sourced from a developing country. This could result in the Annex 1 Party benefiting from negative emissions against its commitments, whereas positive emissions go unreported in the source country.

While implementation of the current LULUCF Decision (529/2013/EU) is under way with the aim of delivering improved accounting systems by 2020, without a legal framework consolidating this implementation and defining the applicable rules for the period post-2020, the way in which LULUCF would be included in the overall framework could continue to be heterogeneous across the EU. In July 2016, the European Commission published a proposal for a regulation on the inclusion of GHG emissions and removals from LULUCF into the 2030 climate and energy framework and amending Regulation No. 525/2013 on a mechanism for monitoring and reporting GHG emissions. The objective of the proposal is to determine how the governance of the LULUCF sector can be further developed within the EU climate-policy framework as of 2021. It has been argued that the UNFCCC has had little real impact on emissions so far (Jacoby and Chen, 2015), and improved emissions accounting systems and legally binding agreements could go some way towards rectifying this situation.

11.3.2.1.2 EU Emissions Trading System

At the level of industrial installations and other high-emitting industries (such as aviation) in the EU, the emissions trading system (ETS), launched in 2005, is the key mechanism for reducing GHG emissions. The ETS embodies a flexible 'cap-and-trade' system whereby a cap is set on the total amount of certain GHGs that can be emitted by companies covered by the system, reduced over time so that total emissions fall. The current cap is set to fall by 1.74% annually to achieve a target of reducing emissions in 2020 to 21% lower than in 2005. Within the cap, companies can trade their allowances, providing an incentive to reduce their emissions. Each year a company must surrender enough allowances to cover all its emissions, otherwise heavy fines are imposed. In its current and third phase (2013–2020), the ETS has set aside 300 million allowances in the New Entrants Reserve to fund the deployment of innovative renewable energy and carbon capture and storage demonstration projects. The Commission's proposal for the revision of the ETS after 2020 proposes an Innovation Fund endowed with 450 million emissions allowances to support innovative technologies in carbon capture and storage, renewable energy and energy-intensive industry, although no CCS projects have yet received any support from this scheme.

11.3.2.1.3 Renewable Energy Directive and Fuel Quality Directive

Across the EU, the Renewable Energy Directive (RED) (Directive 2009/28/EC) supported by Article 7a of the Fuel Quality Directive (FQD) (Directive 2009/30/EC) requires 20% of total energy consumption to be based on renewable sources, including biomass, bioliquids and biogas, by the year 2020. A specific target of 10% for renewable transport fuels is expected to be met largely by biofuels. To reach these goals, legally binding (though different) national targets have been imposed on Member States. Only biofuels certified as sustainable can contribute to this policy target. The limited environmental sustainability criteria relate primarily to the GHG reduction requirements of biofuels and place restrictions on the types of land that can be used to grow biofuel feedstock. There are no social criteria, although the EC is required to monitor the social impacts of demand for biofuels, including the effects on commodity and food prices (Helliwell and Tomei, 2016). It has been suggested that the RED is focused on sustainability issues associated with first-generation biofuels and is not prepared to account for emissions from second-generation biofuels (Whittaker et al., 2011). Widespread concern that these targets cannot be sustainably met in Europe and will instead require significant imports, which will be difficult to monitor with regard to sustainability, has led to amendments to the RED to establish new rules to reduce the risk of iLUC from biomass production (European Commission, 2016). Limiting the share of biofuels from crops grown on agricultural land, the new rules require that biofuels produced in new installations emit at least 60% less GHG than fossil fuels.

In November 2016, the Commission published a proposal for a revised RED with a target of at least 27% renewable energy sources in the final energy consumption by 2030 (European Commission, 2016) to be met through individual Member States' contributions. To address concerns around iLUC emissions, Article 7 introduces a decreasing maximum share of first-generation biofuels and bioliquids starting from 2021, offset by an increase annually towards a total 3.6% share for second-generation biofuels in 2030. The RED will also introduce stricter sustainability criterion for peatland protection and new risk-based sustainability criterion for forest biomass, as well as a LULUCF requirement for ensuring proper carbon accounting of carbon impacts of forest biomass used in energy generation. New biofuel plants will attract a 70% GHG-saving performance requirement while a new 80% saving requirement is applied to biomass-based heating/cooling and electricity with a fuel capacity above 20 MW. The revised RED adopts for the first time a common GHG accounting methodology for biomass fuels for heat and power.

As the United Kingdom prepares to leave the EU and the single EU energy market, it is difficult to predict what the effects on national climate-change and energy policies will be until the government has determined its negotiating position. However, energy imports, whether biofuels or biomass, are likely to become more expensive if the exchange rate of the pound against the currencies of trading partners is adversely affected. Balancing economic interests with the public interest, and the relative security of the single energy market with leaving the EU, will undoubtedly be politically challenging. Yet, it could be viewed as a fortuitous opportunity to bring UK energy policies into better alignment with agricultural and industrial policies to help tackle the wicked problems of climate change and food security while supporting economic growth. Nevertheless, the United Kingdom's physical proximity to the EU and the extent of shared energy resources and infrastructures means that future energy policies will need to consider this persevering interface. Either way, the future UK energy system should be one focused on decarbonisation involving a suite of low-carbon technology options. Of these, bioenergy and CCS are considered as the most critical (Energy Technologies Institute, 2015).

11.3.2.2 National Policy Frameworks in the United Kingdom
11.3.2.2.1 Renewables Obligation and Contracts for Difference
The United Kingdom has incentivised renewable electricity, including electricity from bioenergy, for over 25 years. Early schemes included the Non-Fossil Fuel Obligation (NFFO), established in the 1990s, which contracted with particular developers to purchase renewable electricity on a long-term contract. These had limited success in incentivising bioenergy because of its higher market cost, but did manage to secure some landfill gas and waste-to-energy capacity (Thornley, 2006). A key part of this scheme was that obligations were placed on supply companies to produce a certain proportion of their electricity from renewables. It did not fix the price of the renewable electricity, but awarded green certificates for every unit supplied, which could then be traded to obtain revenue, the market value for such certificates being determined by the 'buy-out price' that suppliers who failed to produce sufficient certificates had to make. The principle was that this market-based system would encourage competition and the lowest-cost renewable solutions. Eventually, this differential was recognised and the Renewables Obligation was 'banded' to facilitate a range of payment levels, with different technologies recognised as being in need of different levels of support to maintain a competitive marketplace and awarded different numbers of renewable obligation certificates (ROCs). Subsequently, feed-in-tariffs (FITs) were introduced in April 2010 for small-scale renewable generators and the scheme makes guaranteed payments to all eligible installations for electricity generated for a period of 20 years, giving market-price stability to generators, which supports financing and investment. Larger-scale bioelectricity systems enter directly into a limited number of contracts for differences (CfDs), which facilitate fixed payments for electricity delivered.

11.3.2.2.2 Renewable Transport Fuel Obligation

The success of the Renewables Obligation led to a significant increase in UK electricity production from renewables and so a similar scheme was crafted to address a similar need in the transport fuel sector. The Renewable Transport Fuel Obligation (RTFO) was implemented in 2008 to support the UK government's policy on reducing GHS emissions from vehicles by encouraging the production of renewable biofuels with lower carbon intensity. Under the RTFO, suppliers of transport and mobile machinery fuel in the United Kingdom must be able to show that a certain percentage of the fuel they supply comes from certified renewable and sustainable sources. In 2012 GHS emissions reporting was introduced so that suppliers had to demonstrate how they had achieved at least 35% GHG emissions saving, with this saving rising over time.

The first phase of the Renewable Heat Incentive (RHI) was introduced in November 2011, as part of UK efforts to expand energy-sector RE targets and GHG reductions. It is an environmental programme designed to increase the uptake of renewable heat technologies by providing incentive payments to eligible generators of renewable heat for commercial, industrial, not-for-profit and public-sector purposes and to producers of biomethane. It provides a subsidy, payable for 20 years, to eligible, non-domestic renewable heat generators (e.g. biogas CHP) and biomethane to grid projects.

11.4 Ensuring Environmental, Economic and Social Sustainability of a BECCS System

11.4.1 Environmental Sustainability and System Scope

A bioenergy system must include a conversion device that converts biomass into a required energy vector. However, if there is a desire to take into account wider environmental impacts of bioenergy, the wider system beyond this conversion process must also be considered, particularly production of the biomass. Figure 11.1 shows the many different aspects that could be included in a bio-CCS system: land preparation/change, growth of the biomass, harvesting, transport, processing, haulage and CO_2 sequestration. A key decision that must be taken at the very outset is defining the scope and objectives of the study. It is critical for LCA studies to consider the LCA question being addressed at the outset of the study, as this should actually inform the methodological choices, particularly the scope of system included in the calculation. Therefore, the scope of system chosen should be appropriate to the LCA 'question' being asked. Variations in this scope of system are a major cause of variation in published LCA results for bioenergy systems. For example, previous work has shown that including the GHS impact of land-use change into bioenergy system LCA calculations can make the difference between a bioenergy system achieving significant GHS reductions and actually increasing emissions compared to a fossil fuel system (Upham et al., 2009).

Indeed, the underpinning rationale for the negative emissions characteristic of a bio-CCS system is the choice of scope of system. Essentially, the significant sequestration impacts of the 'biomass growth' stage more than offsets the emissions incurred in the land preparation, harvesting, processing, transport and haulage stages. If the carbon dioxide emissions associated with the conversion process were then released to the atmosphere, the result would be an energy system with a net GHS balance where the sequestration from the atmosphere offset a significant amount of the emissions released to the atmosphere, resulting in a lower carbon footprint than a fossil fuel system, where there is no offset from sequestration during the growth stage. However, if we then add in a final carbon sequestration (geological storage) stage, we end up with two negative numbers in the overall equation (one from natural carbon sequestration from the atmosphere during biomass growth and the other from forced sequestration of carbon in appropriate reservoirs) and that is sufficient, for many systems studied, to tip the balance to result in net negative emissions of carbon dioxide across the whole system.

As discussed in Section 11.2.2, the concept of 'net negative' is therefore entirely predicated on the scope of the system and in particular the inclusion of negative credits for carbon dioxide removal from the atmosphere during plant growth, which is subsequently stored after energy has been extracted from the biomass. The scope of the system must be related to the study objective or LCA question being asked, and this often involves a critique of the rationale or philosophy behind the choice of the system, e.g. the causal links that justify relating a particular quantity of carbon sequestered during growth to generation of a particular energy output. It was suggested above that afforestation and carbon dioxide storage could have the same net climate impact as a BECCS system and could be much more manageable to achieve, since it does not require any of the engineering challenges associated with the development of advanced-technology BECCS systems. Ideologically, though, it is less satisfying as there is no causal link between the forest carbon sequestration and the energy provision and therefore it would not be plausible to attribute any negative emissions to the energy system in an energy model or other energy-system accounting framework. Also there is no explicit rationale for maintaining the permanency of the sequestration or storage in forest systems. These alternative paradigms raise very significant questions about the integrity/governance of what we call a BECCS 'system'. There is a natural inclination to consider a system as the physical alignment of material and energy flows and so it is not reasonable to credit a 'net negative' result to a forest and fossil fuel CCS system; yet it is reasonable to provide such credit when a proportion of the forest is used in the conversion system.

Putting aside ideological concerns, the key issues raised by this focus around whether it makes sense to incentivise and manage BECCS systems as systems or as discrete components are: is it necessary or desirable to have policies or provisions in place that will optimise the net negative effect of the system around which we choose to draw a physical boundary or is it enough to separately increase afforestation and sequestration, with a similar net effect? These issues become more complex when linked with the complexities posed by trade and globalisation. In one sense, it could be argued that most industrialised nations effectively outsource their carbon reduction commitments by being net importers of goods manufactured in (and incurring emissions in) poorer countries in the global South (often ones that do not have legally binding emissions reduction commitments) and so the issues raised by BECCS governance are no different from those raised by the general phenomena of global trade, which encourages richer countries to effectively export their carbon commitments rather than actually act to deliver the reductions intended. A consumption-based approach to emissions is often cited as a potential solution to this problem, and using LCA to evaluate net GHS balances along a bioenergy (or any other fuel) supply chain is effectively a step in that direction.

However, normally this extension of supply-chain/consumption approach is intended to attribute additional emissions to consuming countries, whereas in the case of BECCS we are effectively transferring the credit associated with carbon sequestration/biomass production to the consuming country, which also benefits from the energy produced. This raises issues of equity and governance: for example, in what context can or should a producer country effectively transfer its ownership of the carbon sequestration credits associated with BECCS in order to deliver the desired negative emissions?

11.4.2 Economic Sustainability and System Scope

Energy-system modelling is widely used to assess whether a sufficient level of BECCS can be deployed in the United Kingdom to support cost-effective decarbonisation pathways for the United Kingdom out to 2050. But assessments of financial sustainability and cost-effectiveness at the level of different BECCS pathways alone are not enough to determine the long-term economic sustainability of BECCS implementation. Sustainable systems-wide change is dependent on economic viability for all stakeholders along the BECCS supply chain. Although these networks may be focused around national pathways, in a globalised context national energy visions often rely on resources from wider networks and markets (Schot and Geels, 2008). In the case of BECCS, some models estimate the need for both imported and domestic biomass, each at levels of more than double the amount currently being used for bioenergy in the United Kingdom (Energy Technologies Institute, 2016).

In a global bioenergy system, the economic case for importing biomass for use in the United Kingdom can only be made if the environmental or social sustainability credentials of producing the biomass also stack up. Globally, the markets that regulate biomass supply and demand are driven almost entirely by political subsidies enacted at different territorial scales (Raman and Mohr, 2014). This politics of scale can be used both to obscure the socio-ecological consequences of emissions reduction strategies, which are often felt in distant places, and to justify the need to develop 'locally' sustainable bioenergy supply chains. Where bioenergy supply chains have extended across national borders, some of these have been linked to 'land grabs' leading to dispossession of local people and loss of livelihoods, often enabled by weak governance regimes (van der Horst and Vermeylen, 2011). One of the drivers posited for growing biomass in developing countries for export to the world bioenergy market is 'rural development'. However, such countries remain entrenched as suppliers of raw materials as opposed to higher 'value-added' end products (van Gelder and German, 2011).

Domestic sourcing of biomass would not in itself address all the potential negative impacts of imports nor necessarily deliver the rural development opportunities envisioned in national bioenergy policies. Some key assumptions embedded in visions for use of UK marginal land for bioenergy crops so as not to compete with food crops grown on arable land do not reflect real-life farming conditions (Mohr et al., 2016). Farmers' decision-making regarding whether to diversify activities to include bioenergy crops or to redeploy residues from food or feed crops involves numerous challenges. A survey of 249 arable farms found that over 80% of farmers would not consider switching to dedicated energy crops (Glithero et al., 2013). A key factor influencing farmer decision-making is that they consider food production to be the moral purpose of farming. Elsewhere research has shown that bioenergy crops such as SRC willow are perceived as wood and, therefore, contrary to the purpose of arable farming (van der Horst and Evans, 2010).

Arable farmers' decision-making is further complicated by the fact that they are themselves locked into complex supply chains involving agronomists, supermarkets and agricultural policy-makers over which they have little control. Decisions over the end use for agricultural residues such as baled straw are generally taken by merchants who have an interest in preserving their existing customer base in non-bioenergy sectors (Helliwell, 2016). In the case of livestock farmers, they do not consider their own land to be of marginal quality and, in this sense, available for bioenergy crops (Helliwell, 2016). Such crops are regarded as alien by farmers who lack the necessary experience and equipment and whose business models and existing cropping and harvesting patterns may not be compatible with bioenergy crops. While market/profit incentives are important to farm-level decision-making, farming/rural cultures and the legacy of existing norms, practices and infrastructures, including relationships and advice from other supply-chain intermediaries, mean that building new domestic supply chains will be challenging. The extent to which there is sufficient marginal land in the United Kingdom on which to sustainably cultivate bioenergy crops for a domestic supply chain is also in question.

The economic viability of BECCS systems faces a further potential challenge posed by trade rules. Bioenergy crops are seen as part of a global agricultural economy and biomass is a globally traded commodity. Subsidies for creating domestic biomass markets in place of foreign-sourced biomass or bioenergy may be deemed to have a global-trade-distorting effect prohibited by World Trade Organization (WTO) rules, although opinions on this matter are contradictory (Raman et al., 2015). Bioenergy trade networks have become spatially organised around North–South configurations whereby the benefits of consumption are largely experienced in the global North while the burdens of production are unjustly transferred to the global South. The scale of such networks has been linked to socio-economic and environmental harms in the global South where weak governance regimes have aided the establishment of monoculture plantations where once subsistence farming was the norm. EU bioenergy policies have been implicated in this shift in economic circumstance through their setting of bioenergy consumption targets (Raman and Mohr, 2014). National policies such as the RTFO have further exacerbated the situation by allowing decisions regarding feedstock, technology and vector type, as well as their place of production and consumption, to be determined by the market.

BECCS deployment will require an established CCS infrastructure to be in place. However, significant uncertainties about the economic viability of CCS led the UK parliament to cancel a proposed £1 billion CCS demonstration competition. In response, a recent report by the Parliamentary Advisory Group on Carbon Capture and Storage chaired by Lord Oxburgh sets out a case for continuing CCS investment, arguing it has a central role to play across the UK economy if we are to deliver on our emissions reduction commitments at the lowest possible cost to the United Kingdom (Oxburgh, 2016).

11.4.3 Social Sustainability and System Scope

In the real-world contexts in which BECCS systems will operate, social impacts are intrinsically interconnected with economic and environmental impacts, such that their artificial separation in assessments of their sustainability will be problematic (Mohr and Raman, 2013). Moreover, restricting social sustainability to questions of impacts on society risks ignoring important social and value judgements involved in choices over which impacts to include and how to assess them (Palmer, 2012). For example, while some environmental impacts such as GHG emissions represent gains/losses to the planet as a whole, others such as eutrophication are location specific (Thornley and Gilbert, 2013). Included in what initially appears to be a specific environmental criterion for sustainability assessment are impacts arising from socio-environmental change that place disproportionate burdens on some social groups, places or ecologies, while elsewhere their sustainability might be enhanced (Swyngedouw, 2007).

Many of the social impacts of BECCS deployment will be experienced at the local level, especially where biomass is cultivated in poorer Southern regions for bioenergy use elsewhere (van der Horst and Vermeylen, 2011) or where bioenergy production or carbon capture and storage facilities may impact on local air quality or on the aesthetics of the local landscape. Local contextual factors are not usually captured by models, such as integrated assessment models, which often focus on the global scale and where the need to simplify the representation of complex systems obfuscates the inclusion of local heterogeneity (Geels et al., 2016). As discussed above, the interaction of sustainability certification schemes with the heterogeneity of local contexts has similarly highlighted their limited scope in dealing with social impacts. Nor are the direct and indirect human and social impacts caused by land-use change, particularly in the poorer regions of the global South, alleviated by iLUC factors, which are instead directed towards measuring carbon emissions to the exclusion of local impacts. This tendency to fetishise carbon as something that is 'placeless' and that exists independently of socio-economic considerations has been recognised as symptomatic of climate-change debates generally (Swyngedouw, 2007).

BECCS highlights the fact that the physical elements of the supply chain have to be located somewhere with a distinct sociocultural heritage and a distinct physical environment. Governance mechanisms must therefore be sensitive to these nuances of place and context and the fact that even local/national BECCS systems may interface with global networks and other resource systems and have impacts beyond the material boundaries of the supply chain. How these interfaces and the spatially sensitive questions they raise are governed will be crucial for the future development of BECCS systems.

Imported energy crops are expected to play a significant role in the UK energy sector by 2050. A techno-economic study commissioned by the Energy Technologies Institute (ETI) to inform the commercial deployment of biomass CCS concluded that large volumes of biomass supplies will be available for all combustion and gasification technologies such that feedstock availability or suitability will not be a deciding factor in the choice of biomass CCS options (Mac Dowell, 2014). Narrowly framing the issue as one of sufficient supply of biomass regardless of its provenance, however, neglects important social sustainability issues, which we have already raised in this chapter. Beyond issues of specific local impacts, social sustainability is also dependent on broad societal support. The strength of public feeling against the use of imported feedstock over domestic feedstocks, as long as the United Kingdom maintains at least current levels of food self-sufficiency, is informative in this regard. A 2016 YouGov survey conducted by the ETI on public perceptions of bioenergy in the United Kingdom found concerns over imported biomass came second only to competition for land (ETI, 2016). The intended location of BECCS plants on the coast may be logical to take advantage of biomass imports and access offshore storage sites, but it could prove counter-productive to the development of local BECCS supply chains beyond just those few coastal locations. If the socio-economic and environmental benefits of siting BECCS facilities are perceived to be benefiting a select few, then public support beyond those locations may be limited.

The scale at which BECCS systems will be organised will also be critical in determining not only their carbon sustainability but also the spatial and temporal distribution of impacts associated with the systems. As we are talking about BECCS systems, which are yet to be implemented at a significant scale, there is a need to make explicit expectations of intended benefits as well as key concerns about the distribution of impacts that might arise. Expectations of benefits from BECCS systems for delivering negative carbon emissions are based on the assumption that biomass production can be done sustainably without clashing with food, resource and livelihood security. However, positive expectations are tempered by concerns that, in practice, bioenergy development in a global economy can clash with food and livelihood security with incentives favouring sourcing of feedstocks from poorer developing countries. Thus, there is a critical need to address these direct and indirect social sustainability issues across the full scope of BECCS systems.

11.4.4 Trade-Offs Between Different Sustainability Components

Of course, the GHS balance is only one impact of a BECCS system. There are many others: primary energy consumption (and associated fuel/resource depletion); airborne emissions (even when flue gases are captured there remain airborne emissions that have air-quality impacts associated with the supply chain, e.g. for vehicle transport or wood drying); eutrophication and acidification impacts on land; jobs created; materials consumed *etc.* Increasingly bioenergy systems are being evaluated not just for their GHS or environmental impacts (where the aim is often to ensure that there is not significant harm being done) but also for their wider social and economic impacts. While these are more challenging to quantify, they are dependent on the feedstock, technology and system design, are part of the plant impact and so are important considerations during project development.

One way of evaluating these impacts is to consider the 'environmental risk' associated with implementation of a bioenergy system compared to the counterfactual. Figure 11.2 shows the environmental impact profile of a typical bioenergy system (without CCS) (for more details see Thornley and Gilbert, 2013). Environmental risk is a product of the probability of the risk event occurring and the severity of impact if it does occur. This can be expressed as a product on semi-quantitative scales (high, medium or low) producing numerical indicators for a range of different (environmental, economic and social) impacts of bioenergy systems.

Figure 11.2, therefore, is a composite display of the sustainability of a bioenergy system compared to a fossil fuel system. The fossil fuel system is the circular reference level in the diagram, and if the bioenergy system is more sustainable, then it will score higher for that particular attribute: making the overall system more sustainable can be visualised as stretching the circle to make its area larger.

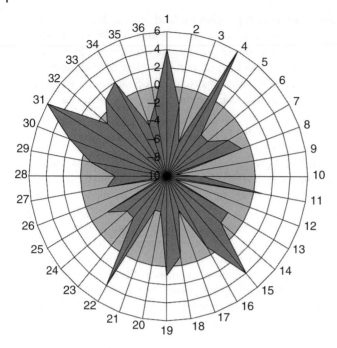

Figure 11.2 Radar diagram showing overall sustainability assessment of biodiesel from Argentinean soy compared with mineral diesel (light grey = reference level; dark grey = scores for Argentinean soy system).

Imagine, therefore, if we tried to improve the GHG performance of a particular bio-energy system. We might do so by increasing fertiliser input and thereby yield. This would cause the attribute related to GHGs per unit of energy produced to decrease, and, therefore, the sustainability of the system with respect to that attribute would improve. But the additional use of phosphate would cause the attribute related to natural resource depletion to decrease. Alternatively, we could switch to a larger, more efficient system, when energy efficiency and airborne pollution would both improve, but job creation for local communities and ecosystem impacts on soils would most likely get worse (depending, of course, on how agronomy and supply chains changed in response to this conversion plant shift). So, there is no hard-and-fast boundary between a system that is 'sustainable' and one that is 'unsustainable'; changes made to improve one component of sustainability will inevitably affect other components as well (negatively and positively) and there are constantly evolving trade-offs to consider between the myriad of different impacts of bioenergy systems.

It is a basic principle of risk management that, wherever possible, risk should be placed with the party best able to mitigate and deal with it. When we look at Figure 11.2, it is clear that the interrelationship between different sustainability impacts challenges that notion significantly. How can the most appropriate risk management party be identified if their system intervention might result in a change in a different system impact which may be beyond their control? Overall optimisation of such a system (or perhaps more realistically, minimisation of the system's cumulative negative impacts) can only be achieved if there is simultaneous oversight, cognisance and control of the different impacts.

The big challenge presented by BECCS is that the most significant desired impact (minimisation of GHG emissions) is actually a composite output of the actions of several different actors, and it is difficult to conceive of a governance framework that would effectively enable minimisation of the net GHG output, while taking into account other impacts that may be beyond the control and perhaps even the sight of the party being incentivised by claiming the GHG reductions.

Sustainability criteria have been developed as a partial response to this problem – preventing the worst damage from occurring upstream of the conversion plant – but they are inherently limited in their ability to offer a more holistic solution.

11.5   Governance of BECCS Systems

A recurrent criticism aimed at bioenergy governance has been its inability to respond consistently and comprehensively to the full range of externalities linked to bioenergy production. This is in part because EU policy has delimited bioenergy specifically as a low-carbon energy technology and not a process or alignment of interconnected processes. Accordingly, governance of bioenergy has focused on delivering end products (e.g. bioelectricity, biofuels, *etc.*) without providing steering on preferences for feedstocks, places and practices of production, *etc.*, in other words, without assessing the complex global supply chains that constitute the final product (Helliwell, 2016). In the United Kingdom, the Renewables Obligation (RO: overseen by Ofgem, the government regulator for gas and electricity markets) and the RTFO (overseen by the Department for Transport) are the only mechanisms governing bioenergy production. The numerous interfaces with other systems along the length of biofuel supply chains, including and especially agriculture, are not envisaged by the specific policy focus on bioenergy/biofuel as a low-carbon energy technology.

It can be argued that the sustainability certification criteria established in the EU RED and subsequently imposed on the RO and RTFO encompass sites and modes of biomass production, but these are quite narrowly focused and many aspects of agricultural practices, conversion, distribution and end use are not included. However, many of the bioenergy controversies and negative socio-economic and environmental externalities identified throughout this chapter arise at the interfaces with adjacent systems and are the products of the complex relations between the heterogeneous human and non-human elements of bioenergy supply chains. Consequently, a whole-systems governance of BECCS is about moving beyond carbon myopia to develop hybrid policy solutions to match the complexity and breadth of sustainability issues raised. Bringing the whole supply chain (as opposed to just the end product) into the purview of governance opens up opportunities to not only regulate the techno-economic components of the supply chain but also to be responsive to specific ways in which these components interact with humans in different contexts and at different times. But can this be achieved in practice?

The rationale for a comprehensive systems-level evaluation of BECCS to assess not only its capacity to provide net removal of GHGs from the atmosphere but the full extent and range of its impacts appears sound. However, such a comprehensive scope of the system also poses some distinctive challenges for its governance. From a governance perspective, the question is not so much whether there is a process or physical link between the different system elements, but how the system (and its myriad human and technical elements) behaves as an entity. As we have noted, even though GHG emissions are a global issue, the effects of interventions (e.g. growing, transporting and processing biomass, siting of bioenergy and CCS facilities) are often context specific. Many emissions reduction choices depend on specific aspects of particular supply-chain situations, including potential direct and indirect impacts on interfacing systems. For this reason, policy and fiscal interventions such as regulations, prices and other incentives may need to be somewhat individualised according to the particular part of the supply chain and its interfaces with other parts of the supply chains and allied systems. But policy and fiscal incentives alone will be insufficient drivers to establish new BECCS systems.

Pursuing a negative emissions agenda will require economies to be nudged down more sustainable paths of carbon-negative development. But governance strategies that focus on techno-economic advances in the processes of energy production to cut emissions often overlook the role of people in shaping energy and climate systems. Human interactions with low-carbon systems are more than just fiscal in nature. An emphasis on supply-side technical fixes obfuscates that it is often broader social, economic and environmental concerns that determine whether energy and climate policies are effective. Because BECCS systems are characterised by complexity and uncertainty, their governance requires a certain level and breadth of understanding to help facilitate transitions that meet the targets for limiting GHG emissions. Modelling based on narrow techno-economic assumptions will need to be supplemented with assumptions based on empirical analysis of the specific contexts under investigation. In designing governance policies and programmes, it is important to recognise that judgements of the risks of climate change have both objective and subjective elements involving both facts and values and that trade-offs between some elements of security or sustainability with others will be required (Stern et al., 2016).

Governance will be central to the successful societal embedding of BECCS, yet little attention is paid to this synergistic relationship. The tensions between reductions in global GHG emissions and local impacts such as pollution from transporting biomass are social as well political in nature. Governing systems-level changes in a context where power is distributed across diverse societal subsystems and among many societal actors calls for governance to encompass public debate, political decision-making, policy formation and complex interactions among public authorities, business and civil society (Meadowcroft, 2007). This characterisation of governance is broader in scope than the dominant techno-economic rationality that tends to drive policy decision-making. But to understand the different impacts that can be promoted *via* different governance and policy approaches, we must take into account the specific contexts of BECCS deployment, that is, the different system elements (including humans) that comprise a particular supply chain.

What, then, should the scope of BECCS system governance look like? At a global level, countries vary greatly in their GHG emissions levels as well as in their levels of integration in global supply chains and global governance systems of treaties and protocols. Thus, countries are differently engaged in global emissions governance. Some countries exert their values and economic power to influence others. While this can lead to convergent trajectories of economic development and environmental impact, it can also lead to sharply divergent paths of development and the offshoring of adverse environmental impacts to less powerful nations (Jorgenson, 2012). However, treaties among nations, such as the COP21 Paris Agreement, that place demands on all signatories are unlikely to be both achievable and strongly implementable and, therefore, have less practical potential for reducing carbon emissions than the aggregate of less ambitious treaties involving fewer nations (Keohane and Victor, 2016). While deep and lasting cooperation across territorial, sectoral or supply-chain boundaries will be challenging, many partial efforts could build confidence and lead to larger cuts in emissions. This strategy of decentralised governance could lead incrementally to deeper cooperation. Policy and governance mechanisms for developing BECCS systems at different scales and that are sensitive to different spatial impacts therefore need to be examined.

11.6   Conclusions: The Future of BECCS Policy and Governance

Uptake of BECCS in integrated assessment models is predicated on an assumed scope of system that combines biological carbon dioxide sequestration from the atmosphere with energy production and physical sequestration of the CO_2 even if these activities take place in different countries. This alignment is appropriate if trying to establish the net impact of the supply chain, but is not consistent with the principles of operation of the UNFCCC agreement, which focuses on national, territorial responsibility for GHGs emissions. There have been many challenges to this established method of accounting in recent years, with some claiming that a consumption-based emissions accounting system would be more appropriate and equitable. This would require countries to account for the GHG emissions associated with their imports, which would include wood products and therefore would address some of the issues associated with management and responsibility for emissions. In other words, it would be more legitimate for a country to claim sequestration credits in relation to biomass they import and utilise if they were also accounting for GHG emissions associated with other materials they import and utilise. The present framework does not facilitate that and so claiming the sequestration credit seems perverse.

Using a supply-chain framing allows consideration of how to minimise emissions along that supply chain, but that does not necessarily minimise global emissions, which is the key objective of the UNFCCC and international climate policy. Evaluating the net global impact of BECCS systems requires a consequential LCA approach and wider perspective on emissions trade-offs between different technology and development pathways. It is difficult to envisage how best practice could be managed, governed or incentivised within existing governance frameworks, and this should be a focus of future evaluation. In fact, a set or suite of overlapping policy and governance mechanisms for developing BECCS systems at different scales and that are sensitive to different spatial impacts need to be explored. Such a multilevel governance approach would help to focus attention on the local nodes of global supply chains where issues such as public support for BECCS facilities, managing local impacts and strengthening local supply chains are often beneath the scope of national and global frameworks.

A multilevel governance approach might also help to bridge the value–action gap that exists between global strategies and national policies for emissions reductions on the one hand and their connection to everyday life practices on the other. While people in general are supportive of reducing emissions, this support is not matched by the equivalent level of action. Often this has to do with the way in which the policy or governance problem is framed in national and global discourses and how this connects, or not, to the ways in which people are experiencing and valuing practices related to emissions reductions in everyday life. In this context, existing governance frameworks are not fit for purpose. Integrated assessment models privilege simplified representations of the global/national scale over the dynamics and scope of local heterogeneity. Emissions reduction along supply chains is not a simple case of technological improvements and economic incentives or deterrents but also requires us to understand the complex ways in which the heterogeneous supply-chain elements interact with one another over space and time. Alongside techno-economic innovation, understanding of the role of institutional and socio-economic innovation, human–technology interactions and social practices and routines in effecting lasting system change will be just as important for BECCS governance.

References

Committee on Climate Change (2015). *The Fifth Carbon Budget: The Next Step Towards a Low-Carbon Economy*. Committee on Climate Change.

Di Lucia, L. (2010). External governance and the EU policy for sustainable biofuels, the case of Mozambique. *Energy Policy* **38** (11): 7395–7403.

Energy Technologies Institute (2015). *Options, Choices, Actions: UK Scenarios for a Low Carbon Energy System Transition*. Loughborough: ETI.

Energy Technologies Institute (2016). *The Evidence for Deploying Bioenergy with CCS (BECCS) in the UK*. Loughborough: ETI.

European Commission (2016). Proposal for a Directive of the European Parliament and of the Council on the promotion of the use of energy from renewable sources (recast). Brussels, COM/2016/0767 final/2–2016/0382 (COD).

Geels, F.W., Berkhout, F., and Van Vuuren, D. (2016). Bridging analytical approaches for low-carbon transitions. *Nature Climate Change* **6** (6): 576–583.

van Gelder, J.W. and German, L. (2011). *Biofuel Finance*. Centre for International Forestry Research (CIFOR) Brief http://www.cifor.org/publications/pdf_files/infobrief/3340-infobrief.pdf.

Glithero, N.J., Wilson, P., and Ramsden, S.J. (2013). Straw use and availability for second generation biofuels in England. *Biomass and Bioenergy* **55**: 311–321.

Helliwell, R. (2016). Imagining biofuels – building agricultural supply chains in the UK: a comparison of UK policy expectations with on-farm perspectives. Unpublished thesis. University of Nottingham.

Helliwell, R. and Tomei, J. (2016). Practising stewardship: EU biofuel policy and certification in the UK and Guatemala, agriculture and human values. *First Online*, 17 October 2016.

Hornborg, A. (2008). Environmental load displacement in world history. In: *Sustainable Development in a Globalized World* (ed. B. Hettne), 94–116. Basingstoke: Palgrave Macmillan.

van der Horst, D. and Evans, J. (2010). Carbon claims and energy landscapes: exploring the political ecology of biomass. *Landscape Research* **35** (2): 173–193.

van der Horst, D. and Vermeylen, S. (2011). Spatial scale and social impacts of biofuel production. *Biomass and Bioenergy* **35** (6): 2435–2444. doi: 10.1016/j. biombioe.2010.11.029.

Jacoby, H.D. and Chen, Y.H.-H. (2015). *Launching a New Climate Regime*. MIT Joint Program on the Science and Policy of Global Change.

Jorgenson, A.K. (2012). The sociology of ecologically unequal exchange and carbon dioxide emissions, 1960–2005. *Social Science Research* **41**: 242–252.

Keohane, R.O. and Victor, D.G. (2016). Cooperation and discord in global climate policy. *Nature Climate Change* **6**: 570–575.

Mac Dowell, N. (2014). Power generation in the UK: carbon source or carbon sink? Presentation given at the UKCCSRC Direct Air Capture/Negative Emissions Workshop, 18 March.

Mander, S., Polson, D., Roberts, T., and Curtis, A. (2010). Risk from CO_2 storage in saline aquifers: a comparison of lay and expert perceptions of risk. *Energy Procedia* **4**: 6360–6367.

Markusson, N., Kern, F., Watson, J. et al. (2012). A socio-technical framework for assessing the viability of carbon capture and storage technology. *Technological Forecasting and Social Change* **79** (5): 903–918.

Meadowcroft, J. (2007). "Who is in charge here?" Governance for sustainable development in a complex world. *Journal of Environmental Policy & Planning* **9**: 299–314.

Mohr, A. and Raman, S. (2013). Lessons from first generation biofuels and implications for the sustainability appraisal of second generation biofuels. *Energy Policy* **63**: 114–122.

Mohr, A., Shortall, O., Helliwell, R., and Raman, S. (2016). How should land be used? Bioenergy and responsible innovation in agricultural systems. In: *Food Production and Nature Conservation: Conflicts and Solutions* (ed. I. Gordon and H. Prins). London: Earthscan (Routledge).

OECD/IEA (2016). *20 Years of Carbon Capture and Storage. Accelerating Future Deployment*. Paris: International Energy Agency.

Oxburgh (2016). Lowest cost decarbonisation for the UK: the critical role of CCS. Report to the Secretary of State for Business, Energy and Industrial Strategy from the Parliamentary Advisory Group on Carbon Capture and Storage (CCS).

Palmer, J. (2012). Risk governance in an age of wicked problems: lessons from the European approach to indirect land use change. *Journal of Risk Research* **15** (5): 495–513.

Parkus, A., Röder, M., Gawel, E. et al. (2015). Handling uncertainty in bioenergy policy design – a case study analysis of UK and German bioelectricity policy instruments. *Biomass and Bioenergy* **79** (August): 64–79.

Raman, S. and Mohr, A. (2014). Biofuels and the role of space in sustainable innovation journeys. *Journal of Cleaner Production* **65**: 224–233. doi: 10.1016/j.jclepro.2013.07.057.

Raman, S., Mohr, A., Helliwell, R. et al. (2015). Integrating social and value dimensions into sustainability assessment of lignocellulosic biofuels. *Biomass and Bioenergy* **82**: 49–62.

Schot, J. and Geels, F. (2008). Strategic niche management and sustainable innovation journeys: theory, findings, research agenda and policy. *Technology Analysis and Strategic Management* **20** (5): 537–554.

Stern, P.C., Sovacool, B.K., and Dietz, T. (2016). Towards a science of climate and energy choices. *Nature Climate Change* **6**: 547–555.

Swyngedouw, E. (2007). Impossible "sustainability" and the Postpolitical condition. In: *The Sustainable Development Paradox: Urban Political Economy in the United State and Europe* (ed. R. Krueger and D. Gibbs), 13–40. New York: Guildford Press.

Thornley, P. (2006). Increasing biomass based power generation in the UK. *Energy Policy* **34** (15): 2087–2099.

Thornley, P. and Gilbert, P. (2013). Biofuels: balancing risks and rewards. *Interface Focus* **3** (1): 2042–8901.

Tomei, J. (2015). The sustainability of sugarcame-ethanol systems in Guatemala: land, labour and law. *Biomass and Bioenergy* **82** (November): 94–100.

Upham, P., Thornley, P., Tomei, J., and Boucher, P. (2009). Substitutable biodiesel feedstocks for the UK: a review of sustainability issues with reference to the UK RTFO. *Journal of Cleaner Production* **17**: S37–S45.

Vaughan, N. and Gough, C. (2016). Expert assessment concludes negative emissions scenarios may not deliver. *Environmental Research Letters* **11** (9): 095003. doi: 10.1088/1748-9326/11/9/095003.

Whittaker, C., McManus, M., and Hammond, G.P. (2011). Greenhouse gas reporting for biofuels: a comparison between the RED, RTFO and PAS2050 methodologies. *Energy Policy* **39**: 5950–5960.

12

Social and Ethical Dimensions of BECCS

Clair Gough[1], Leslie Mabon[2] and Sarah Mander[1]

[1] Tyndall Centre for Climate Change Research, School of Mechanical Aerospace and Civil Engineering, University of Manchester, UK
[2] Robert Gordon University, Aberdeen, UK

12.1   Introduction

Previous chapters have explored the possible role for BECCS in achieving climate-change targets, technical issues associated with its implementation and the means to verify that it can deliver its assumed negative emissions potential. This chapter considers the social and ethical factors associated with its use. With lack of progress on halting the global increase in greenhouse gas emissions, the challenge of meeting agreed targets for limiting global average temperature increases has become huge (see Chapter 9); a reliance on BECCS to deliver 'negative emissions' is a key feature of the majority of scenarios likely to deliver carbon budgets compatible with these targets (Fuss et al., 2014; Vaughan and Gough, 2016). The large-scale implementation of carbon capture and storage (CCS) infrastructure connected to unprecedented amounts of bioenergy production implied by the assumed contribution from BECCS is likely to present significant social and ethical challenges. Thus, understanding the potential for BECCS means understanding the social and ethical context in which BECCS approaches must operate, with implications spanning a variety of scales – from the individual components that make up a BECCS chain, ranging from small-scale projects to plantation level, right up to global supply chains and policy frameworks. Furthermore, relying on BECCS to deliver global net negative emissions as a means of offsetting emissions, delaying ambitious mitigation measures or overshooting emissions targets, brings additional layers of complexity to the potential social and ethical considerations.

Although there is a substantial body of research forming around the possible societal responses to CCS (see overviews in Ashworth et al. (2013, 2015) and L'Orange Seigo et al. (2014)) and similarly for the use of biomass energy (Dale et al., 2013; German et al., 2011; Raman et al., 2015; Ribeiro and Quintanilla, 2015), there is very little research exploring the impact that using CCS technologies with biomass rather than fossil feedstocks might have on how it will be received across society. The bioenergy requirement sufficient to deliver net negative emissions from BECCS will mean increasing demands

Biomass Energy with Carbon Capture and Storage (BECCS): Unlocking Negative Emissions, First Edition.
Edited by Clair Gough, Patricia Thornley, Sarah Mander, Naomi Vaughan and Amanda Lea-Langton.
© 2018 John Wiley & Sons Ltd. Published 2018 by John Wiley & Sons Ltd.

on available land for bioenergy production, both in socio-economic terms (such as food production, housing etc.) and environmental implications (biodiversity, water resources, fertiliser use etc.) (Tilman et al., 2009; Slade et al., 2014; Powell and Lenton, 2012). Furthermore, various ethical issues associated with the use of CCS as a mitigation option have been discussed (Boucher and Gough, 2012; Gough and Boucher, 2013; Mabon and Shackley, 2015; McLaren, 2012a; Medvecky et al., 2014; Spreng et al., 2007). Dowd et al. (2015) consider insights from social science relevant to BECCS, and there is a small number of studies which explicitly explore specific social impacts of BECCS technologies (for example, Dütschke et al., 2014; Feldpausch-Parker et al., 2015).

CCS activities to date have been concentrated in certain key regions with existing fossil fuel infrastructure and storage potential (e.g. Europe, the United States, China, Australia *inter alia*), but interest in the technology extends beyond these areas (see for example Roman, 2011). Introducing BECCS on large scales suggested in climate mitigation scenarios will necessarily expand the number of nations and regions involved in CCS mitigation, particularly those with the greatest potential biomass resource (including many developing countries). This, in turn, will have implications for climate and energy policies, regulation and international regulatory frameworks that can accommodate the international dimensions of the BECCS supply chain – there is very little research on the global governance of CCS (Bäckstrand et al., 2011), let alone BECCS. Meeting society's energy needs while delivering strict climate constraints in this complex context suggests policy intervention at state level and above will be required (see Chapter 11); these challenges are described in detail and explored with explicit reference to CCS by Torvanger and Meadowcroft (2011).

This chapter explores some of the principal social and ethical issues associated with the use of BECCS technologies in pursuit of global net negative emissions: the relationship between CCS and fossil fuels and how it relates to other mitigation options and other negative emissions technologies; the notion of sustainable decarbonisation and the potential social responses to the technology. The final section of the chapter explores various aspects of justice (distributional, procedural, financial and intergenerational) in relation to BECCS. Each section draws on published literature, supplemented with material drawn from a specialist technical workshop of invited experts on governance and ethics of CCS and BECCS, convened as part of a series of technical workshops supported by the UKCCSRC.[1] The workshop provided a mix of presentations covering the themes of ethical dimensions and governance issues associated with CCS, BECCS and negative emissions, followed by facilitated discussions in small groups to identify key ethical principles and governance issues.

12.2 Fossil Fuels and BECCS

The implementation of CCS with biomass energy should be considered in the context of current fossil fuel use; whether it is deployed as part of a system co-firing biomass and fossil fuels (and co-firing with a proportion of 30% (see Chapter 6) which may deliver

---

1 UKCCSRC (2015). Issues of governance and ethics in CCS. Specialist technical workshop. https://ukccsrc. ac.uk/news-events/events/ccs-issues-governance-and-ethics.

negative emissions) or as a dedicated system, the demonstration and deployment of CCS is strongly tied to fossil fuel energy systems. Here, we reflect on the possible consequences of the relationship between biomass energy, CCS and fossil fuel energy. Indeed, participants at the workshop highlighted the linking of BECCS to fossil fuels and the associated ethical and social implications. Important concerns were identified as the potential for carbon and infrastructural lock-in, the perceived 'techno-fix' nature of BECCS and the ethical implications of burning fossil fuels; we expand on these issues in the following text.

Typically, optimistic assessments of the potential for BECCS to deliver negative emissions on a global scale are analysed in isolation from the need for supporting CCS infrastructure. The Representative Concentration Pathways for limiting temperature increases to 2 °C presented in the IPCC Fifth Assessment Report (IPCC, 2014) are based on Integrated Assessment Modelling (IAM) scenarios, the majority of which rely on negative emissions from BECCS (Fuss et al., 2014) (see Chapter 9 for more details). Within the IAM scenarios, there is a widespread assumption that CCS is deployed at commercial scale, along with infrastructure to transport and store CO_2, which supports the deployment of BECCS (Koelbl et al., 2014). Looking more widely at studies focused specifically on BECCS, either the assumption of the availability of CCS infrastructure, such as CO_2 pipelines and appropriate storage sites, is common (see, for example, Koornneef et al., 2012; Ricci and Selosse, 2013) or studies do not consider the costs of CCS pipelines within their assessment (see, for example, Akgul et al., 2014). Many authors (see for example Ricci and Selosse, 2013; Akgul et al., 2014) further reinforce the link between fossil fuels and BECCS as they identify co-firing as a route to BECCS. Akgul et al. (2014) argue that the conversion of existing coal-fired power stations to co-fire biomass provides a relatively low-cost route to BECCS, and their modelling work highlights co-firing as an inexpensive route to negative emissions (at a project level).

Furthermore, it has been suggested that deploying CCS in conjunction with biomass energy could provide a route out of a fossil fuel lock-in effect associated with CCS, particularly if initially it is introduced *via* a co-firing approach (Vergragt et al., 2011). The concept of lock-in describes how particular technologies become so strongly connected to their supporting physical, social, institutional and political infrastructures that their dominance of use persists even when 'better' technological options are available (Unruh, 2000). Using a social-technical transition approach to compare fossil CCS with BECCS Vergragt et al. (2011), caution that the fossil CCS 'niche' is far stronger than the BECCS 'niche', and although much of the research produced for fossil CCS is also applicable to BECCS, the current structures for the diffusion of that knowledge are geared towards fossil fuel. Shackley and Thompson (2012) suggest that carbon lock-in associated with CCS could be avoided. They conclude that if CCS is developed flexibly so that its use is associated with other low-carbon vectors such as hydrogen, any initial lock-in may be shallow. Furthermore, the widespread cancellation of CCS demonstration projects suggests that the fossil CCS niche is not established, and, with a central role for BECCS within modelled scenarios presented in the IPCC Fifth Assessment Report (AR5) (Fuss et al., 2014), it is possible to see how political emphasis on fossil CCS could shift. Although the demonstration of fossil CCS currently is further advanced, and at a larger scale, than BECCS plants, fossil CCS is far from widely established and it remains to be seen whether or not it provides the route for establishing BECCS.

Whether or not BECCS is part of a co-fired or dedicated bioenergy system, it can be seen as a 'techno-fix', a technological solution to a problem created by modern society.

So, the argument goes, rather than changing or cutting back on the activities that caused the problem of climate change, BECCS is proposed (and still not proven) as a means of delivering negative emissions on a scale that can reverse the trend of ever-increasing atmospheric CO_2, providing a 'technological rescue' (Garvey, 2008) enabling us to continue as we are and, conveniently, remove damaging levels of CO_2 from the atmosphere (Anderson, 2015a). By using the same centralised fossil fuel combustion systems that initiated the process of climate change, it becomes harder to disentangle the problem from the solution (see, for example, Ehrlich, 2011) and does nothing to change the practices that created the problem in the first place. Moreover, CCS and BECCS are not 'open-ended' solutions – eventually suitable storage sites will be filled and/or resources will be depleted (Haszeldine, 2016). In identifying and discussing BECCS as a 'techno-fix', workshop participants also noted that its requirement for continued use of large-scale centralised infrastructure takes us further away from opportunities for smaller-scale or decentralised approaches.

Whilst there are currently few policy incentives for BECCS, and for negative emissions, a strong case can be made for looking to develop CCS, not for new applications within the power sector, but instead close to areas with existing large emissions for CO_2 from biomass sources. Sweden, with a large paper and pulp industry, and Brazil, with a large bio-ethanol industry, have both been identified as good potential locations for BECCS, due to the presence of both CO_2 sources and sinks (Gough and Upham, 2011; Vergragt et al., 2011). Similarly, Arasto et al. (2014) suggest that cost-effective reductions in Finland could come from iron and steel production and paper and pulp production.

12.3 Alternative Approaches

12.3.1 Negative Emissions Approaches and CDR

Negative emissions approaches provide permanent or long-term carbon dioxide removal (CDR) from the atmosphere. Chapter 1 describes some of the alternative approaches that also have the potential to remove CO_2 from the atmosphere. Along with afforestation and reforestation, BECCS is the most commonly cited approach to CDR, featuring prominently in policy discourses and emissions modelling. Other approaches include various techniques for direct air capture, ocean fertilisation, enhanced weathering, soil and habitat management techniques and sequestration in materials (for use in construction, for example).

None of the proposed approaches has been implemented or tested on a large scale: all remain somewhat tentative propositions that, in order to deliver sufficient negative emissions to benefit carbon budgets, would be required to function on a global scale (see Chapter 9). Some of the techniques require significant energy input, for example, to regenerate solvents/sorbents (e.g. BECCS, lime/soda process) or to produce and transport large quantities of raw materials (e.g. ocean liming), which may at best reduce the efficacy of the process and at worst lead to greenhouse gas emissions that must be abated by other means (such as CCS). McGlashan et al. (2012) present a techno-economic assessment of negative emissions technologies, which compares the economic and energy costs of different approaches, based on published data. Furthermore,

McLaren (2012b) provides an assessment of the socio-technical performance (including controllability, accountability, side-effects and costs) of 14 negative emissions technologies, including BECCS, noting the uncertainty associated with such an assessment, given that the majority are not commercially proven. Although forest and habitat restoration and sequestration in construction materials are identified as being at a high level of technology readiness, uncertainties remain over their potential to deliver significant net negative emissions, alongside concerns over their land-use implications. At a mid-range technology readiness, while the deployment of BECCS remains limited at a large scale, it is nevertheless currently at a more advanced stage of technology development than many of the alternative negative emissions technologies – many of which remain experimental and potentially not ready for use before the 2050s (McLaren, 2012b).

All of the options that rely on biological productivity for carbon removal (including BECCS) present challenges in relation to land-use change, with implications for food production, water availability and accountability. Some social and governance issues are specific to BECCS, such as the long-term monitoring and liability associated with geological storage (although direct air capture approaches also depend on long-term storage of CO_2), and some apply to all negative emission technologies – for example, how they are assumed to contribute to long-term climate policy and the implications of not meeting expectations in terms of delivering net negative emissions (Fuss et al., 2014). Furthermore, many aspects of CDR approaches hold implications related to notions of nature and the natural world, and how human society might be interfering with it. The complex relationship between how people think about CDR techniques in relation to human influence on nature and the natural world has been explored by Corner et al. (2013). The positive influence on public perceptions of technologies framed as depending on processes analogous to natural processes, for example 'artificial trees', has been identified (Corner and Pidgeon, 2014) and may also apply to geological storage if it is framed as 'putting CO_2 back into the ground' (e.g. Davison et al., 2001).

There is clearly a compelling case for establishing responsible governance frameworks, within the principles of responsible research and innovation (RRI) for negative emissions approaches (and indeed all forms of geoengineering) whereby the governance of research processes, as well as deployment, is anticipatory, reflective, deliberative and responsive (McLaren, 2012b; Parkhill et al., 2013). Some of the features of how such an approach might look in practice for CCS/BECCS were presented during the specialist technical workshop (Bickerstaff, 2014). More specifically, the issue of proportionality of governance was raised during workshop discussions and could be of particular relevance to BECCS (and other NETs) given the temporal and geographical dispersal of potential impacts; achieving fairness in the governance process depends on those who will be affected by the consequences of decisions having power to influence the decisions (Brighouse and Fleurbaey, 2010). Using an RRI approach, Raman et al. (2015) identify distinct value-based visions for development of biofuel production and propose ways in which future development might reconcile such different priorities. Other issues raised during the workshop related to how to evaluate between competing technologies (whether negative or otherwise) and the ethical implications for BECCS if an environmentally and economically sustainable option for direct air capture was available, suggesting that this would make BECCS a less attractive proposition.

12.3.2 Different Mitigation Approaches

As a negative emissions technology, BECCS can be seen as a different type of approach to other mitigation options. Conventional mitigation approaches reduce or avoid atmospheric emissions of greenhouse gases, whereas negative emissions technologies are categorised as a family of geoengineering approaches that remove carbon dioxide from the atmosphere, at a scale sufficient to reduce CO_2 concentrations. What sets BECCS apart from the majority of other negative emissions technologies (with the exception of biochar) is that the technology is associated with the provision of energy services (e.g. electricity generation or biofuel production) as well as CDR. How BECCS fits within the wider mitigation context is partly a question of scale: moving from negative emissions at a project level to delivering global net negative emissions to achieve atmospheric CO_2 concentration targets shifts the negative emissions discourse from 'offsetting hard to abate sectors', or a spatial decoupling of abatement from emissions (Lomax et al., 2015), to a more fundamental pillar of managing the future climate by removing CO_2 from the atmosphere, temporally decoupling emissions from their abatement (Lomax et al., 2015). Thus, BECCS can be seen as a means of enhancing existing emissions efforts, potentially reducing the costs of delivering atmospheric concentration targets and providing 'head room' for some continued fossil fuel use (e.g. mobile sources); it could also be used as a way of 'buying time' as social practices adjust and mitigation technologies are advanced (allowing an 'overshoot' of carbon budgets in the near term) or as a 'climate recovery' strategy; each of these framings brings with them their own risks (Meadowcroft, 2013).

This latter strategy of allowing an overshoot of concentration targets introduces a potential 'moral hazard' whereby the promise of negative emissions in the future leads to a reduction in mitigation efforts in the near term (Preston, 2013). It is by no means proven that BECCS can deliver global net negative emissions; it is not yet an economically viable approach that can be seen as an *alternative* to other forms of mitigation and is unlikely to be ready for large-scale commercial deployment for another two decades or more (McLaren, 2012b; McGlashan et al., 2012). If BECCS proves to be no more than wishful thinking on a global scale, relying on large-scale negative emissions to meet urgent climate targets could be seen as 'moral recklessness' (Garvey, 2008) or even a form of 'moral corruption' whereby rich nations find excuses to avoid climate action (Gardiner, 2006). Limitations to the speed and scale with which the negative emissions benefits of BECCS could be realised, and the uncertain social and environmental impacts of doing so, suggest that it should not be seen as an alternative to radical emissions reductions (Tavoni and Socolow, 2013; Anderson, 2015b). Indeed, by avoiding the framing of BECCS, or negative emissions, as an alternative to mitigation, it could be argued that there is a moral imperative to reducing uncertainties and pursuing the technology; the potential for such a 'reverse moral hazard' requires further research (Preston, 2013).

Furthermore, Preston (2014) argues that, by seeing CDR approaches, such as BECCS, as a combination of emissions mitigation and enhancing and managing carbon sinks (i.e. not an alternative to reducing emissions) and framing it as a means of reducing anthropogenic climate forcing, some of the challenges posed by these types of moral arguments may be avoided.

Challenges specific to BECCS (and other NETs) include the quantitative verification and accounting of negative emissions; ensuring secure long-term storage and complexities associated with managing dispersed social and environmental impacts across heterogeneous technology chains (Meadowcroft, 2013). Any mitigation and abatement technology can be associated with a variety of known and unknown social consequences and responses; it is thus imperative that they are evaluated relative to each other, in the context of their wider implications and as part of a strategic approach to climate action.

Workshop participants raised issues around the potential for BECCS to lead to a reduction in alternative mitigation efforts, questioning whether it should be viewed as a mitigation 'last resort'. Views ranged from seeing its value as a bridging technology to seeing it as an approach that could ultimately be better than renewable energy.

12.4   Sustainable Decarbonisation

While decarbonisation of the energy system (and further reducing atmospheric carbon) represents the overarching goal of BECCS technologies as part of a climate-change mitigation strategy, here we consider the contexts in which BECCS might represent a sustainable as well as a just solution. This depends on the wider impacts of the approach across its expansive supply chain, considering the full life cycle of the processes feeding into the BECCS chain (as discussed in Chapter 11). A review of literature on life-cycle emissions of different CCS technologies highlights the importance of the non-CO_2 impacts of CCS. These may include upstream issues associated with the production and transport of feedstocks, whether mined hydrocarbons, agricultural production of biomass, chemical emissions from CO_2 capture processes using amines or downstream impacts associated with CO_2 transport and storage, for example (Corsten et al., 2013). A review of the sustainability implications of bioenergy options can be found in Thornley et al. (2009).

The feasibility and sustainability of large-scale bioenergy use can be significantly impeded by various, often interconnected, socio-economic and environmental factors. The so-called *food energy environment trilemma* encapsulates the potential conflicts between bioenergy production and food supply on biodiversity, water resources and fertiliser use (Powell and Lenton, 2013; Slade et al., 2014; Tilman et al., 2009). The future availability of land for dedicated bioenergy crops is an important factor governing the potential magnitude of bioenergy resources: competing with other land uses and attendant ecosystem services and relying on economic drivers can cause deforestation of primary forests (Wise et al., 2009). Growing dedicated energy crops on abandoned agricultural land seeks to circumvent some of these issues (Tilman et al., 2009). However, there is a trade-off between assumptions about yields and available land area, with higher yields requiring less land area to achieve the same bioenergy potential (Bonsch et al., 2014). Bioenergy production at the scale required to deliver global net negative emissions will put significant pressure on water resources; higher yields can be achieved through irrigation than with rain-fed agriculture. However, irrigation may lead to conflict with other potential users and degradation of freshwater ecosystems but requires less land area (Bonsch et al., 2014). Assumptions about crop yield and agricultural productivity improvements also rely upon the use of fertilisers to improve yields, which impact on greenhouse gas balances. In addition to the direct biodiversity impacts of deforestation, for example, indirect biodiversity impacts may also occur if bioenergy crops replace food crops in one area but lead to food crop expansion in undisturbed areas (Immerzeel et al., 2014). A recent review of the biodiversity impacts of bioenergy production concludes that land-use change is a key driver leading to habitat loss and changes in species richness and abundance (Immerzeel et al., 2014).

These themes of competing land uses and environmental impacts were taken up during workshop discussions, noting detailed concerns such as the waste disposal requirements for solvents used in the capture processes. But more general questions of sustainability were also introduced – such as the wider human rights issues associated with maintaining a healthy environment and the technology's compatibility with the precautionary principle in environmental law. As environmental pressures drive systems towards efficiency improvements, CCS essentially reverses this trend by introducing inefficiencies into the energy production process, bringing a significant energy penalty (see Chapter 7).

12.5 Societal Responses

There are, as yet, few examples of operational BECCS plants worldwide from which to gauge the potential societal response to the approach, and, whilst a few planned schemes have been identified (Kemper, 2015), further details of these were scant at the time of writing. CO_2 from a bio-ethanol plant has been injected into the Mount Simon sandstone aquifer, as part of the Illinois Basin Decatur project which captured and stored 1 M tonnes of CO_2 over a 3-year period from 2011 to 2014.[2] The bio-ethanol plant providing the CO_2 for the project commenced operation in 1978 and was therefore established and known within the local community. Follow-on funding has been secured for the Illinois Industrial CCS project, which is intended to scale up this demonstration project to commercial scale and inject in the region of 1 M tonnes of CO_2 a year from early 2016; it is not clear, however, how this scheme is progressing and when injection will commence.[3] The Illinois Basin projects have been developed under the umbrella of the Midwest Geological Sequestration Consortium (MGSC), one of the seven regional partnerships intended to support the deployment of CCS in the United States. Engagement with publics and stakeholders has been a key part of the work of the MGSC with the successful implementation of onshore storage, an indication of the success of their approach.[4] The focus of the MGSC has been on proving the storage of CO_2, thus, whilst the aquifer is storing biogenic CO_2, the purpose of the project was not BECCS as such. Another scheme to capture CO_2 from an ethanol plant in Greenville, Ohio (within a different regional partnership, the Midwest Regional Carbon Sequestration Partnership (MRCSP)) was abandoned due to local opposition and concerns over the storage. When learning from these early demonstration plants for the deployment of BECCS in the longer term, it is important to consider that schemes which can appear at the outset to be quite similar can have different outcomes; social responses to BECCS need to be

---

2 https://sequestration.mit.edu/tools/projects/decatur.html.
3 https://sequestration.mit.edu/tools/projects/illinois_industrial_ccs.html.
4 http://ac.els-cdn.com/S1876610209009370/1-s2.0-S1876610209009370-main.pdf?_tid=bed27126-1390-11e6-a7ad-00000aab0f6b&acdnat=1462542488_7b419438f877ac38129399332d9f76b7.

considered on a case-by-case basis and with an approach that is sensitive to the social contexts in each case.

Moving from this operational scheme to consider research into public perceptions of BECCS more generally, to date little work has been undertaken to explore and understand social responses to the technology specifically, and, thus, there remains limited evidence for how the introduction of biomass into the CCS system could change public responses. This issue of awareness of BECCS was raised at the workshop, participants highlighting that BECCS was 'too abstract a concept for people to care', and questioning the level of awareness that would be necessary to deploy BECCS. In work focusing on geoengineering, Parkhill et al. (2013) suggest that BECCS is not perceived as a progressive technology for the long term. They found that perceptions of biomass energy are more complex than for other renewables, suggesting that responses may be contingent on the fuel type, application and context of their use. A stated-choice questionnaire study, conducted in Switzerland, explored the preferences of the sample for different CCS 'settings'; one of the conclusions of the study was that biogas-fuelled CCS plants were perceived more positively than those fuelled by natural gas (Wallquist et al., 2012). Similar findings were reported by Dütschke et al. (2014), with CCS rated less positively by a representative sample of German citizens when the source of the CO_2 was a coal-fired power plant than when the source was either from biomass or industry; with some large industrial sources of biogenic CO_2 such as the paper and pulp industry, this may present an effective route for early deployment of BECCS in terms of social responses.

BECCS was included in research in six European countries by Upham and Roberts (2011) exploring public acceptance of CCS and the role of communication. A post-focus group questionnaire suggested that acceptance of BECCS had shifted from a 'non-opinion' before the focus groups to more negative after the focus groups, during which a film on CCS had been shown. The authors, however, conclude that this was due to a negative association with CCS, rather than being a considered opinion of BECCS. Concerns over the possible negative impacts of association with other technologies, such as hydraulic fracturing (fracking) or underground coal gasification (UCG), for example, were raised by workshop participants; a link with induced seismicity associated with fracking was reported by Gough et al. (2014) during focus groups discussion on CO_2 transport. To date, however, there has not been a study which seeks to explore in more detail the reasons behind the indications for a more positive reaction to bioenergy than fossil CCS.

From a social perspective, a key challenge of deploying BECCS is that it is not a single technology, but instead requires the integration of a biomass supply chain with an energy-conversion technology and CO_2 capture, transport and storage infrastructure. As highlighted by Parkhill et al. (2013), social responses to biomass energy may depend on the fuel type, energy-conversion technology and context. In the example of the Illinois Basin project, the plant is an existing ethanol plant, using a corn feedstock, which could be domestic or imported; CO_2 is transported a short distance for onshore storage. From a social perspective, issues of concern to stakeholders and publics would depend very much on the plant, with a new-build power plant, using imported feedstock and connecting into a CO_2 pipeline network for offshore storage, potentially raising very different issues amongst the different affected stakeholders and communities. Workshop participants raised the issue of whose opinion counts in cases where separate and disparate communities may be affected. The issues of acceptance are also likely to vary depending on whether a scheme is in a developed, industrialising or developing country (Dowd et al., 2015).

As discussed in Chapter 9, BECCS offers a cost-effective way of mitigating emissions across a number of sectors and decreasing atmospheric concentrations of CO_2 in the event of an overshoot within climate modelling. Gough and Upham (2011) identify that these are benefits of BECCS that may contribute to a positive social response to the technology. That said, awareness of BECCS amongst publics is likely to be low; public awareness of BECCS has yet to be surveyed at a large scale and while bioenergy is widely known in most countries, few members of the public have heard of CCS or know what it is (European Commission, 2011).

Pidgeon et al. (2012) have surveyed awareness and responses to geoengineering (including negative emissions or CDR) in the United Kingdom and found it to be extremely low. The media are key sources of information for publics and stakeholders; studies exploring how CCS is reported in the media suggest that, in the absence of many examples of operational schemes, reporting tends to focus on policy and funding announcements and be somewhat abstract and lacking in technical detail (Mander et al., 2012). A lack of journalistic understanding may also contribute to a questioning of whether use of the media is a good route for people to learn about CCS. Nerlich and Jaspal (2013) show how reporting on CCS in the United Kingdom has dropped from a high in 2009. Looking at media reporting in the United States in areas with bioenergy and CCS projects, Feldpausch-Parker et al. (2015) found that the two technologies tended to be reported separately, with little mention of the two combined. Whilst we are not aware of similar analysis for other countries, this separation in reporting may be repeated elsewhere.

Social responses to BECCS can thus be considered across two levels. The first level is support, or otherwise, for BECCS as a mitigation option and its role in CDR and the second level is the social responses to the different elements of the CCS chain at a community level. A UK-focused study exploring public values and energy-system change highlights that use of waste biomass was viewed more favourably than dedicated energy crops, which were linked to potential impacts on food prices and availability, particularly for those living in countries where the crops may be grown (Demski et al., 2015). Such issues of justice are discussed in Section 12.6. In the Demski et al. study, whilst biofuels and CCS appear not to have been connected, both were judged separately by participants to be 'non-transitions' associated with a continued dependence on finite resources. At the second level, deployment of BECCS requires facilities for each element of the BECCS chain, which is likely to affect several different communities, in distant areas, and with different facilities, including variances in feedstock, energy-conversion facility, CO_2 transport method and storage site (which may be onshore, as in the Illinois Basin, or offshore). In each instance, social responses are likely to be multidimensional, going beyond straight opposition or support (Roberts and Mander, 2011), with each stage potentially raising specific concerns. For example, bioenergy supply issues may relate to water use, or competition for alternative land uses may be important (Tilman et al., 2009), while the energy-conversion element may prompt local concerns related to air pollution, transport of fuel or the visual impact of the facility (Thornley et al., 2009). Thus, social responses are influenced by myriad factors, including perceived effectiveness and impact of technologies, which may be judged differently by different communities and subject to different decision-making processes, both at the level of choices of technology type and in deployment and engagement processes themselves. Thus, the benefits and impacts of BECCS will vary along the elements of its supply chain and must be considered across local and global scales.

The novelty and unfamiliarity of the storage of CO_2 has also been shown to be a primary focus of concern about CCS amongst members of the lay public (Eurobarometer, 2011; Feenstra et al., 2010). Few studies have specifically focused on perceptions of geological storage of CO_2, although a study of social responses at five onshore storage sites in Europe demonstrated the importance of non-technical factors in site selection, with outcomes highly dependent on local and temporal contingencies, the trust between communities and developers and the actions of other stakeholders (such as NGOs, local policy makers, media, etc.), which may influence trust (Oltra et al., 2012). Early empirical research suggests that offshore storage, as is proposed in the United Kingdom, is not necessarily likely to meet less opposition (Mabon et al., 2014a, b), given the importance of the marine environment to publics and stakeholders (Roberts et al., 2013). There is a high level of uncertainty and a limited scientific knowledge base, regarding impacts on marine ecosystems, and the more alien nature of the marine environment can make working offshore highly challenging (Roberts and Upham, 2012). Offshore development can impact on livelihoods as well as on the marine environment. These issues, along with the need to consider how the values associated with the sea relate to offshore storage, and the naturalness of the marine environment when compared to many terrestrial environments, were raised during the Edinburgh workshop. Other issues raised with respect to societal responses included how CO_2 storage may be viewed from an ethical perspective, by leaving a monitoring and liability legacy for future generations, as well as hiding CO_2 'out of sight and out of mind', alongside wider concerns about the importance of procedural fairness and justice, as discussed below.

Originating in the mining industry but now becoming more widely used in other sectors (Hall et al., 2015; Moffat et al., 2015), the Social Licence to Operate (SLO) concept offers a useful framing for future deployment of BECCS and was proposed as an approach relevant to BECCS deployment by workshop participants. The concept can be broadly defined as informal permission given by the local community and broader society to industry to pursue technical work (Thomson and Boutilier, 2011, quoted in Dowd and James (2014)). Although the use and meaning of the term varies across different industries (Hall et al., 2015), important elements of an SLO include acceptance, desire, beneficial relationships or mutual benefit and lack of opposition (Dowd and James, 2014). The essential characteristic of the SLO is partnership between communities, operators and government, and it is thus dependent on a variety of factors that contribute to building trust between the stakeholders.

The SLO concept can exist at multiple levels; Hall et al. (2015) suggest that the local level is key for the SLO. When considering BECCS, this leads us to the question of what is local for a BECCS scheme – where is the fuel grown, the CO_2 captured or the CO_2 stored? Trust is key to maintaining an SLO, and in the case of BECCS, this will be contingent on the regulatory aspects of the chain, which are as yet undeveloped, though some of the key questions are clear: how do you ensure the sustainability of the biomass source, and the extent of any land-use change-related emissions? Where are the negative emissions allocated, and which nation gets the benefit of reductions which occur in cross-boundary ways?

12.6 Justice

Justice has emerged as a key concept in thinking on the societal implications of climate change. In addition to the now well-established field of climate ethics, which we refer to in this chapter, climate justice has found its way into international climate negotiations. The Paris Agreement, signed at COP21 in December 2015, calls for 'Noting the importance of ensuring the integrity of all ecosystems, including oceans, and the protection of biodiversity, recognised by some cultures as Mother Earth, and noting the importance for some of the concept of 'climate justice', when taking action to address climate change' (UNFCCC, 2015). Similarly, the Intended Nationally Determined Contributions (INDCs) which formed the backbone of the COP21 negotiations made an explicit mention of 'fair' contributions to climate-change mitigation and adaptation from each nation.

'Fairness' and justice are therefore not only issues of scholarly concern but also crucial dimensions of climate policy requiring attention if agreement on action is to be achieved. Justice dimensions also come into play when we reflect on what the objective of climate-change mitigation is. Mabon and Shackley (2015) consider the question of the kind of society that may be produced through different forms of climate mitigation. McLaren (2015) goes further to question whether we are aiming for carbon abatement at all costs or a just and sustainable society, arguing in the context of BECCS that if the objective is the latter of these two, then how climate stability is delivered matters a lot.

This section therefore reviews current thinking on justice in the context of energy and climate change, paying particular attention to issues that are relevant to CCS and in particular where BECCS may add additional complexity to justice issues already raised for conventional CCS. We look in turn at distributional, procedural, financial and intergenerational aspects of justice.

12.6.1 Distributional Justice

Distributional justice is relatively well covered in environmental social science literature (Shrader-Frechette, 2002; Adams et al., 2013). It relates to the distribution of impacts, responsibility, and costs and benefits – meaning claims to distributional injustice may arise if such impacts, risks or positive benefits are not distributed fairly across society. The potential for distributional justice to be an issue for CCS has been identified (de Groot and Steg, 2011). However, in much of the CCS-specific literature, this consideration of distributional justice centres on the physical location and characteristics of key parts of the conventional CCS 'chain' – capture plants, pipelines and storage sites. A recurring theme in this is the question of whether communities may be exposed to potential risks associated with CO_2 storage and/or transport without receiving energy or economic benefits (Gough et al., 2014). Building on Blowers' (1999) idea of landscapes of dependence, Mabon (2016) similarly argues that CCS infrastructure may inadvertently perpetuate unequal relationships of dependency between host communities and the energy industries they already host. Conversely, Reiner and Nuttall (2011) suggest benefits from CCS can accrue within host communities to the extent that a PIMBY (please in my back yard) attitude may emerge. ter Mors et al. (2012) argue that host community compensation may give one mechanism for ensuring exposure to risk and negative effects is balanced out for host communities, and Boyd (2015) and Mabon and Littlecott (2016) in the contexts of west Canada and north-east Scotland, respectively, suggest CCS infrastructure could help to sustain employment within communities dependent on declining extractive industries. These types of social exchange, in which communities balance the risks and negative impacts of a technology with the benefits it brings, such as jobs, may be significant in determining how well communities tolerate technology developments such as CCS (Zhang and Moffat, 2015).

By contrast, workshop participants raised a much wider range of issues around distributional justice, many of which link back to the broader ethical and moral questions raised earlier in this chapter and centre on the 'responsibility' dimension identified by Adams et al. (2013). One cluster of arguments focused on the responsibility of developed nations to make CCS work, due not only to their access to the infrastructure, knowledge and wealth necessary for research and development but also to their ultimate responsibility for emissions arising from goods produced in developing countries for developed-nation consumption. Another cluster of arguments similarly centred on inequalities within and between countries. These included more economically developed countries gaining financially from selling CCS technology (see, for example, Kapila and Haszeldine (2009) on CCS in India); differences in access to storage capacity between countries due to varying geological conditions and questions of power relations between actors in the exchange of intellectual property, knowledge and technology. Participants identified an additional suite of distributional justice issues to do with the practical and physical realities of governing the capture, transport and storage of CO_2 across national boundaries. Within this third group of arguments were questions over whether or not storage is fungible; how to manage liabilities across national boundaries and how to deal with one nation having access to a reservoir with the potential to address a global issue with global effects.

It is perhaps these more wide-ranging points raised in the workshop that highlight the additional distributional justice complexities BECCS has over conventional CCS. Extant thinking on distributional injustice in CCS at the host community level arguably fails to capture the possibility for bioenergy production in developing countries to exacerbate competing land uses and food poverty (Gamborg et al., 2012). However, thinking in terms of carbon dioxide capture, transport and sequestration in the context of a global system can draw attention to the possibility for BECCS to magnify existing distributional injustices between developed and less developed nations. Moreover, giving over large tracts of land to producing fuel for BECCS may mirror the ethical fault line of environmental justice (justice for animals and ecosystems) identified by Gough and Boucher (2013) if subsequent ecosystem changes disproportionally expose non-humans to the negative environmental effects of production for BECCS. Indeed, McLaren (2015) concludes that mitigation in the rich world in the short term may be considered more 'ethical' than the exploitation of less economically developed countries' biomass potential.

12.6.2 Procedural Justice

Like distributional justice, there is relatively good awareness of what procedural justice means in an energy and environmental context: justice in the processes through which decisions are reached (Shrader-Frechette, 2002; Gross, 2007). Claims to procedural injustice could hence arise if the processes employed to reach decisions about energy or the environment are not perceived as being fair. This is another aspect of justice that has received emergent attention within the CCS sphere, McLaren (2012a) advocating greater consideration of the procedural justice dimensions of CCS.

Indeed, issues of fairness and propriety in decision-making procedures were argued to be central to the failures of the Dutch Barendrecht project, Feenstra et al. (2010) criticising the 'decide–announce–defend' nature of public consultation and Terwel et al. (2012) noting a much wider range of factors to do with trust in the developer as influencing local opinion rather than the techno-scientific risks of the project itself. In work with citizens in Poland and Scotland on hypothetical CCS projects, Brunsting et al. (2012) likewise identified the perception of CCS plans as a 'done deal' (i.e. giving the impression at the consultation stage of being confirmed and/or imminent regardless of public input) as being a major source of concern. In these cases, as argued by Mabon and Shackley (2015), claims to procedural injustice could arise if citizens attend or get involved in 'consultation' events believing they may be able to substantially alter the nature of project plans, when in reality many of the major details have already been set.

More critical thought has linked engagement of this nature to a drive to engender acceptance of CCS projects – that is, to 'convince' publics and stakeholders to accept the end goal of CCS deployment (McLaren, 2012a; Hansson, 2012). Mabon et al. (2015) suggest that whilst this is logical and to be expected (i.e. that a developer will, of course, want a project to proceed as smoothly as possible), subsequent claims to injustice may arise if public engagement is closed down to focus on narrowly technical or risk-based arguments, leaving no room for more values-based arguments. Bielicki and Stephens (2008) too believe that technical arguments may be of little consequence if it is issues of value that are at stake. Procedural justice in the context of CCS thus entails sensitivity towards the grounds on which different actors actually form their views on the technology, and the fora and processes through which they expect to be able to negotiate these viewpoints.

Such concerns over the nature of decision-making processes for CCS emerged in the workshop. Some of these had clear connections to the distributional justice issues raised above and reflected the processes through which siting outcomes may be reached: namely, how to seek consent from communities for siting and to ensure they have an ability to refuse compensation (i.e. that vulnerable communities are not 'targeted'); how to avoid falling into the trap of assuming existing industrialised areas will consent to new industrial infrastructure and how to assess the extent to which a developer can legitimately claim to have a SLO. Other concerns raised related to the extent to which developers could be trusted to implement fair processes themselves versus the need for regulations and policies to ensure 'fair' processes for governing CCS deployment. Potential developer claims to social licences and the possibility of building on past positive experience with infrastructure were set against workshop participant concerns over citizens' and communities' lack of ability to engage in local or regional debate and/or to have any meaningful influence on national energy policy decisions. In particular, participants noted the following: the possibility for consequentialist decision-making processes that could override local concerns in the national interest; whether the state should assume project liability as an incentive to CCS and (after Proelss and Gussow, 2011) whether existing waste legislation would be sufficient to encompass the complexities of captured carbon dioxide.

A third cluster of concerns extended beyond procedural justice to encompass questions of epistemic justice, defined in the CCS context by Mabon et al. (2015) as justice in the whole process of using knowledge to frame CCS as a potential 'solution' to the 'problem' of climate change. Issues identified by participants included how to include different forms of knowledge and value systems in governance processes, and the importance of taking seriously emotions in governance processes (Cass and Walker, 2009). Questions were also asked over whether the use of metrics (e.g. 450 ppm, 2 °C, 80% reduction) were understandable and transparent to everyone and how to integrate non-numeric environmental values with this. Lastly, the need was stressed for procedures that allow different claims to knowledge to be discussed in situations such as those around CCS and climate change where even scientists cannot provide 100%-certain 'proof'.

As with distributional justice, it may be the case that BECCS serves only to make the procedural justice challenges that arise with conventional CCS even more pronounced and complicated. For instance, following McLaren (2015), questions may be asked over the processes through which land use is decided and allocated. Just as there is already a small field of literature on how conventional CCS has the potential to exacerbate the deleterious effects of fossil fuel extraction (Ha-Duong and Loisel, 2011; Bower-Bir, 2014), so it is the case that consideration of where the feedstock for biomass comes from adds a whole new set of concerns to the CCS decision-making process. Particularly, if produce is grown in less economically developed countries, the points made above about power imbalances in decision-making processes and the right for communities to refuse/agree to siting in their own terms, how to account for different knowledge and conceptualisations of land ownership (e.g. indigenous knowledge) and cultural differences in what is considered a fair and appropriate decision-making process become all the more acute. Claims to epistemic injustice may likewise be further complicated by competing knowledge claims over the extent to which BECCS really can be considered a 'negative' emissions technology, given the emissions associated with production and transportation of biomass (Smolker and Ernsting, 2012).

12.6.3 Financial Justice

Extant concerns over the financial 'fairness' or otherwise of CCS have much to do with the perception of its close association with the fossil fuel industries, in particular, coal. That is, support for CCS research and development may come to be perceived as subsidising – or at least giving licence for continued production – to fossil fuel producers (Stephens, 2014). This may be especially true if there is a perception that 'captured' CO_2 will then be used for enhanced oil recovery processes without critical reflection on the need to move away from fossil fuels (Melzer, 2012; Mabon and Littlecott, 2016). This in itself may not be a justice issue unless one believes fossil fuels are intrinsically 'bad'; however, claims to environmental injustices have been raised against many of the companies also viewed as being involved in CCS processes, and questions may also be asked as to private operators' sincerity in tackling climate change (de Coninck, 2010). There is hence potential for other forms of CCS (industrial CCS, BECCS) to be viewed by publics and opinion-influencers as being complicit in funding and perpetuating the same kind of injustices as conventional power-sector CCS.

This in turn leads into a wider financial justice question – who benefits financially from CCS research, development and deployment? Or, to put it differently, is investment in CCS worthwhile to society as a whole in terms of the climate-change mitigation potential it offers (Gough and Boucher, 2013)? Brown (2008) argues the burden of proof is on governments and developers to demonstrate that investment in CCS is a worthwhile and necessary part of climate-change mitigation that cannot be achieved through renewable energy sources alone. There are, of course, links back to distributional justice here, in that claims to financial injustice may arise if the 'benefits' from deployment accrue to the profits of private developers and/or those in rich nations rather than providing a cost-efficient means of mitigating climate change for society as a whole.

In the workshop, this issue of CCS diverting attention away from other means of attaining a sustainable low-carbon society achieved some prominence, participants picking up on the diversion of attention and resources away from renewables; the risk of importing technology from abroad versus building up a domestic technology base; and also the fact that CCS could perpetuate the process of uncritically connecting economic growth and energy consumption (which others, e.g. Anderson (2015a), have argued to be incompatible with the level and nature of mitigation required). The question of who ought to 'pay' for CCS also rose to prominence, in particular, whether funding ought to come *via* state aid or polluting industries themselves under some kind of 'polluter pays' system; whether the costs of rising fuel bills for low-carbon electricity will result in increased fuel poverty; how to balance 'costs' of energy with the 'costs' of climate change and the financial implications of long-term storage.

It could be argued that many of the issues identified in the workshop are relevant to all energy technologies, not just CCS. This is especially true in a justice context, where a number of low-carbon energy forms may arguably increase fuel costs in a way likely to impact the least well-off disproportionately (Ekins and Lockwood, 2011). Where BECCS brings further complexity for financial justice is in ensuring fair payment for the use of land and purchasing of fuel – especially given wealthier nations' (and the multinational corporations headquartered within them) less-than-impressive record to date in ensuring financial benefits from resource extraction are spread out fairly between countries. This would add an additional layer of complexity to the already challenging question of what are 'fair' levels and terms of financing richer nations ought to provide to their poorer counterparts for climate mitigation and adaptation (Light, 2013). Moreover, in the light of the land and vegetation requirements of biofuel production, one may also question how the use and appropriation of land for fuel production could sit alongside some other financially driven climate-change and environmental initiatives such as payments for ecosystem services. That is, could attempts to ensure fair payment for use of land in economically developing nations unintentionally hinder other attempts at financial justice in the context of climate change?

A critical governance issue that makes BECCS uniquely challenging, and one with potentially significant distributional justice implications, was also raised during the workshop (and expanded in Chapter 11) – that of how to account for, value and certify negative emissions compared to carbon reduction. With elements of the BECCS supply chain potentially located across a vast geopolitical range, who gets the credits for CO_2 reduction or removal from BECCS?

12.6.4 Intergenerational Justice

The very nature of CO_2 storage ought to force us to think over generational timescales. The whole basis of CCS as a climate-change mitigation technology rests on the premise of us being able to use geological formations to store CO_2 and prevent it entering the atmosphere for thousands of years. Debates on leaving a burden of responsibility and/or negative effects for future generations are well covered in the climate ethics literature (Gardiner, 2006; Jamieson, 2010) and resonate well with intergenerational justice concerns relevant to CCS. The most obvious of these is whether or not it could be considered fair to leave a legacy of sequestered CO_2 – which requires particular knowledge, geopolitical formations and associated financial input to manage and maintain – for future generations who may not have consented to CCS processes in the first place. Likewise, if the security and viability of storage cannot be proven in the very long term, then it may be the case that sequestering CO_2 now unfairly exposes future generations to climate risks if stored CO_2 ultimately finds its way back into the atmosphere (after Brown, 2011). This, in turn, raises questions over how the geological timescales required for consideration of the propriety of CO_2 storage sit with the short-term cycles over which most decision-making is currently undertaken, and, in particular, what happens if our regulations and governance regimes are not up to the challenges of CO_2 storage (Gough and Boucher, 2013).

Whilst most of the intergenerational justice issues around CCS understandably focus on storage, claims to injustice could also arise if the development of CCS infrastructure (to which BECCS could contribute) leads to society being 'locked in' to a fossil-fuel-intensive economy (Markusson et al., 2012) and/or diverts attention away from research, development and investment in renewable energy sources, demand reduction and behaviour change. What makes this a justice issue in particular is that, following Gardiner's 'moral corruption' argument, it shifts the burden of responsibility for taking such actions onto future generations when present generations may know such an action is already necessary. At a more philosophical level, it could even be argued that by relying on CCS as a 'technological fix', current generations are shifting the burden to undertake much deeper and more painstaking reflection on societal organisation and society's use of resources onto future generations (Hale and Grundy, 2009; Borgmann, 2012; Mabon and Shackley, 2015) rather than addressing such questions in the present.

Much of what was picked up on in the workshop concerned the question of how well the time frames for CO_2 storage – which are, of course, relevant whatever the fuel – fit with our existing policies, regulations and governance arrangements more generally. Building on the issues already mentioned in the preceding paragraphs, participants questioned who ought to be responsible for storage in the long term – especially for shared reservoirs; who ought to be responsible if CO_2 leaks after storage; and a lack of certainty about the time frame for how long a reservoir should operate (i.e. how long one should continue to inject CO_2 and subsequently monitor). The intergenerational justice dimension coming out of this is again that storing CO_2 now without appropriate consideration as to its long-term management and governance may unfairly impose a financial, political and knowledge burden on future generations who had not consented to storage in the first instance.

Where BECCS perhaps moves this on, however, is in relation to the climate ethics thinking on moral corruption and responsibility for taking action. Paralleling the 'moral hazard' arguments raised by Preston (2013) regarding the promise of negative emissions leading to a reduction in near-term mitigation efforts, or 'mitigation obstruction' (Morrow, 2014) or 'deterrence' (McLaren et al., 2016), Anderson (2015b) expresses serious concern at the prominence given to negative emissions technologies such as BECCS in the Paris Agreement. Anderson argues that 'betting' on negative emissions technologies becoming viable in the longer term has taken preference over, and distracts from, shorter-term but more challenging mitigation measures (see also Fuss et al., 2014). Furthermore, even using rapid power-generation build rates in China and in the United Kingdom as a precedent, McGlashan et al. (2012) suggest it could take up to several hundred years to achieve BECCS capacity sufficient to remove 1 ppm of CO_2 from the atmosphere.

Quite apart from the distributional and procedural concerns raised by allocating land for biofuel production, such reliance on longer-term negative emissions technologies may give rise to serious claims to injustice by (a) shifting responsibility for taking action onto future generations and (b) exposing future generations to even greater climate risks should technologies such as BECCS fail to come to fruition. However, others argue that negative emissions technologies such as BECCS could provide a crucial level of flexibility in tacking climate change and that consequently investing in their development is necessary to avoid missing out on innovation opportunities (Lomax et al., 2015). In other words, it may be that investing now in BECCS leads to innovation opportunities which may prevent greater claims to injustice arising in the future.

12.6.5 Summary

Within the CCS community, BECCS has thus far tended to be presented as a somehow 'greener' and 'better' version of CCS given its negative emissions potential – see, for example, the positive publicity generated around the possibility for the UK CCS demonstrator candidate White Rose project to be co-fuelled with biomass. This perception is in a large part dependent on BECCS being positioned as an alternative to coal and hence sidestepping the justice criticisms associated with fossil fuels which have thus far dogged conventional CCS. Nonetheless, many of the distributional, procedural, financial and intergenerational justice concerns raised for coal or gas CCS remain relevant to BECCS. As discussed in the reviewed literature and reinforced by the workshop contributions, fairness in host community relations with sites of CCS infrastructure, intergenerational questions raised by leaving a legacy of stored CO_2 and the potential for CCS to benefit private developers through processes such as CO_2-EOR remain relevant regardless of fuel. BECCS is also not exempt from justice questions that affect the energy sector more widely, such as whether the costs of low-carbon energy will ultimately have to be met by consumers with knock-on effects on pricing and fuel poverty and whether 'technological fixes' merely stall deeper reflection on societal patterns of resource consumption. Moreover, BECCS introduces new justice questions of its own around the use of poorer nations' land and ecosystems in the name of rich-world consumption and has the potential to complicate – or even hinder – already challenging debates on what 'fairness' between rich and poor nations means in the context of climate-change mitigation and adaptation. Whilst many of the justice issues raised for BECCS are common to more conventional CCS, then, BECCS arguably introduces a number of additional justice issues which serve only to attract even greater scrutiny of the question in whose interests – and to what end – CCS and decarbonisation of the energy sector are undertaken.

12.7   Summary

For BECCS to work as a mitigation option contributing to the avoidance of dangerous climate change, it must work at an ethical and societal level. Linking biomass energy sources with CCS brings together any moral and social issues already associated with the approaches individually, as well as introducing new challenges when the two disparate and complex actor networks connect. Added to this is the approach's distinct characteristic, that of the promise of delivering negative emissions, as society struggles to address the consequences of a rapidly dwindling carbon budget. In this context, the need to understand the social and ethical implications of BECCS becomes as urgent and challenging as the climate crisis.

BECCS *could* enable us to get closer to carbon budget constraints and thus reduce the impacts of climate change, if other mitigation measures cannot be implemented quickly and deeply enough. The introduction of biomass co-firing with CCS *could* get us on the pathway to deploying BECCS more quickly and get us out of a lock-in to fossil fuels. Or, it could be argued that, by not moving us to new ways of generating and using energy more efficiently, it locks us in to similar ways of thinking to those that created the climate-change problem. How the role of BECCS in climate-change mitigation is both framed and pursued in relation to carbon budgets in the coming decades may be key to whether it will be viewed ultimately as a 'gamble' (see Anderson, 2015b; Fuss et al., 2014) or an 'investment' (see, for example, Lomax et al., 2015). Whether it is used to justify an overshoot of the budget, to buy time to establish deep enough carbon reductions or to offset sectors that are particularly challenging to decarbonise, over-optimistic assessments of the potential for BECCS to deliver global net negative emissions could lead to a reduced mitigation effort in the near term (the so-called moral hazard).

Thus, BECCS raises questions not only in terms of whether it has a role to play in achieving climate stability but also in the debate about how society should achieve that goal. These are big moral questions but the approach is far from established in the public consciousness. How it will be received at a societal level remains highly uncertain; the large variety of approaches covered by the term BECCS, and the variety of social contexts in which its presence may be felt, will be matched by the variety of social responses with which it will be met. The geographical spread of its potential impacts and hence the challenge of establishing a social licence for it, in both broad terms as a response to climate change and on the ground along its various supply chains, should not be underestimated. The potential consequences of BECCS extend from possible impacts on biodiversity, competing land-use requirements and the marine environment to direct local impacts of its component techniques.

Attention to issues of justice will be critical in how these non-technical questions about BECCS are to be answered: distributional justice in the context of the siting of facilities that may bring both risks and/or benefits to host communities and avoiding perpetuating inequalities within and between countries; procedural justice to ensure fairness and propriety in the decision-making process in the context of competing knowledge claims associated with emissions; financial justice, in particular how we account for, value and certify negative emissions and manage the use of land and eco-system services; and intergenerational justice and how we manage liability and allocate responsibility for storage. What makes these challenges unique to BECCS compared to CCS, or indeed any other low-carbon energy technology, is its potential to deliver negative emissions; this is the critical governance issue at the heart of BECCS.

Perhaps the key litmus test for BECCS is whether it is able to fit within what Garvey (2008) calls a 'morally adequate' response to climate change. We have argued in this chapter that for this to happen, the challenge is to see BECCS not as an alternative to wholesale changes in our levels and mode of energy consumption, but as an addition to ensure that the energy we do use is safe for the climate.

References

Adams, C., Bell, S., Taylor, P. et al. (2013). Equity across borders: a whole-systems approach to micro-generation. In: *Energy Justice in a Changing Climate: Social Equity and Low-Carbon Energy* (ed. K. Bickerstaff, G. Walker and H. Bulkeley), 91–115. London: Zed Books.

Akgul, O., Mac Dowell, N., Papageorgiou, L.G., and Shah, N. (2014). A mixed integer nonlinear programming (MINLP) supply chain optimisation framework for carbon negative electricity generation using biomass to energy with CCS (BECCS) in the UK. *International Journal of Greenhouse Gas Control* **28**: 189–202.

Anderson, K. (2015a). Duality in climate science. *Nature Geoscience* **8**: 898–900.

Anderson, K. (2015b). Talks in the city of light generate more heat. *Nature* **528**: 437.

Arasto, A., Onarheim, K., Tsupari, E., and Kärki, J. (2014). Bio-CCS: feasibility comparison of large scale carbon-negative solutions. *Energy Procedia* **63**: 6756–6769.

Ashworth, P., Dowd, A.M., Rodriguez, M. et al. (2013). *Synthesis of CCS Social Research: Reflections and Current State of Play in 2013.* Global CCS Institute.

Ashworth, P., Wade, S., Reiner, D., and Liang, X. (2015). Developments in public communications on CCS. *International Journal of Greenhouse Gas Control* **40**: 449–458.

Bäckstrand, K., Meadowcroft, J., and Oppenheimer, M. (2011). The politics and policy of carbon capture and storage: framing an emergent technology. *Global Environmental Change* **21**: 275–281.

Bickerstaff, K. (2014). Innovation, equity and energy system transformation: implications for CCS. Presentation to the UKCCSRC Specialist Technical Workshop CCS: Issues in Governance and Ethics, Edinburgh (23 September 2014). Slides available at: http://www.slideshare.net/UKCCSRC/karen-bickerstaff-govethicssept14.

Bielicki, J. and Stephens, J.C. (2008). Public perception of carbon capture and storage technology workshop report. In: *Energy Technology Innovation Policy Group Workshop Series.* Cambridge, MA: Harvard Kennedy School.

Blowers, A. (1999). Nuclear waste and landscapes of risk. *Landscape Research* **24** (3): 241–264.

Bonsch, M., Humpenoder, F., Popp, A. et al. (2014). Trade-offs between land and water requirements for large-scale bioenergy production. *Global Change Biology Bioenergy* doi: 10.1111/gcbb.12226.

Borgmann, A. (2012). The setting of the scene: technological fixes and the design of the good life. In: *Engineering the Climate: The Ethics of Solar Radiation Management* (ed. C. Preston), 189–200. Lanham: Rowman & Littlefield.

Boucher, P. and Gough, C. (2012). Mapping the ethical landscape of carbon capture and storage technology. *Poiesis and Praxis* **9**: 249–270.

Bower-Bir, N. (2014). The hidden costs of hiding carbon. Unpublished MSc thesis. University of Edinburgh.

Boyd, A. (2015). Connections between community and emerging technology: support for enhanced oil recovery in the Weyburn, Saskatchewan area. *International Journal of Greenhouse Gas Control* **32**: 81–89.

Brighouse, H. and Fleurbaey, M. (2010). Democracy and proportionality. *Journal of Political Philosophy, pp.* 137–155.

Brown, D. (2008). The Ethics of Allocating Public Research Funds for Carbon Capture and Storage. Widener Law Blog, 16 October 2008. http://blogs.law.widener.edu/climate/2008/10/16/the-ethics-of-allocating-public-research-funds-for-carbon-capture-and-storage/#sthash.n2kKsXnR.dpuf.

Brown, D. (2011). Comparative ethical issues entailed in the geological disposal of radioactive waste and carbon dioxide in the light of climate change. In: *Geological Disposal of Carbon Dioxide and Radioactive Waste: A Comparative Assessment* (ed. F. Toth), 317–337. Paris: IAEA.

Brunsting, S., Mastop, J., Pol, M. et al. (2012). *SiteChar Deliverable 8.2: Trust Building and Raising Public Awareness*. Amsterdam: ECN http://www.sitechar-co2.eu/FileDownload.aspx?IdFile=590&From=Publications (accessed 11 February 2014).

Cass, N. and Walker, G. (2009). Emotion and rationality: the characterization and evaluation of opposition to renewable energy projects. *Emotion, Space and Society* **2**: 62–69.

de Coninck, H. (2010). Advocacy for carbon capture and storage could arouse distrust. *Nature* **463**: 293.

Corner, A., Parkhill, K., Pidgeon, N., and Vaughan, N.E. (2013). Messing with nature? Exploring public perceptions of geoengineering in the UK. *Global Environmental Change* **23** (5): 938–947.

Corner, A. and Pidgeon, N. (2014). Like artificial trees? The effect of framing by natural analogy on public perceptions of geoengineering. *Climatic Change* 1–14.

Corsten, M., Ramirez, A., Shen, L. et al. (2013). Environmental impact assessment of CCS chains – lessons learned and limitations from LCA literature. *International Journal of Greenhouse Gas Control* **13**: 59–71.

Dale, V.H., Efroymson, R.A., Kline, K.L. et al. (2013). Indicators for assessing socioeconomic sustainability of bioenergy systems: a short list of practical measures. *Ecological Indicators* **26**: 87–102.

Davison, J., Freund, P., and Smith, A. (2001). *Putting Carbon Back into the Ground*. IEA Greenhouse Gas R&D Programme.

Demski, C., Butler, C., Parkhill, K.A. et al. (2015). Public values for energy system change. *Global Environmental Change* **34**: 59–69.

Dowd, A.-M. and James, M. (2014). A social licence for carbon dioxide capture and storage: how engineers and managers describe community relations. *Social Epistemology* **28**: 364–384.

Dowd, A.-M., Rodriguez, M., and Jeanneret, T. (2015). Social science insights for the BioCCS industry. *Energies* **8**: 4024–4042.

Dütschke, E., Schumann, D., Pietzner, K. et al. (2014). Does it make a difference to the public where CO_2 comes from and where it is stored? An experimental approach to enhance understanding of CCS perceptions. *Energy Procedia* **63**: 6999–7010.

Ehrlich, P.R.E.A.A.H. (2011). Technology: not a panacea, maybe a poison pill. In: *TechnoFixes: Why Technology Won't Save us or the Environment* (ed. M.H.A.J. Huesemann). Canada: New Society Publishers.

Ekins, P. and Lockwood, M. (2011). *Tackling Fuel Poverty During the Transition to a Low-Carbon Economy*. York: Joseph Rowntree Foundation http://www.jrf.org.uk/publications/tackling-fuel-poverty-low-carbon-economy (accessed 11 February 2014).

Eurobarometer (2011). *Public Awareness and Acceptance of CO_2 Capture and Storage*. Brussels: European Commission.

European Commission (2011). *Special Eurobarometer 364: Public Awareness and Acceptance of CO_2 Capture and Storage*. Brussels: European Commission.

Feenstra, C.F.J., Mikunda, T., and Brunsting, S. (2010). *What Happened in Barendrecht? Case Study on the Planned Onshore Carbon Dioxide Storage in Barendrecht, the Netherlands*. Amsterdam: Energy Research Centre for the Netherlands http://www.csiro.au/files/files/pybx.pdf.

Feldpausch-Parker, A., Burnham, M., Melnik, M. et al. (2015). News media analysis of carbon capture and storage and biomass: perceptions and possibilities. *Energies* **8**: 3058–3074.

Fuss, S., Canadell, J.G., Peters, G.P. et al. (2014). Betting on negative emissions. *Nature Climate Change* **4**: 850–853.

Gamborg, C., Millar, K., Shortall, O., and Sandøe, P. (2012). Bioenergy and land use: framing the ethical debate. *Journal of Agricultural and Environmental Ethics* **25**: 909–925.

Gardiner, S.M. (2006). A perfect moral storm: climate change, intergenerational ethics and the problem of moral corruption. *Environmental Values* **15**: 397–413.

Garvey, J. (2008). *The Ethics of Climate Change: Right and Wrong in a Warming World*. London: Continuum Books.

German, L., Schoneveld, G., and Pacheco, P. (2011). The social and environmental impacts of biofuel feedstock cultivation: evidence from multi-site research in the forest frontier. *Ecology and Society* **16** (3): 24.

Gough, C. and Boucher, P. (2013). Ethical attitudes to underground CO_2 storage: points of convergence and potential faultlines. *International Journal of Greenhouse Gas Control* **13**: 156–167.

Gough, C., O, Keefe, L., and Mander, S. (2014). Public perceptions of CO_2 transportation in pipelines. *Energy Policy* **70**: 106–114.

Gough, C. and Upham, P. (2011). Biomass energy with carbon capture and storage (BECCS or bio-CCS). *Greenhouse Gases: Science and Technology* **1**: 324–334.

de Groot, J.I.M. and Steg, L. (2011). Psychological perspectives on the geological disposal of radioactive waste and carbon dioxide. In: *Geological Disposal of Carbon Dioxide and Radioactive Waste: A Comparative Assessment* (ed. F.L. Toth), 339–363. Paris: IAEA.

Gross, C. (2007). Community perspectives of wind energy in Australia: the application of a justice and community fairness framework to increase social acceptance. *Energy Policy* **35** (5): 2727–2736.

Ha-Duong, M. and Loisel, R. (2011). Actuarial risk assessment of expected fatalities attributable to carbon capture and storage in 2050. *International Journal of Greenhouse Gas Control* **5**: 1346–1358.

Hale, B. and Grundy, W. (2009). Remediation and respect: do remediation technologies alter our responsibility? *Environmental Values* **18**: 397–415.

Hall, N., Lacey, J., Carr-Cornish, S., and Dowd, A.-M. (2015). Social licence to operate: understanding how a concept has been translated into practice in energy industries. *Journal of Cleaner Production* **86**: 301–310.

Hansson, A. (2012). Colonizing the future: the case of CCS. In: *The Social Dynamics of Carbon Capture and Storage* (ed. N. Markusson, S. Shackley and B. Evar), 74–90. London: Earthscan.

Haszeldine, S. (2016). Can CCS and NET enable the continued use of fossil carbon fuels after CoP21? *Oxford Review of Economic Policy* **32**: 304–322.

Immerzeel, D.J., Verweij, P.A., van der Hilst, F., and Faaji, A.C.P. (2014). Biodiversity impacts of bioenergy crop production: a state-of-the-art review. *Global Change Biology Bioenergy* **6**: 183–209.

IPCC (2014). *Climate Change 2014: Mitigation of Climate Change* (eds. O. Edenhofer, R. Pichs-Madruga, Y. Sokona, et al.). Cambridge University Press.

Jamieson, D. (2010). Climate change, responsibility and justice. *Science and Engineering Ethics* **16**: 431–445.

Kapila, R. and Haszeldine, R.S. (2009). Opportunities in India for carbon capture and storage as a form of climate change mitigation. *Energy Procedia* **1** (1): 4527–4534.

Kemper, J. (2015). Biomass and carbon dioxide capture and storage: a review. *International Journal of Greenhouse Gas Control* **40**: 401–430.

Koelbl, B.S., van den Broek, M.A., Faaij, A.P.C., and van Vuuren, D.P. (2014). Uncertainty in carbon capture and storage (CCS) deployment projections: a cross-model comparison exercise. *Climatic Change* **123**: 461–476.

Koornneef, J., van Breevoort, P., Hamelinck, C. et al. (2012). Global potential for biomass and carbon dioxide capture, transport and storage up to 2050. *International Journal of Greenhouse Gas Control* **11**: 117–132.

Light, A. (2013). An equity hurdle in international climate negotiations. *Philosophy and Public Policy Quarterly* **31** (1): 27–34.

Lomax, G., Lenton, T.M., Adeosun, A., and Workman, M. (2015). Investing in negative emissions. *Nature Climate Change* **5**: 498–500.

L'Orange Seigo, S., Dohle, S., and Siegrist, M. (2014). Public perception of carbon capture and storage (CCS): a review. *Renewable and Sustainable Energy Reviews* **38**: 848–863.

Mabon, L. (2016). Responsible risk-taking, or – how might CSR be responsive to the nature of contemporary risks? Reflections on sub-seabed carbon dioxide storage in Scotland and marine radioactive contamination in Fukushima prefecture, Japan. In: *Corporate Social Responsibility: Academic Insights and Impacts* (ed. S. Vertigans and S. Idowu). New York: Springer.

Mabon, L. and Littlecott, C. (2016). Stakeholder and public perceptions of CO_2-EOR in the context of CCS – results from UK focus groups and implications for policy. *International Journal of Greenhouse Gas Control* **49**: 128–137.

Mabon, L. and Shackley, S. (2015). Meeting the targets or re-imagining society? An empirical study into the ethical landscape of carbon dioxide capture and storage in Scotland. *Environmental Values* **24**: 465–482.

Mabon, L., Shackley, S., and Bower-Bir, N. (2014). Perceptions of sub-seabed carbon dioxide storage in Scotland and implications for policy: a qualitative study. *Marine Policy* **45**: 9–15.

Mabon, L., Shackley, S., Vercelli, S. et al. (2015). Deliberative decarbonisation? Assessing the potential of an ethical governance framework for low-carbon energy through the case of carbon dioxide capture and storage. *Environment and Planning C: Politics and Space* doi: 10.1068/c12133.

Mander, S., Gough, C., Wood, R. et al. (2012). New energy technologies in the media: a case study of carbon capture and storage. In: *Low-Carbon Energy Controversies* (ed. T. Roberts, P. Upham, S. Mander, et al.). London: Routledge.

Markusson, N., Shackley, S., and Evar, B. (2012). *The Social Dynamics of Carbon Capture and Storage*. London: Earthscan.

McGlashan, N., Shah, N., Caldecott, B., and Workman, M. (2012). High-level techno-economic assessment of negative emissions technologies. *Process Safety and Environmental Protection* **90** (6): 501–510.

McLaren, D. (2012a). Procedural justice in carbon capture and storage: a review. *Energy & Environment* **23**: 345–365.

McLaren, D. (2012b). A comparative global assessment of potential negative emissions technologies. *Process Safety and Environmental Protection* **90** (6): 489–500.

McLaren, D. (2015). Ethical considerations of BECCS. Presentation to the UKCCSRC Specialist Technical Workshop CCS: Issues in Governance and Ethics, Edinburgh (23 September 2014). Slides available at: http://www.slideshare.net/UKCCSRC/duncan-mclaren-govethicssept14.

McLaren, D., Parkhill, K.A., Corner, A. et al. (2016). Public conceptions of justice in climate engineering: evidence from secondary analysis of public deliberation. *Global Environmental Change* **41**: 64–73.

Meadowcroft, J. (2013). Exploring negative territory. Carbon dioxide removal and climate policy initiatives. *Climatic Change* **118** (1): 137–149.

Medvecky, F., Lacey, J., and Ashworth, P. (2014). Examining the role of carbon capture and storage through an ethical lens. *Science and Engineering Ethics* **20**: 1111–1128.

Melzer, L.S. (2012). *Carbon Dioxide Enhanced Oil Recovery (CO_2 EOR): Factors Involved in Adding Carbon Capture, Utilization and Storage (CCUS) to Enhanced Oil Recovery*. Melzer Consulting: Midland, TX.

Moffat, K., Lacey, J., Zhang, A., and Leipold, S. (2015). The social licence to operate: a critical review. *Forestry* **89**: 477–488.

Morrow, D.R. (2014). Ethical aspects of the mitigation obstruction argument against climate engineering research. *Philosophical Transactions of the Royal Society A: Mathematical, Physical and Engineering Sciences* **372**.

ter Mors, E., Terwel, B.W., and Daamen, D.D.L. (2012). Potential of host community compensation in facility siting. *International Journal of Greenhouse Gas Control* **11** (Supplement): S130–S138.

Nerlich, B. and Jaspal, R. (2013). UK media representations of carbon capture and storage, actors, frames and metaphors. *Metaphor and the Social World* **3** (1): 35–53.

Oltra, C., Upham, P., Riesch, H. et al. (2012). Public responses to CO_2 storage sites: lessons from five european cases. *Energy & Environment* **23** (2): 227–248.

Parkhill, K., Pidgeon, N., Corner, A., and Vaughan, N. (2013). Deliberation and responsible innovation: a geoengineering case study. In: *Responsible Innovation: Managing the Responsible Emergence of Science and Innovation in Society* (ed. R. Owen, J. Bessant and M. Heintz), 219–241. Chichester: Wiley.

Pidgeon, N., Corner, A., Parkhill, K. et al. (2012). Exploring early public responses to geoengineering. *Philosophical Transactions of the Royal Society* **370**: 4176–4196.

Powell, T.W.R. and Lenton, T.M. (2012). Future carbon dioxide removal *via* biomass energy constrained by agricultural efficiency and dietary trends. *Energy and Environmental Science* **5**: 8116.

Powell, T. and Lenton, T.M. (2013). Scenarios of global agricultural biomass harvest reveal conflicts and trade-offs for bioenergy with CCS. *Geophysical Research Abstracts* **15**: (EGU2013-13338).

Preston, C.J. (2013). Ethics and geoengineering: reviewing the moral issues raised by solar radiation management and carbon dioxide removal. *Wiley Interdisciplinary Reviews: Climate Change* **4** (1): 23–37.

Preston, C. (2014). Framing an ethics of climate management for the anthropocene. *Climatic Change* **130**: 359–369.

Proelss, A. and Gussow, K. (2011). Carbon capture and storage from the perspective of international law. In: *European Yearbook of International Economic Law 2011* (ed. C. Hermann and J.C. Terhechte), 151–169. New York: Springer.

Raman, S., Mohr, A., Helliwell, R. et al. (2015). Integrating social and value dimensions into sustainability assessment of lignocellulosic biofuels. *Biomass and Bioenergy* **82**: 49–62.

Reiner, D. and Nuttall, W. (2011). Public acceptance of geological disposal of carbon dioxide and radioactive waste: similarities and differences. In: *Geological Disposal of Carbon Dioxide and Radioactive Waste: A Comparative Assessment* (ed. F.L. Toth), 295–216. Paris: IAEA.

Ribeiro, B.E. and Quintanilla, M.A. (2015). Transitions in biofuel technologies: an appraisal of the social impacts of cellulosic ethanol using the Delphi method. *Technological Forecasting and Social Change* **92**: 53–68.

Ricci, O. and Selosse, S. (2013). Global and regional potential for bioelectricity with carbon capture and storage. *Energy Policy* **52**: 689–698.

Roberts, T. and Mander, S. (2011). Assessing public perceptions of CCS: benefits, challenges and methods. *Energy Procedia* **4**: 6307–6314.

Roberts, T. and Upham, P. (2012). Prospects for the use of macro-algae for fuel in Ireland and the UK: An overview of marine management issues. *Marine Policy* **36**: 1047–1053.

Roberts, T., Upham, P., Mander, S. et al. ed. (2013). *Low-Carbon Energy Controversies*. Routledge.

Roman, M. (2011). Carbon capture and storage in developing countries: a comparison of Brazil, South Africa and India. *Global Environmental Change* **21**: 391–401.

Shackley, S. and Thompson, M. (2012). Lost in the mix: will the technologies of carbon dioxide capture and storage provide us with a breathing space as we strive to make the transition from fossil fuels to renewables? *Climatic Change* **110**: 101–121.

Shrader-Frechette, K. (2002). *Environmental Justice: Creating Equality, Reclaiming Democracy*. New York: Oxford University Press.

Slade, R., Bauen, A., and Gross, R. (2014). Global bioenergy resources. *Nature Climate Change* **4**: 99–105.

Smolker, R. and Ernsting, A. (2012). BECCS (bioenergy with carbon capture and storage): climate saviour or dangerous hype? BioFuelWatch http://www.biofuelwatch.org.uk/files/BECCS-report.pdf.

Spreng, D., Marland, G., and Weinberg, A.M. (2007). CO_2 capture and storage: another faustian bargain? *Energy Policy* **35**: 850–854.

Stephens, J.C. (2014). Time to stop CCS investments and end government subsidies of fossil fuels. *Wiley Interdisciplinary Reviews: Climate Change* **5**: 169–173.

Tavoni, M. and Socolow, R. (2013). Modeling meets science and technology: an introduction to a special issue on negative emissions. *Climatic Change* **118** (1): 1–14.

Terwel, B.W., ter Mors, E., and Daamen, D.D.L. (2012). It's not only about safety: beliefs and attitudes of 811 local residents regarding a CCS project in Barendrecht. *International Journal of Greenhouse Gas Control* **9**: 41–51.

Thomson, I. and Boutilier, R. (2011). The social licence to operate. In: *SME Mining Engineering Handbook* (ed. P. Darling), 673–690. Littleton, Colorado: Society for Mining, Metallurgy, and Exploration.

Thornley, P., Upham, P., Huang, Y. et al. (2009). Integrated assessment of bioelectricity technology options. *Energy Policy* **37**: 890–903.

Tilman, D., Socolow, R., Foley, J.A. et al. (2009). Beneficial biofuels – the food, energy, and environment trilemma. *Science* **325**: 270–271.

Torvanger, A. and Meadowcroft, J. (2011). The political economy of technology support: making decisions about carbon capture and storage and low carbon energy technologies. *Global Environmental Change* **21**: 303–312.

United Nations Framework Convention on Climate Change (2015). Adoption of the Paris agreement. https://unfccc.int/resource/docs/2015/cop21/eng/l09r01.pdf (accessed 13 December 2017)

Unruh, G.C. (2000). Understanding carbon lock-in. *Energy Policy* **28**: 817–830.

Upham, P. and Roberts, T. (2011). Public perceptions of CCS: emergent themes in pan-European focus groups and implications for communications. *International Journal of Greenhouse Gas Control* **5**: 1359–1367.

Vaughan, N.E. and Gough, C. (2016). Expert assessment concludes negative emissions scenarios may not deliver. *Environmental Research Letters* **11**: 095003.

Vergragt, P.J., Markusson, N., and Karlsson, H. (2011). Carbon capture and storage, bio-energy with carbon capture and storage, and the escape from the fossil-fuel lock-in. *Global Environmental Change* **21**: 282–292.

Wallquist, L., L'Orange-Seigo, S., Visschers, V.H.M., and Siegrist, M. (2012). Public acceptance of CCS system elements: a conjoint measurement. *International Journal of Greenhouse Gas Control* **6**: 77–83.

Wise, M., Calvin, K., Thomson, A. et al. (2009). Implications of limiting CO_2 concentrations for land use and energy. *Science* **324**: 1183–1186.

Zhang, A. and Moffat, K. (2015). A balancing act: the role of benefits, impacts and confidence in governance in predicting acceptance of mining in Australia. *Resources Policy* **44**: 25–34.

13

Unlocking Negative Emissions

Clair Gough[1], Patricia Thornley[1], Sarah Mander[1], Naomi Vaughan[2] and Amanda Lea-Langton[1]

[1] *Tyndall Centre for Climate Change Research, School of Mechanical Aerospace and Civil Engineering, University of Manchester, UK*
[2] *School of Environmental Sciences, University of East Anglia, Norwich, UK*

13.1   Introduction

This book presents an integrated and holistic analysis of BECCS technologies and systems, and it is summarised in this chapter in two sections. The first section provides an overview of the key conclusions presented in each of the preceding chapters individually. The book introduces the reader to the detailed technical and engineering characteristics of the components of BECCS technologies and builds up through the assessment of integrated BECCS processes before placing the technology in the wider contexts of climate mitigation, policy and society.

To limit future global warming and meet the aspirations of the Paris Agreement (UNFCCC, 2015), modelled future emission scenarios rely heavily on BECCS to deliver global net negative emissions. In the second section of this summary chapter, the editors draw together the insights from across the book to critique key questions implicit in this assumption.

13.2   Summary of Chapters

Section 1 describes the component technologies of BECCS and the specific challenges of introducing them into a BECCS system. Thus, the book begins with consideration of the sustainable production of biomass energy feedstocks (Chapter 2), followed by the implications of using biomass energy in conjunction with pre- and post-combustion capture technologies (Chapters 3 and 4) and the techno-economic implications of different combinations of the component technologies (Chapter 5).

In *Chapter 2*, Welfle and Slade explore the issues associated with estimating and delivering bioenergy at the scales required for global negative emissions delivered through BECCS. Many countries are likely to increase bioenergy to meet energy demands; if BECCS deployment increases, even more biomass will be needed. European

Biomass Energy with Carbon Capture and Storage (BECCS): Unlocking Negative Emissions, First Edition.
Edited by Clair Gough, Patricia Thornley, Sarah Mander, Naomi Vaughan and Amanda Lea-Langton.
© 2018 John Wiley & Sons Ltd. Published 2018 by John Wiley & Sons Ltd.

renewable energy targets have rendered Europe the world's primary biomass trading hub and demand region for biomass energy. Estimating the resources is complex and uncertain, with a variety of methods and huge number of assumptions resulting in a large range of estimates; central forecasts fall between 50 and 300 EJ/year by 2050. Expansion is likely to be in the form of large-scale dedicated or co-fired (with fossil fuels) power generation, with or without CCS. The availability of resources for BECCS is inextricably linked to this co-evolving demand. A key concern is diversion of land from food production, although this is just one aspect of sustainability, which also encompasses water, conservation, biodiversity and social impacts. Conversely, there are significant benefits that could be harnessed using perennial crops, for example, to increase soil carbon and improve fertility.

In *Chapter 3*, Finney et al. discuss post-combustion and oxy-fuel combustion capture technologies; both are well understood – the extensive research and development already conducted renders them suitable for current and imminent deployment on a global scale. Our knowledge of these processes can be adapted to suit biomass fuels and thus BECCS-specific applications, although there is scope for further work to address technical challenges, which would be expected to improve technical performance and reduce costs. Specific challenges regarding BECCS operating conditions include various forms of solvent degradation, slagging, fouling, deposition and corrosion, but there are options for minimising and mitigating these. Pretreatments of fuels, through washing, hydrothermal carbonisation and torrefaction can remove problematic species (e.g. particulate matter) before combustion; and additional stages in the emissions-control systems are able to eliminate or minimise these before the flue gases enter the carbon capture plant.

In *Chapter 4*, Lea-Langton and Andrews describe how pre-combustion technologies, as part of an integrated gasification process, are also a technically feasible method for carbon capture. However, the application of these processes to power systems is less established compared to post-combustion techniques, and there have been few trials with biomass feedstocks, although existing techniques could be adapted to BECCS. Enabling technologies such as syngas-compatible energy systems and fuel-cell technologies are not yet established and so demand for pre-combustion CCS technology is more limited than that for post-combustion CCS technology. Specific challenges are associated with the gasification step of the process and include avoidance of tar build-up, which could be exacerbated by variable fuel quality and the presence of trace elements. There is also substantial research activity ongoing regarding the CO_2 separation stage of the process, especially in the areas of identifying novel or modified materials as adsorbents, solvents or separation membranes. The deployment potential of all capture options needs to be considered in terms of applications of these technologies for biomass fuels, and there are clear openings for much additional work in this area to be undertaken to make the potential of net negative emissions from BECCS a global reality.

In *Chapter 5*, Bhave et al. present a techno-economic analysis based on detailed process modelling conducted on eight BECCS technologies considered to be the most promising for large-scale deployment by 2050. The technologies covered were co-fired power generation with amine scrubbing, oxy-fuel combustion, carbonate looping or utilising integrated gasification combined cycle (IGCC);and dedicated fully biomass-fired power stations utilising amine scrubbing, oxy-fuel, chemical looping or IGCC.

Their analysis attempts to quantify the uncertainties associated with different parameters and consider how far away from market they are.

The results show that each technology has its benefits and risks; for example, technologies that have higher potential efficiency generally have lower technology readiness, making cost assessment much harder. The eight technology combinations considered here vary in their technology readiness level (TRL) over a wide range from TRL 4 (components tested in the laboratory) to TRL 7 (pilot plant >5% of commercial scale). Wherever direct comparison is feasible from a similar unabated plant, it is observed that net efficiency decrease due to carbon capture varied from 6% to 13%, specific investment costs increased significantly by between 45% and 130% and annualised operating and maintenance costs were increased by 4–58%. The high cost of large-scale demonstration projects on fossil-fuel-based plants, and the need for government subsidies, means that the pace of deployment is slow, particularly in light of the cancellation of the UK CCS demonstration programmes.

Section 2 assesses the wider characteristics of these component technologies operating as an integrated BECCS system. This section begins by setting out the challenges and opportunities for carrying out such an assessment across the full supply chain of a BECCS system (Chapter 6). Detailed performance parameters are presented for a full fossil-fuel-based CCS supply chain (Chapter 7) and another in conjunction with biomass feedstocks (Chapter 8).

In *Chapter 6*, Falano and Thornley discuss the rationale for supply-chain assessment to compare the performance of BECCS systems and the useful role that life-cycle assessment (LCA) can play in this context. The importance of framing an appropriate LCA question is emphasised (e.g. absolute emissions, CO_2/MWh, etc.) since it dictates the scope of the system and the calculation framework. Common sources of variability in LCA results are examined, and it is stressed that extreme caution is needed when comparing different studies, which may have different system scopes, methodological approaches and assumptions. The relatively small number of BECCS LCAs that have been published in the scientific literature are discussed, reaching the conclusion that defined BECCS systems can deliver negative emissions. However, if these are to be relied upon they must be robustly demonstrated with adequate attention to both the naturally variable biomass supply chain and the unknown and unproven components of the engineering system.

In *Chapter 7*, Hammond presents a detailed assessment of the sustainability-related performance characteristics of contemporary fossil CCS, which will form the basis for BECCS technologies. These assessments are based on well-established, often quantifiable, economic, energy-related and environmental (including climate change) indicators. Capturing carbon from a power plant incurs an energy penalty, which lowers the system (thermodynamic) efficiency and hence impacts the plant's economics. CCS is currently more expensive than its competitors and will rely on cost reduction to become commercially viable. In the case of post-combustion plants, CCS with enhanced oil recovery (EOR) results in a 62–106% increase in the cost of electricity (COE) on a whole CCS chain basis compared to the reference system, whereas storage in gas wells leads to a 93–97% increase. For natural gas combined cycle (NGCC) technologies, CCS with EOR results in an increase of 91–142% and CCS with gas well storage a 118–122% increase. Integrated gasification combined cycle (IGCC) plants with EOR result in an increase of 27–60% and CCS with gas well storage a 45–48% increase. The average price increase for all scenarios is about 84%. Greater financial incentives for carbon abatement could, in principle, be secured through a higher carbon price from the EU ETS, although in its first 10 years of operation this fell from about €20 per tonne in 2005 to around €5 per tonne in 2015. The environmental performance of CCS development is assessed in terms of climate-change impacts (including parameters such as carbon intensity and the cost of carbon avoided or captured), as well as the effects on biodiversity, land use and water resources.

In *Chapter 8*, Newton-Cross and Gammer explore the technical potential for BECCS to play a significant role in the United Kingdom. The authors conclude that barriers to the deployment of BECCS technologies are unlikely to be technical but that the most significant barriers will be the scale of investment required, the limited price of carbon and hence the limited value of negative emissions. Furthermore, most cost savings in the next two decades will be delivered through reducing costs by deployment, rather than fundamental technology breakthroughs. The chapter describes numerous bioenergy value chains with the potential to deliver significant carbon savings and sizeable negative emissions from BECCS, based on certain feedstocks. The analysis suggests that the United Kingdom is exceptionally well placed to exploit the benefits of BECCS, given its vast offshore storage opportunities; experience in bioenergy deployment and UK academic and industrial research and development strength across bioenergy and CCS.

Section 3 extends the scope of assessment to consider the role of BECCS within global energy systems as we respond to the challenge of climate-change mitigation. This section begins by setting out the scale of BECCS required to deliver global net negative emissions in relation to atmospheric carbon budgets and considers the implications of the required levels of deployment (Chapter 9). The role of BECCS in emission scenarios directed at delivering specific climate targets is then analysed in an integrated assessment model (IAM) (Chapter 10). The final chapters discuss the policy and accounting implications of delivering negative emissions through BECCS (Chapter 11) and the associated social and ethical implications of doing so (Chapter 12).

In *Chapter 9*, Mander et al. demonstrate how the emissions budgets associated with meeting the ambitions of the UNFCCC Paris Agreement (UNFCCC, 2015) are dwindling rapidly in the face of continuation of current levels of emissions. BECCS is not explicitly mentioned within the agreement, but integrated assessment model results suggest that atmospheric concentrations cannot be kept within remaining carbon budgets without significant levels of negative emissions, primarily provided by BECCS in the modelled scenarios. This chapter describes how, in order to understand the implications of these findings, it is necessary to place them within the context of the underpinning IAM assumptions, and whilst some of these are explicit, others are less obvious. For example, there are assumptions concerning the presence of a CO_2 transport and storage infrastructure and the sustainability of biomass supply chains, which will be challenging to deliver without appropriate global oversight and international governance arrangements. IAMs typically assume that BECCS deployment occurs at scale from 2020 onwards, with IAM results suggesting that up to 10 Gt CO_2 negative emissions are required by 2050 (van Vuuren et al., 2013); this is in contrast to bottom-up assessments for the deployment of BECCS, which estimate an economic potential for the storage of 3.5 $GtCO_2$ by 2050 (IEAGHG, 2011). This mismatch between the outputs from top-down and bottom-up modelling approaches is also evident when considering requirements for both CO_2 transport and storage infrastructure (Whiriskey, 2014; Stewart et al., 2014). Whilst acknowledging that BECCS is not a blue-skies mitigation option, the authors stress that it is in early stages of development. It is also clear that in order to address the climate-mitigation challenge a wholesale and rapid transition to a new low-carbon energy system is required. The key questions that IAM models cannot answer are whether BECCS can deliver what is anticipated in an adequate timescale and whether it should be preferred to other ways of achieving the same net emissions when environmental, social and economic perspectives are taken into account.

In *Chapter 10*, Anandarajah et al. used the TIAM-UCL global energy-system model to explore the extent to which BECCS is critical for meeting the 2 and 1.5 °C limits of the Paris Agreement (UNFCCC, 2015). Several scenarios were developed that varied the availability of biomass, the year in which a peak in greenhouse gas emissions occurs, and the availability of BECCS. The results show that the availability of BECCS can reduce the pressure on near-term mitigation requirements under a 2 °C scenario, allowing a longer use of coal and fewer stranded assets. Alternatively, if there is no BECCS available, the carbon price in 2050 is doubled from 150 to 300 $/tCO$_2$ to meet the 2 °C target, requiring almost complete removal of coal from electricity production. The results suggest that the 1.5 °C target is impossible to achieve without BECCS or other negative emissions processes. The pathways explored in this chapter involve extremely rapid reductions in global emissions, implying unprecedented global cooperation and dramatic energy-system transformations. The authors note that these results be regarded as techno-economically optimal and highlight that the socio-technical plausibility remains open to question. The availability of BECCS provides some 'breathing space' to enable mitigation efforts to ramp up but the authors caution that this should not be interpreted as reducing the need for near-term action. They highlight the optimistic technology deployment rates in the model that do not account for the behavioural, institutional or political factors that slow technology adoption in the real world.

In *Chapter 11*, Thornley and Mohr consider the governance challenges associated with BECCS systems. As described in Chapter 6, the ideological rationale for a system-level evaluation is clear, but there is no enforcement or incentive framework that operates at the spatial, temporal and sectoral levels that are required to shape and govern such a system. Additionally, the frameworks and assumptions used in most common assessment and regulatory approaches carry with them implicit value judgements, which are rarely transparent, explicit or tested. This is problematic when considering the ethical justification of the assessments, and any associated policy initiatives, which combine carbon sequestration and emissions incurred in different countries and sectors. There is also the challenge of large-scale system implementation coupling to local supply and impacts. The chapter concludes that a multilevel governance approach might help to bridge the value–action gap that exists between global strategies, national policies and everyday actions, but that there is a need for understanding socio-economic innovation and human-technology interactions to inform the development of BECCS in tandem with technology and engineering advances.

In *Chapter 12*, Gough et al. explore the social and ethical implications of using BECCS to deliver negative emissions. BECCS raises questions about the potential role of negative emissions in achieving climate stability, but also about how society should achieve that goal, in the context of carbon budgets and a broad mitigation portfolio. Although there is a large body of literature looking at possible social responses to CCS, there is very little experience of the technology in practice, beyond demonstration-scale installations, and it remains a largely hypothetical technology at commercial scale. The geographical spread of its potential impacts will be felt in broad terms as a response to climate change but also along its various supply chains. While biomass energy is currently very extensively used at small scales, the use of BECCS for achieving global negative emissions would instead introduce a smaller number of very-large-scale facilities. The potential for BECCS to deliver negative emissions is also critical to its governance and introduces unique challenges to ensuring distributional, procedural, financial and intergenerational justice in its deployment. Particular care is needed to ensure that it does not perpetuate inequalities within and between countries. It presents new and complex governance challenges, not least in how negative emissions can be accounted, verified and certified; in how we prioritise competing land-use needs and ecosystem services and in how we manage liability and allocate responsibility for storage many generations in to the future.

The summary of chapters in this book shows that BECCS is a workable technology that could contribute to climate-change mitigation efforts. However, uncertainties over technology development, enabling frameworks and policy development make it much harder to determine the precise role and magnitude of the contribution that BECCS can make to delivering global net negative emissions and limiting future global warming. In the following section of this summary chapter, we reflect on the key issues associated with growing BECCS from the project level to strategically providing atmospheric carbon dioxide removal at a planetary global scale.

13.3 Unlocking Negative Emissions: System-Level Challenges

13.3.1 Terminology, Scale and Quantification

In Chapter 1 we defined the terms 'negative emissions' and 'global net negative emissions', as they are used in the negative emissions literature, and illustrated them with a schematic (Figure 13.1). Here, we continue to use those definitions and present the schematic again for the reader.

The terminology illustrated in Figure 13.1 describes how negative emissions are conceptualised in the IAMs. In IAM scenarios, BECCS systems that deliver negative emissions, Figure 13.1b,c, are included from around 2020, but global net negative emissions (Figure 13.1d) begin around 2070 (Figure 1a in Fuss et al., 2014). Global net negative emissions (Figure 13.1d) are achieved when global negative emissions from BECCS (or other negative emissions technologies, e.g. direct air capture and storage or enhanced weathering) outweigh emissions from all other human sources.

The key to moving from being a technology that delivers 'negative emissions' at a project level (Figure 13.1a) to one that delivers global net negative emissions (Figure 13.1d) is scale (and, ultimately, the scope of how we account for those emissions). This pervades both our understanding of BECCS and the issues associated with its potential use and implementation. Ensuring that BECCS (whether as part of a wider portfolio of negative emissions approaches or not) removes more CO_2 from the atmosphere than the combined emissions of all other anthropogenic sources at a global level

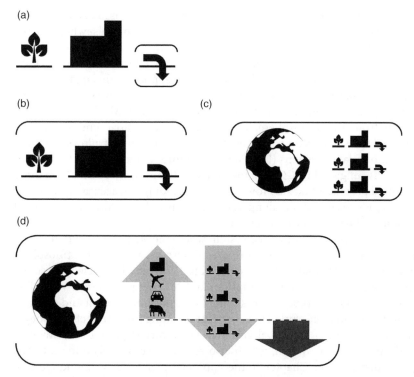

Figure 13.1 Schematic illustrating terminology for negative emissions from BECCS. (a) *CO₂ stored*, which can be calculated at a project, national or global level. (b) *Negative emissions*, which can be calculated at a project or national level and are achieved when the CO_2 removed from the atmosphere during biomass growth is greater than the CO_2 emissions from all the processes in the supply chain. (c) *Global negative emissions*, the global sum of all negative emissions activity; this removes carbon dioxide from the atmosphere and can be used to 'offset' other anthropogenic CO_2 emissions. (d) *Global net negative emissions* occur when the global amount of negative emissions exceeds the CO_2 emissions from all *other* human sources, e.g. energy, transport and agriculture (Gough et al., 2018).

presents scalar challenges across every aspect of its implementation; it requires massive ramping-up of the technical, biological, logistical, governance and social processes that constitute the BECCS system. Here, we explore the implications of advancing BECCS from the project scale to a means of unlocking net negative emissions on a global scale.

The BECCS supply chain is extensive and is interconnected with other human systems, such as food production, water resources and land-use change. This creates challenges for where to delineate the boundary around a BECCS supply chain when calculating its negative emissions at a project level (Figure 13.1b). Spatial and temporal scales become increasingly important factors when determining the negative emissions at a national (Figure 13.1b) and global level (Figure 13.1c) and whether BECCS can deliver global net negative emissions (Figure 13.1d). We argue that <u>global</u> net negative emissions, as conceptualised in Figure 13.1d, are not meaningful at sub-global scales due to the global distribution of biomass production, its associated implications (e.g. on land-use change), the structure of associated accounting frameworks and allocation of anthropogenic emissions from elsewhere in the socio-technical system.

Accounting frameworks have been designed to allow countries to take control of their own emissions reduction efforts rather than accurately apportion the split of greenhouse gas emissions along a supply chain. Consumption-based emissions may offer one route to addressing this for biomass and other traded products which use land and sequester carbon from the atmosphere. Consideration of negative emissions on a national or regional basis requires a consequential LCA approach that allows examination of the impact of related systems (land and agriculture, water, construction, waste etc.) on the feasibility of global BECCS deployment and the impact of BECCS on those systems. This would provide a first step towards understanding and accurately accounting for the greenhouse gas emissions along a supply chain and therefore the delivery of negative emissions at a national or regional level.

'Negative emissions' describe a bounded system where the elements included within the specified system boundary remove more greenhouse gases from the atmosphere than they emit under the assumed process conditions. With any life-cycle assessment framing, there is always a reliance on assumptions and key data parameters. Variability in these may affect results, and practitioners should take this into account and examine appropriate sensitivities in their work. However, extrapolating from project-level negative emissions (Figure 13.1b) to global negative emissions (Figure 13.1c) involves changes in scale and scope of system, which also require a completely different LCA methodology. As discussed in Chapter 6, the approach that needs to be adopted to evaluate the greenhouse gas balance of a single BECCS project is very different from that which needs to be used to evaluate the consequences for implementing BECCS globally. The latter requires consideration of the feedstock requirements, technologies likely to be used (scale, efficiency and greenhouse gas balance), displaced land use, changes in global wood production and use patterns, effects on wider energy systems to assess counterfactuals, etc. This is a complex calculation requiring interdisciplinary assessment of the possibility of scaling up BECCS sufficiently to meet IAM projections; completing such work on a global scale, while taking into account competing uses of biomass and co-evolution of the energy system, is a very significant undertaking, and one fraught with uncertainty.

13.3.2 Non-Technological Challenges

There is a number of key governance challenges associated with BECCS development (as detailed in Chapter 11). Negative emissions are not currently valued under any existing international framework and the accounting mentality that underlies the delivery of negative emissions (i.e. supply chains across national boundaries) does not sit easily alongside the existing UNFCCC territorially based accounting systems. Significant ethical issues arise over 'ownership' and 'accountability' of the different components of emissions along this supply chain, not least when they are conflated to produce a single negative emissions value that is calculated from composite contributions with different provenances. This is compounded by concerns that there may be accompanying increases in greenhouse gas emissions in other areas that are not being included in the assessment, such as through indirect land-use change or indirect consequences on the energy system through changes to other sectors. To what extent should these be considered?

It is often argued that energy producers, and others, should only be accountable for changes they have caused or where a causal link can at least be partly demonstrated, but geographic spread and market effects can make physical demonstration very challenging. For example, we have to make assumptions about what the biomass used in a BECCS supply chain would otherwise have been used for, e.g. fuel wood or in construction, and this choice would affect the calculation of greenhouse gas emissions for the BECCS system (i.e. the greenhouse gas benefit of BECCS compared to no BECCS). The calculations would also be affected by assumptions about whether the land used would otherwise have reverted to natural forest or been used for agricultural production. There is significant uncertainty attached to such assumptions, particularly when future long-term projections are involved.

In reality, this issue of 'counterfactuals', or alternative uses, depends on market evolution, which can change in unexpected ways. For example, forests established for pulp and paper production may see their global markets and prices fall with social and cultural changes (such as reading habits and electronic media) to the extent that forest owners may abandon plantations that are no longer generating sufficient income. This could be offset and forest land cover retained if BECCS demand were to provide additional wood-product markets. However, the plausibility of this is far beyond the domain of energy-system modelling and demonstrating causal effects is extremely challenging. For example, establishment of supply chains from particular countries to BECCS facilities may require the establishment of new transport, ports and other infrastructures. Once established, these may open up other trading opportunities for that region, which could conceivably result in local communities changing their current production patterns to take advantage of new export opportunities. Thus, it may become viable to export horticultural produce, previously not possible without those transport links, resulting in changes to local agricultural and land-use patterns. To what extent has that change been caused by and to what extent should it be attributed to the BECCS system? There is no direct link and no easy answer.

There are also very significant challenges associated with incentivising negative emissions. First, there is no explicit value on carbon removed. National and international policy incentives that recognise and reward renewable or low-carbon energy provision have been developed. It is conceivable that some of these could be adapted to suit negative emissions delivered by BECCS systems, may be even tiered in a way that incentivised greater carbon reductions. However, the valued commodity is the energy supplied, and there are trade-offs between delivering negative emissions and efficiency. A system that removes more carbon from the atmosphere is often achieved at the expense of efficiency, meaning that while more greenhouse gas emissions reductions are achieved, less of the primary product (energy) is produced and more of the valuable biomass resource utilised. Without any contrary price signals, it is inevitable that power plants will seek to maximise their useful energy output to existing markets rather than maximising the carbon sequestered in the overall system.

The second challenge to incentivising negative emissions is that, as described, the existing accounting frameworks (UNFCCC and the associated national inventories) do not sum emissions in a way that reflects a net removal of carbon from the atmosphere across the entire BECCS system. Instead, the negative emissions are split across different accounting sectors e.g. energy and land use, in different countries, and there is no overall management incentive or responsibility corresponding to the whole BECCS system. This provides a very significant governance challenge. The beneficial outcome of negative emissions from BECCS is not recognised in a single framework but split across different stakeholders operating in different regimes, meaning that there is currently no consolidated entity that could be manipulated to incentivise BECCS deployment.

An assessment of the global biomass resource potential was presented in Chapter 2 using different analyses and techniques. The authors identified that there were many forecasts of national and global biomass potential with significant degrees of variability, related to methodology and scope, but most point to a significant resource with energy crops, wastes, agricultural residues and forestry products all key elements. However, BECCS development at a scale that delivers global net negative emissions (Figure 13.1d) is likely to require deployment of very large power generation and CCS facilities. There is a very significant challenge associated with matching a dispersed set of distinct biomass resources to large consolidated central conversion facilities, and differentiation with particular facilities focusing on particular resource categories is perhaps most likely. Ensuring sustainability of these supply chains is essential if the negative emissions are to be achieved and is extremely challenging to put into practice. Some aspects such as forest carbon stock or impacts on long-term soil carbon are most appropriately viewed on a large and long-term basis to be assured of capturing all relevant impacts. However, practically, mobilisation of biomass resources needs to take place on a local or regional basis and for which risks of land-use change, water impacts, social impacts, conservation and biodiversity must be addressed and managed. Sustainable supply chains require appropriate management of the local environmental issues in pursuit of consolidated systems that satisfy global sustainability objectives. Simultaneous effective governance at both of these scales is required; present systems do not provide an adequate framework for progressing this.

Social and ethical research has to date focused on fossil CCS as an incremental technology that can be easily identified as part of existing energy system, whether it is seen as a bridge towards a fully decarbonised system or as a component in a broad portfolio of options. While introducing biomass energy as feedstock at a project or even regional level may change the perceptions and acceptance of the basic technology, moving to global net negative emissions adds an extra dimension to the discourse. Once BECCS is proposed as a means of changing the shape of the carbon budget over the next century and beyond (see Chapter 9), it opens up some broader ethical questions. For example, is it morally defensible to delay mitigation action in the near term, leading to a temporary overshoot of the carbon budget and committing to future negative emissions to remove CO_2 from the atmosphere and rebalance the carbon budget? Or, can it be argued that it is simply not feasible to meet our carbon budgets without negative emissions, lending a moral imperative to pursuing approaches such as BECCS? There remain scientific uncertainties related to global carbon cycle–climate feedbacks and our understanding of the relationship between future reductions in atmospheric CO_2 concentrations and climatic responses. A mitigation strategy that depends on negative emissions may place an additional burden on future generations to implement a radical carbon dioxide removal programme and/or bear the risks of greater climate change, in the event of passing tipping points or the failure to deliver assumed atmospheric CO_2 concentration reductions.

13.3.3 Technical Challenges

As discussed earlier, a variety of biomass resources would be needed to deliver global net negative emissions, each with different management requirements and production timescales. Establishing large-scale forestry plantations for energy production has the advantage of accelerating sequestration of CO_2 from the atmosphere, with the uptake of CO_2 fastest during the early years of tree growth. The intention with BECCS is that this sequestered carbon will be captured and stored in the future. If the BECCS system were to fail in any way, there is a risk that years of sequestered CO_2 will be released back to the atmosphere over a very short time and at a point when the carbon budget has already been exceeded.

Generating electricity from biomass is a well-established technology; however, it has a thermal efficiency that is usually lower than for conventional fossil-fuel-powered facilities. In addition, if we are considering post- or oxy-combustion capture (the most mature BECCS technologies), then there is a significant and explicit energy penalty associated with capture and removal of carbon dioxide. So, it must be clearly understood that implementing BECCS as a negative emissions technology will actually increase global energy demand. Essentially, there is an energy penalty to pay for the negative emissions, which needs to be appropriately accounted for in IAMs, but can really only be fully verified with progression of practical technology deployment. One way to address this is development of alternative, more efficient technologies and some of these (pre-combustion capture and gasification or chemical looping combustion) have been discussed in this book (see Chapters 4 and 5). Technological innovation could reduce the impact of this energy penalty, but needs to be incentivised and initiated.

The role of BECCS has evolved from a process to deliver enhanced mitigation at project level to widespread incorporation in IAMs and, as a result, has become a central element in the projected pathways to limiting global warming to 2 °C. To a large extent, the current emphasis on the use of BECCS to deliver negative emissions derives from its inclusion in IAMs, which, basing results on cost optimisations, typically conclude that BECCS provides either a more cost–effective route to delivering atmospheric concentration targets or, as is the case for 1.5 °C pathways, an essential component in reaching those targets. Thus, estimates of the costs of implementing the technology are central to the models' conclusions. There is a risk that if individual projects underperform when deployed and inadequate time is allowed for learning and improvement, this could result in an over-optimistic global assessment.

13.4 Can Negative Emissions be Unlocked?

This book has provided an overview of the technologies, challenges and implications of using biomass energy and CCS in the context of negative emissions. Reflecting on the prospects for its potential to be realised, the problem of unlocking negative emissions can be encapsulated in four overarching questions, summarised in the following sections.

13.4.1   Do we Need This Technology?

Chapters 9 and 10 provide clear evidence that, given our current understanding of carbon budgets and the consequential climate responses, and even with very ambitious conventional mitigation, global net negative emissions are a necessary component of strategies to meet current mitigation targets. Of the available greenhouse gas removal technologies, BECCS is probably the most technologically advanced and fits most comfortably with the current energy-system infrastructure. However, to have the impact required and assumed within integrated assessment models, there is a need to progress technology and project development as a matter of urgency.

13.4.2   Can it Work?

Historical experience of thermal power technology deployment shows that it takes a significant length of time for technologies, even at high TRLs, to become fully commercial to the extent that financiers and investors are comfortable with the associated economic risks. Examples of previous experience in the deployment of (low-carbon) energy technologies in the United Kingdom illustrate how problems with technology designs are often not exposed until commercial deployment begins on the ground. Despite a large national technology base and research programme, the development of nuclear power in the mid-twentieth century involved substantial adaptation of engineering designs, leading to timescales from inception to delivery in excess of 10 years. Similarly, in a variety of contexts during the 1990s, practical issues were encountered that were not fully apparent until operation and deployment was progressed, for example, combined-cycle gas turbines, waste and biomass plants and the wave of renewable energy power plant construction under the Non-Fossil Fuel Obligation (NFFO) programme. In the latter case, the lack of a single technology base impeded learning across projects, and, in the early years, many failed to meet the performance objectives that had underpinned their financial viability. The experience with the technologies described above suggests that it would be reasonable to assume a time horizon of at least 10 years between implementation of an incentivisation policy and delivery of an engineered solution that actually delivers what was originally anticipated. This would be further extended by consent, permission and other issues related to the significant infrastructure requirements associated with the carbon-capture and storage elements of proposed BECCS projects. So, extreme caution needs to be exercised in making projections about likely timescales for delivery of BECCS plants; in this context, the deployment represented in IAMs by 2020 seems, at best, optimistic – practical deployment to affirm the engineering issues would need to begin now to establish projects by 2030.

13.4.3   Does the Focus on BECCS Distract From the Imperative to Radically Reduce Demand and Transform the Global Energy System?

Although many modelling assessments highlight a role for BECCS, do these lead to an underestimation of the scale of mitigation challenge because of emphasis on BECCS in climate scenarios? Global emissions have continued to rise, despite the fact that for a number of decades there has been a strong scientific case for the need to reduce emissions and an understanding of the mitigation options available (including decarbonisation and reducing energy consumption). There is a mismatch between the understanding of what needs to be done and the reality of deploying technologies on the ground. It could be argued that the prospect of negative emissions allowing an overshoot of budgets relieves some of the pressure relating to the urgency and ambitiousness of mitigation required (mitigation deterrence) or that relying on an emerging technology as a significant component of mitigation efforts risks falling short of future climate goals (moral hazard). However, it could also be argued that because conventional mitigation efforts have consistently failed to keep pace with the challenge of dwindling carbon budgets, it is now more urgent than ever that we pursue negative emissions technologies. If we need both radical transformation of energy systems and negative emissions in order to constrain climate change, what is the risk of not developing BECCS (reverse moral hazard)?

13.4.4 How Can BECCS Unlock Negative Emissions?

This book has attempted to provide readers with the information to reflect on their own answers to the above questions. The United Kingdom, through its academic research councils, the Committee on Climate Change, and the Department of Business, Energy and Industrial Strategy (BEIS), for example, is exploring the feasibility and implications of incorporating negative emissions into national climate change plans. If it can be shown that BECCS is needed to avoid temperature rises beyond 1.5 or 2 °C, that it can work and that its deployment is a necessary addition to our portfolio of options, the final question is: can negative emissions be unlocked with BECCS? The first step should establish a demonstration programme which develops the engineering and design criteria, improves understanding of its performance in terms of negative emissions potential and sustainability parameters and provides the context against which to establish policy and multilevel governance frameworks. That is, a demonstration programme which conceptually goes beyond a straightforward physical construction of BECCS projects to encompass the equally challenging political, social, ethical and governance systems described in this book. Such a demonstration programme must be part of an ambitious drive to reduce carbon emissions, which does not pitch measures against each other but which recognises that some will be more successful and effective than others, and is part of the wider scientific and political debate related to carbon budgets, greenhouse gas concentrations, future climate impacts and adaptation.

Achieving global net negative emissions requires tangible global coordination and governance mechanisms that deliver a global solution that maximises greenhouse gas reductions. Striving for the most ambitious possible reductions in atmospheric CO_2, treating the carbon budget as a maximum which we can underspend, may be what is needed to ensure that the greatest action is taken in the places where it is possible and to support a level and scale of BECCS deployment consistent with global net negative emissions. However, this will require both a scope of governance, oversight and institutional frameworks radically different from those we have today and financing of an unprecedented level of utility and infrastructure projects. Whilst the benefits for individual countries in using BECCS to meet their greenhouse gas targets may be clear, an appropriate global-level policy and governance framework is needed if gains made at national or project level can be scaled up to deliver global net negative emissions.

13.5 Summing Up

Almost all future emissions scenarios that limit future global warming and meet the aspirations of the Paris Agreement rely heavily on BECCS to deliver global net negative emissions. Yet, as BECCS is a technology that increases the cost and reduces the efficiency of power production, it will not be implemented without specific support. Crucially, it will depend on cooperation, incentivisation and a clear international policy drive if it is to be realised at the scale necessary to achieve global net negative emissions. This will require: (1) a high enough carbon price or equivalent fiscal instrument; (2) a mechanism that values and accounts for negative emissions at a global scale and (3) carbon capture and storage infrastructure (which in turn depends on point (1)). The negative emissions from BECCS depend heavily upon the sustainability of the biomass supply, with interconnected issues relating to land use, food, water, biodiversity and the use of biomass elsewhere in the energy sector and therefore impact on other mitigation and adaptation efforts. This presents a significant challenge to governance and regulation when attempting to translate the scale of BECCS deployed in models to the real world. Deploying BECCS as part of a long-term solution to an imminent overspend of the climate budget requires institutions and frameworks to situate negative emissions within mitigation and adaptation contexts that maximise benefits and minimise disadvantages; such a holistic approach to addressing climate change is perhaps the key to unlocking negative emissions.

References

Fuss, S., Canadell, J.G., Peters, G.P. et al. (2014). Betting on negative emissions. *Nature Climate Change* **4**: 850–853.

Gough, C., Garcia Freites, S., Jones, C. et al. (2018). Challenges to the use of BECCS as a keystone technology in pursuit of 1.5°C. *Global Sustainability*, In press.

IEAGHG (2011). Potential for Biomass with Carbon Capture and Storage 2011/06.

Stewart, R.J., Scott, V., Haszeldine, R.S. et al. (2014). The feasibility of a European-wide integrated CO_2 transport network. *Greenhouse Gas Sci Technol* **4**: 481–494. doi: 10.1002/ghg.1410.

UNFCCC (2015). United Nations Framework Convention on Climate Change, Adoption of the Paris Agreement FCCC/CP/2015/10/Add.1.

van Vuuren, D.P., Deetman, S., van Vliet, J. et al. (2013). The role of negative CO_2 emissions for reaching 2°C: insights from integrated assessment modelling. *Climatic Change* **118**: 15–27.

Whiriskey, K., (2014), Scaling the CO_2 storage industry: a study and a tool, Bellona Europa, November 2014, http://bellona.org/publication/scaling-co2-storage-industry-study-tool (accessed 14 December 2017).

Index

bold = extended discussion;   f = figure;   n = footnote or caption;   t = table;
105[–]108 = intermediate pages skipped

Biomass Energy with Carbon Capture and Storage (BECCS): Unlocking Negative Emissions, First Edition.
Edited by Clair Gough, Patricia Thornley, Sarah Mander, Naomi Vaughan and Amanda Lea-Langton.
© 2018 John Wiley & Sons Ltd. Published 2018 by John Wiley & Sons Ltd.